MODELING, CONTROL AND OPTIMIZATION OF COMPLEX SYSTEMS

In Honor of Professor Yu-Chi Ho

THE KLUWER INTERNATIONAL SERIES ON DISCRETE EVENT DYNAMIC SYSTEMS

Series Editor

Yu-Chi Ho
Harvard University

SUPERVISION OF PETRI NETS
Geert Stremersch
ISBN: 0-7923-7486-X

ANALYSIS OF MANUFACTURING ENTERPRISES: ***An Approach to Leveraging Value Delivery Processes for Competitive Advantage***
N. Viswanadham
ISBN: 0-7923-8671-X

INTRODUCTION TO DISCRETE EVENT SYSTEMS
Christos G. Cassandras, Stéphane Lafortune
ISBN: 0-7923-8609-4

OBJECT-ORIENTED COMPUTER SIMULATION OF DISCRETE-EVENT SYSTEMS
Jerzy Tyszer
ISBN: 0-7923-8506-3

TIMED PETRI NETS: ***Theory and Application***
Jiacun Wang
ISBN: 0-7923-8270-6

SUPERVISORY CONTROL OF DISCRETE EVENT SYSTEMS USING PETRI NETS
John O. Moody and Panos J. Antsaklis
ISBN: 0-7923-8199-8

GRADIENT ESTIMATION VIA PERTURBATION ANALYSIS
P. Glasserman
ISBN: 0-7923-9095-4

PERTURBATION ANALYSIS OF DISCRETE EVENT DYNAMIC SYSTEMS
Yu-Chi Ho and Xi-Ren Cao
ISBN: 0-7923-9174-8

PETRI NET SYNTHESIS FOR DISCRETE EVENT CONTROL OF MANUFACTURING SYSTEMS
MengChu Zhou and Frank DiCesare
ISBN: 0-7923-9289-2

MODELING AND CONTROL OF LOGICAL DISCRETE EVENT SYSTEMS
Ratnesh Kumar and Vijay K. Garg
ISBN: 0-7923-9538-7

UNIFORM RANDOM NUMBERS: THEORY AND PRACTICE
Shu Tezuka
ISBN: 0-7923-9572-7

MODELING, CONTROL AND OPTIMIZATION OF COMPLEX SYSTEMS

In Honor of Professor Yu-Chi Ho

edited by

Weibo Gong
University of Massachusetts, Amherst

Leyuan Shi
University of Wisconsin, Madison

SPRINGER SCIENCE+BUSINESS MEDIA, LLC

Library of Congress Cataloging-in-Publication Data

Modeling, control and optimization of complex systems : in honor of Professor Yu-Chi Ho/ edited by Weibo Gong, Leyuan Shi.
p. cm.-- (Kluwer international series on discrete event dynamic systems ; 14)
Includes bibliographical references.
ISBN 978-1-4613-5411-6 ISBN 978-1-4615-1139-7 (eBook)
DOI 10.1007/978-1-4615-1139-7
1. Control theory. 2. System analysis. 3. Mathematical optimization. I. Ho, Yu-Chi, 1934- II. Gong, Weibo. III. Shi, Leyuan. IV. Series.

QA402.3 .M57 2002
003'.5--dc21 2002028784

Printed on acid-free paper.

Contents

List of Figures

List of Tables

Foreword

In a scientific researcher's career, success is measured by the use of words like "pioneering work" and "enduring fundamental contributions" that others use to qualify the accomplishments of the researcher in a specific technical area. In the field of systems and control theory, I can think of few individuals to whom this applies. It makes it all the more remarkable that Larry Ho's name is closely associated with, not one, but three or four important technical areas in our field in which he has been a pioneer and a leader.

Going back to the 1960s, the "golden age" of modern control theory, Larry Ho's name belongs with that of a handful of researchers that laid the foundations for the material now considered the bread-and-butter of standard control engineering textbooks. His name became synonymous with Optimal Control, thanks in part to the landmark book "Applied Optimal Control" he co-authored with Arthur Bryson. In the 1970s, his interests led him to the exploration of facets of game theory such as decision-making for teams and the study of unconventional information structures. His work in this technical area has withstood the test of time, judging from papers he authored during those years which are still widely cited today.

By the late 1970s, it is fair to say that Larry Ho had built a track record that many would envy if they could claim it over an entire lifetime. In his case, it was just a starting point for another great intellectual adventure: Putting aside the comfortable modeling frameworks and well-oiled, control-theoretic machinery for "classical" systems, he looked into the world of not-so-classical systems encountered, for example, in automated manufacturing and communication networks, correctly realizing that the advent of computer technology would make these systems ubiquitous in the years to come. In doing so, he sought to merge ideas and methods from other fields, such as operations research and computer science, with those of systems and control. Anyone who knows about the risks of crossing such frontiers can appreciate what a bold step this can be. As a graduate student joining the Decision and Control group at Harvard in 1978, largely because of a desire to work on game theory, I was presented with the opportunity to play a role in this new direction, did so, and never regretted it. It led to the emergence of yet another technical area associated with Larry Ho's "pioneering work", now well recognized as Discrete Event Systems. The 1980s was the era of Perturbation Analysis for Discrete Event Systems in Larry Ho's research agenda and led to another well-cited book on the subject, co-authored with Xiren Cao.

One would imagine that Larry Ho would be finally done, as far as breaking ground into new technical areas was concerned. After all, there were still a lot of issues and unsolved problems related to Discrete Event Systems that could have occupied his activities in the 1990s. Instead, motivated as always by some "real world" problems, he decided to explore complex systems and ways in which they could be better designed and their performance optimized. He and his most recent students have laid the foundations for what is now known as Ordinal Optimization; the future will show how far the new ideas related to it can take us in our efforts to tackle complexity (admittedly, a difficult concept to properly define to this day).

Many around the world know Larry Ho as the author of seminal papers and widely used books, and as an inspiring speaker at international conferences and seminars. His students have the extra benefit of knowing him as a teacher and a caring mentor. His close colleagues and friends also know him as an individual who has earned their admiration and respect, not just for what he has contributed as a researcher and educator, but also as an individual with seemingly unlimited energy and interests that transcend the narrow bounds of our field, and, perhaps most importantly, as a really "nice guy"...

Christos G. Cassandras

Acknowledgements

This book is a collection of papers presented at the Symposium on Modeling and Optimization of Complex Systems in June 23-24, 2001 and some invited contributions. The symposium was dedicated to Professor Larry Yu-Chi Ho of Harvard University on the occasion of his retirement from teaching. It is our hope that the book reflects the far-reaching influence of Professor Ho's research that is still going on.

We wish to thank all the contributors to the book. Their dedication to scientific research and friendship are deeply appreciated.

We are also very grateful to the Army Research Office, the Air Force Office of Scientific Research, Office of Naval Research, and the National Science Foundation for their generous financial support.

Mr. Songlin Cai, Mr. Yunpeng Pan, and Mr. Mehmet Bozbay's help with the organization of the symposium is crucial for its success. Mr. Cai also did an excellent job compiling the contributed papers in a highly skilled and professional fashion.

Finally we wish to thank Ms. Melissa Fearon of Kluwer for her help in the process of editing the volume.

Editors: Weibo Gong, Leyuan Shi.

Chapter 1

OPTIMAL REPRESENTATION AND VISUALIZATION OF MULTIVARIATE DATA AND FUNCTIONS IN LOW-DIMENSIONAL SPACES

Jian Song
President, Chinese Academy of Engineering

The analysis and processing of massive amount of multivariate data and high-dimensional functions have become a basic need in many areas of science and engineering. To reduce the dimensionality for compact representation and visualization of high-dimensional information appear imperative in exploratory research and engineering modeling. Since D. Hilbert raised the 13th problem in 1900, the study on possibility to express high-dimensional functions via composition of lower-dimensional functions has gained considerable success[1, 2]. Nonetheless, no methods of realization are ever indicated, and not even all integrable functions can be treated this way, a fortiori functions in $L_2(\Omega)$. The common practice is to expand high-dimensional functions into a convergent series in terms of a chosen orthonormal basis with lower dimensional ones. However, the length and rapidity of convergence of the expansion heavily depend upon the choice of basis. In this paper we briefly report some new results of study in seeking an optimal basis for a given function provided with fewest terms and rapidest convergence. All elements of the optimal basis turned out to be products of single-variable functions taken from the unit balls of ingredient spaces. The proposed theorems and schemes may find wide applications in data processing, visualization, computing, engineering simulation and decoupling of nonlinear control systems. The facts established in the theorems may have their own theoretical interests.

Let $F(x) = F(x^1, x^2, ..., x^n)$ be an arbitrarily given function defined on the unit cube Ω in $\mathbb{R}^n$, $F \in L_2(\Omega)$,

$$\int_\Omega (F(x^1, x^2, ..., x^n))^2 dx^1 dx^2 ... dx^n \leq M^2 < \infty, \quad (1)$$

where $x = (x^1, x^2, ..., x^n)$ is a point in Ω.

Since $L_2(\Omega)$ is a separable Hilbert-space, every element F in it can be expressed as a convergent series in terms of chosen basis with respect to the L_2-norm. Such basis may be trigonometric polynomials, Bessel functions, spherical functions, wavelets, etc., considered convenient in each case. However, the choice of particular basis has a great influence on the speed of convergence, necessary space of storage and the workload of processors. It is intended to find a set of one-variable functions $\varphi^\alpha(x^\alpha)$ whose product $\varphi^1(x^1)\varphi^2(x^2)...\varphi^n(x^n) \in L_2(\Omega)$ would best, or optimally, approximate $F(x)$ with the least-square deviation:

$$L = \int_\Omega (F(x^1, x^2, ..., x^n) - \varphi^1(x^1)\varphi^2(x^2)...\varphi^n(x^n))^2 d\Omega = \min, \quad (2)$$

where $d\Omega = dx^1 dx^2 ... dx^n$.

Suppose each $\psi^\alpha(x^\alpha)$ is taken from the unit balls $B_\alpha \in L_2^{(\alpha)}(0,1)$, $B_\alpha = \{\|\psi^\alpha\|_{L_2(0,1)} \leq 1\}, \alpha = 1, 2, ..., n$, the above requirement (2) can be rewritten as

$$L = \inf_{\psi^\alpha \in B_\alpha} \int_\Omega \left(F(x) - \lambda \prod_{\alpha=1}^{n} \psi^\alpha(x^\alpha) \right)^2 d\Omega. \quad (3)$$

Opening up the brackets on the right side we have

$$\int_\Omega \left(F(x) - \lambda \prod_{\alpha=1}^{n} \psi^\alpha(x^\alpha) \right)^2 d\Omega$$

$$= \int_\Omega \left(F^2(x) - 2\lambda F(x) \prod_{\alpha=1}^{n} \psi^\alpha(x^\alpha) + \lambda^2 \prod_{\alpha=1}^{n} (\psi^\alpha(x^\alpha) \right)^2 d\Omega.$$

It is easy to verify that (3) holds if and only if there exists a product of n functions $\prod_\alpha \varphi^\alpha, \varphi^\alpha \in B_\alpha$, and a real $\lambda \in \mathbb{R}, \lambda \neq 0$, which enable the following functional to achieve supremum on all unit balls B_α,

$$\lambda = \sup_{\psi^\alpha \in B_\alpha} \int_\Omega F(x) \prod_{\alpha=1}^{n} \psi^\alpha(x^\alpha) d\Omega = \int_\Omega F(x^1, x^2, ..., x^n)$$

$$\varphi^1(x^1)\varphi^2(x^2)...\varphi^n(x^n) dx^1 dx^2 ... dx^n. \quad (4)$$

For the convenience of discussion in the sequel we distinguish the same spaces $L_2^{(\alpha)}(0,1), \alpha = 1,2,...,n$, and construct a product space $L_2^n(0,1) = L_2^{(1)}(0,1) \bigotimes L_2^{(2)}(0,1) \bigotimes ... \bigotimes L_2^{(n)}(0,1)$, $L_2^n(0,1) = \left\{\sum_\beta a_\beta \prod_{\alpha=1}^n \psi_\beta^\alpha, \psi_\beta^\alpha \in L_2^{(\alpha)}(0,1), \alpha = 1,2,...,n, \beta = 1,2,...,N\right\}$. After introduction of inner product for $\psi, \varphi \in L_2^n(0,1)$, $\psi = \prod_{\alpha=1}^n \psi^\alpha, \varphi = \prod_{\alpha=1}^n \varphi^\alpha$,

$$\langle\psi,\varphi\rangle = \langle\prod_{\alpha=1}^n \psi^\alpha, \prod_{\alpha=1}^n \varphi^\alpha\rangle = \prod_{\alpha=1}^n \langle\psi^\alpha,\varphi^\alpha\rangle = \prod_{\alpha=1}^n \int_0^1 \psi^\alpha(x^\alpha)\varphi^\alpha(x^\alpha)dx^\alpha,$$

with induced natural norm

$$\left\|\prod_{\alpha=1}^n \psi^\alpha\right\|_{L_2^n(0,1)} = \prod_{\alpha=1}^n \|\psi^\alpha\|_{L_2^{(\alpha)}(0,1)},$$

$L_2^n(0,1)$ becomes a linear normed space. The $L_2^n(0,1)$ defined above can be embedded into $L_2(\Omega)$ with preserved norm and becomes a dense subset of the latter[3, 10], while $L_2^{(\alpha)}(0,1)$ is a closed subspace of $L_2(\Omega)$, since $\|\psi^\alpha\|_{L_2^{(\alpha)}(0,1)} = \|\psi^\alpha\|_{L_2(\Omega)}$ always holds on Ω. On the other hand, for the multi-linear functional $f : L_2^n(0,1) \to \mathbb{R}$, defined by

$$f\left(\prod_{\alpha=1}^n \psi^2\right) = \int_\Omega F(x)\prod_\alpha \psi^\alpha(x^\alpha)d\Omega, \tag{5}$$

the following inequality holds for every element of $L_2^n(0,1)$:

$$\left|f\left(\prod_\alpha \psi^\alpha\right)\right| \leq M\prod_\alpha \|\psi^\alpha\|_{L_2^{(\alpha)}(0,1)},$$

where M is the lower bound defined in (1). Hence f is bounded and, by Banach-Steinhaus theorem, is totally continuous in $L_2^n(0,1)$ [6].

Firstly, we need the following Lemma.

LEMMA 1. The n-linear form (5) can achieve its supremum on B^n. Whenever $F(x) \neq 0$, the supremum λ is positive,

$$\lambda = \sup_{\psi^\alpha \in B_\alpha} \int_\Omega F(x)\prod_\alpha \psi^\alpha d\Omega > 0. \tag{6}$$

We list only the primary point of proof bellow. Since all $L_2^{(\alpha)}(0,1)$ are Hilbert-spaces, it is not difficult to show that the combinations of the kind $\prod_{\alpha=1}^n \psi^\alpha \in L_2^n(0,1)$ have the same property: there exists a sequence of combinations $\{\prod_\alpha \psi_k^\alpha(x^\alpha), k=1,2,...\}$ which converges weakly to some element $\prod_\alpha \varphi^\alpha$ in the unit ball $B^n \subset L_2^n(0,1)$ and makes (6) to reach its supremum

$$\lambda = \sup_{B^n} \int_\Omega F(x) \prod \psi^\alpha(x^\alpha) d\Omega = \int_\Omega F(x) \prod \varphi^\alpha(x^\alpha) d\Omega, \qquad (6')$$

even though $L_2^n(0,1)$ is only a linear normed space, but not complete one.

Making use of the compactness of constituent spaces and the Majorized Convergence Theorem [9, 10], it has been shown that there, in fact, exists a sequence $\{\prod \psi^\alpha\}$ which weakly converges to an element $\prod_\alpha \varphi^\alpha \in B^n$ and provides (6') with a positive global supremum [24].

In order to find the extremal element $\prod_\alpha \varphi^\alpha$, we have to apply the necessary conditions which a solution of (4), $\prod_\alpha \varphi^\alpha$, should satisfy. Suppose the expression (4) achieves its supremum at some element $\prod_\alpha \varphi^\alpha \in B^n$, which must be on the boundary. According to Lagrange Principle, $\prod_\alpha \varphi^\alpha$ must satisfy the following conditions with a multiplier λ' [6]:

$$D\left(\int_\Omega F(x^1,...,x^n) \prod_\alpha \varphi^\alpha dx - \lambda' \left(\left\langle \prod_\alpha \varphi^\alpha, \prod_\alpha \varphi^\alpha \right\rangle_{L_2(\Omega)} - 1\right)\right) = 0, \qquad (7)$$

where D denotes the Gateaux directional derivative with respect to all φ^α, λ' is a real to be determined. According to the rules of differentiation, $Df = \sum_{i=1}^n D_i f, D_i f$ is the partial derivative with respect to φ^i. Let h^i be any element in $L_2^{(i)}(0,1), t \in [0,1]$, for the second term f_2 of (7) we have

$$D_i f_2$$
$$= \lim_{t\to 0} \frac{1}{t}\left[\lambda' \left\langle \prod_{\alpha\neq i} \varphi^\alpha(\varphi^i + th^i), \prod_{\alpha\neq i} \varphi^\alpha(\varphi^i + th^i)\right\rangle - \lambda' \left\langle \prod \varphi^\alpha, \prod \varphi^\alpha \right\rangle\right]$$
$$= 2\lambda' \langle \varphi^i, h^i \rangle,$$

while for the first term,

$$D_i\left(\int_\Omega F(x)\prod_\alpha \varphi^\alpha d\Omega\right)$$

$$= \lim_{t\to 0}\frac{1}{t}\left[\int_\Omega F\prod_{\alpha\neq i}\varphi^\alpha(\varphi^i+th^i)d\Omega - \int_\Omega F\prod_\alpha \varphi^\alpha d\Omega\right]$$

$$= \left\langle \Phi_i\left(\prod_{\alpha\neq i}\varphi^\alpha\right), h^i\right\rangle,$$

where Φ_i is an operator generated by G-differentiation, it can be explicitly expressed as $\Phi_i : L_2^{(1)}(0,1)\bigotimes\cdots\bigotimes L_2^{(i)}(0,1)\bigotimes\cdots\bigotimes L_2^{(n)}(0,1) \to L_2^{(i)}(0,1)$. The ith component is

$$\Phi_i\left(\prod_{\alpha\neq i}\varphi^\alpha\right) = \int_{\Omega_i} F(x)\prod_{\alpha\neq i}\varphi^\alpha d\Omega_i. \tag{8}$$

Summing up of the two, with $\lambda = 2\lambda'$, yields

$$\Phi_i\left(\prod_{\alpha\neq i}\varphi^\alpha\right) - \lambda\varphi^i = 0, i = 1,2,,n, \tag{9}$$

or in details,

$$\lambda\varphi^i(x^i) = \Phi_i\left(\prod_{\alpha\neq i}\varphi^\alpha\right) = \int_{\Omega_i} F(x^1,...,x^n)\varphi^1(x^1)\cdots\hat{\varphi^i(x^i)}\cdots\varphi^n(x^n)dx^1 \cdots \hat{dx^i}\cdots dx^n, i = 1,2,\cdots n, \tag{9'}$$

where $\hat{i}$ means absence of ith coordinate, Ω_i is the $(n-1)$ dimensional unit cube without x^i.

The operator $\Phi = (\Phi_1, \Phi_2, ..., \Phi_n)$ generated by G-derivative and defined in (8) is called the gradient operator of functional (5). We have proved the following, the primary theorem as a start-point for further investigation.

THEOREM 2. For any given $F(x) \in L_2(\Omega), F(x) \neq 0$, its gradient operator Φ or, equivalently, the system of homogeneous integral equations (9) possesses at least one positive eigenvalue. The greatest eigenvalue and its associated eigenfunctions satisfy (4), and are a solution of this supremum problem.

The proof of this theorem turned out to be quite lengthy. As for eigenvalue problem of self-adjoint operators in Hilbert Space, it has been shown that there is a sequence of functions $\psi_k^\alpha(x^\alpha), \alpha = 1, 2, \ldots n$, which converges strongly to $\varphi^\alpha, \alpha = 1, \ldots, n$, and indeed satisfies the system of equations (9) with a positive eigenvalue λ as the supremum of (6'). The detailed proof of this Theorem will be presented in a separate paper with focus on the mathematical constructions and elucidations.

In short, it has been proved that there exist n one-variant eigenfunctions $\varphi^\alpha(x^\alpha), \alpha = 1, ..., n$, associated with the eigenvalue λ such that the following equality holds:

$$\sup_{\psi^\alpha \in B_\alpha} \int_\Omega F(x) \prod \psi^\alpha(x^\alpha) d\Omega = \left\langle \Phi_i \left(\prod_{\substack{\alpha=1 \\ \alpha \neq i}} \varphi^\alpha \right), \varphi^i \right\rangle$$

$$= \int_\Omega F(x^1, x^2, \cdots, x^n) \varphi^1(x^1) \varphi^2(x^2) \cdots \varphi^n(x^n) dx^1 dx^2 \cdots dx^n = \lambda, \tag{10}$$

that is, $\prod_\alpha \varphi^\alpha$ is a solution of (9), as claimed in the Theorem.

Let λ_1 and $\prod_{\alpha=1}^n \varphi_1^\alpha$ are the greatest eigenvalue and its associated eigenfunctions, respectively. Construct new function

$$F_1(x^1, x^2, \cdots x^n) = F_0(x^1, x^2, \cdots x^n) - \lambda_1 \prod_{\alpha=1}^n \varphi_1^\alpha(x^\alpha). \tag{11}$$

Let$F_1 \neq 0$. It is easy to check that the norm of F_1 in $L_2(\Omega)$ can be calculated as

$$||F_1||^2 = \left\langle F_0 - \lambda_1 \prod \varphi_1^\alpha, F_0 - \lambda_1 \prod \varphi_1^\alpha \right\rangle$$

$$= ||F_0||^2 - 2\lambda_1 \left\langle F_0, \prod \varphi_1^\alpha \right\rangle + \lambda_1^2 \left\langle \prod \varphi_1^\alpha, \prod \varphi_1^\alpha \right\rangle = ||F_0||^2 - \lambda_1^2. \tag{12}$$

So long as $F_1 \neq 0$, the Lemma 1 and Theorem 2 are applicable for F_1 as well. Having F_0 replaced by F_1 in (8), one gets a new operator Φ_1 and, correspondingly, new system of equations

$$\begin{aligned}
\lambda_2 \varphi_2^i(x^i) &= \Phi_{1i} \left(\textstyle\prod_{\substack{\alpha=1 \\ \alpha \neq i}}^n \varphi_2^\alpha \right) \\
&= \textstyle\int_\Omega \left(F_0(x^1, x^2, , x^n) - \lambda_1 \prod_{\alpha=1}^n \varphi_1^\alpha(x^\alpha) \right) \varphi_2^1(x^1) \\
&\cdots \hat{\varphi_2^i} \cdots \varphi_2^n(x^n) dx^1 \cdots d\hat{x^i} \cdots dx^n, i = 1, 2, \cdots, n.
\end{aligned}$$

By Theorem 2, it possesses the greatest positive eigenvalue λ_2 and associated eigenfunctions $\prod_\alpha \varphi_2^\alpha$. Similarly, we have

$$F_2 = F_1 - \lambda_2 \prod_{\alpha=1}^{n} \varphi_2^\alpha = F_0 - \lambda_1 \prod_{\alpha=1}^{n} \varphi_1^\alpha - \lambda_2 \prod_{\alpha=1}^{n} \varphi_2^\alpha,$$

with its norm

$$||F_2||^2 = ||F_1||^2 - \lambda_2^2 = ||F_0||^2 - \lambda_1^2 - \lambda_2^2.$$

The process can be continued inductively, if $F_N \neq 0$,

$$F_N = F_{N-1} - \lambda_N \sum_{\alpha=1}^{n} \varphi_N^\alpha = F_0 - \sum_{\beta=1}^{N} \lambda_\beta \prod_{\alpha=1}^{n} \varphi_\beta^\alpha, \tag{13}$$

and

$$||F_N||^2 = ||F_{N-1}||^2 - \lambda_N^2 = ||F_0||^2 - \sum_{\beta=1}^{N} \lambda_\beta^2. \tag{14}$$

It is easy to verify, one always has a non-increasing sequence $\lambda_1 \geq \lambda_2 \geq \ldots \geq \lambda_N$. The existence of multiple eigenvalues is possible, but as it will be shown in the sequel, in some spaces the multiplicity is not greater than 1. Generally, each λ_β must be repeated according to its multiplicity. Further, each F_N generates its own gradient operator $\Phi_N = (\Phi_{N1}, \ldots, \Phi_{Nn})$,

$$\Phi_{Ni}\left(\prod_{\alpha\neq i} \varphi_{N+1}^\alpha\right) = \int_{\Omega_i} F_N(x) \prod_{\alpha\neq i} \varphi_{N+1}^\alpha(x^\alpha) d\Omega_i = \lambda_{N+1}\varphi_{N+1}^i(x^i),$$
$$i = 1, 2, \ldots, n. \tag{15}$$

If the process continues infinitely, (13) becomes an infinite series. It turns out that the following equality holds as $N \to \infty$ in the norm of $L_2(\Omega)$,

$$F_0(x^1, x^2, \ldots, x^n) = \sum_{\beta=1}^{\infty} \lambda_\beta \prod_{\alpha=1}^{n} \varphi_\beta^\alpha(x^\alpha). \tag{16}$$

THEOREM 3. Any given $F(x) \in L_2(\Omega), F \neq 0$, can be expressed in the form of series (16) with all positive eignevalues and associated eignefunctions generated by the sequence of gradient operators $\{\Phi_\beta\}$. The series (16) converges exponentially in the sense of the inequality (20) to $F(x)$ in the norm of $L_2(\Omega)$.

An abridged proof goes as following. Indeed, if the process of construction described in (13) terminates at a finite step N, the validity of the first part of Theorem is apparent. Now suppose that (13) becomes an infinite series (16) when $N \to \infty$. It is obvious from (13), however large is N, one always has $||F_N||^2 \geq 0$, and

$$\sum_{\beta=1}^{N} \lambda_\beta^2 \leq ||F_0||^2. \tag{17}$$

The necessary condition of convergence for the series on the left side is $\lambda_N \to 0$ as $N \to \infty$, and (6) implies that the following relationship holds uniformly on B^n,

$$\begin{aligned} \lambda_{N+1} &= \sup_{\psi^\alpha \in B_\alpha} \int_\Omega F_N(x) \prod_\alpha \psi^\alpha(x^\alpha) d\Omega = \left\langle F_N(x), \prod_\alpha \varphi_{N+1}^\alpha \right\rangle_{L_2(\Omega)} \\ &\geq \left| \left\langle F_N(x), \prod_\alpha \psi^\alpha \right\rangle_{L_2(\Omega)} \right|, \forall \psi^\alpha \in B_\alpha, \end{aligned} \tag{18}$$

here $\prod_\alpha \varphi^{\alpha N+1}$ is a solution of (6) and (9) with $F_0(x)$ replaced by $F_N(x)$. Therefore, when $N \to \infty$,

$$\lim_{N\to\infty} \left\langle F_N(x), \prod_\alpha \psi^\alpha \right\rangle_{L_2(\Omega)} \to 0, \forall \psi^\alpha \in B_\alpha, \alpha = 1, 2, \ldots, n.$$

Since B^n is a fundamental set, i.e., its span, $L_2^n(0,1)$, is a dense subset of $L_2(\Omega)$. The above condition suffices for F_N to converge weakly-star to some element F_∞ which is equivalent to 0 in the weak topology [10],

$$\lim_{N\to\infty} \sup_{B^n} \int_\Omega F_N(x) \prod_\alpha \psi^\alpha(x^\alpha) d\Omega = \sup_{B^n} \left\langle F_\infty, \prod_\alpha \psi^\alpha \right\rangle_{L_2(\Omega)} = 0.$$

Recall the fact that the set $E = \{\sum_\alpha e_\alpha^{\gamma_\alpha}, \gamma_\alpha \in \mathbb{N}\}$ consisting of combinations of orthonormal bases of $L_2^{(\alpha)}(0,1)$ is a complete orthonormal basis for $L_2(\Omega)$. Notice $E \subset B^n$, by Lemma 1 and (18), it implies

$$0 = \sup_{B^n} \left\langle F_\infty, \prod_\alpha \psi^\alpha \right\rangle \geq \sup_E \left\langle F_\infty, \prod_\alpha e_\alpha^{\gamma_\alpha} \right\rangle \geq 0.$$

Which means all Fourier coefficients of F_∞ are zero, hence F_N converges to 0 strongly, $F_\infty = 0$. Then (16) holds in the sense

$$||F_0||^2 = \sum_{\beta=1}^{\infty} \lambda_\beta^2. \tag{19}$$

The second statement claimed in the Theorem about the convergence rapidity of the expansion (19) can be justified in the following way. By assumption all $F_\beta \neq 0$, and due to (13) and (14), the following relations and the continuous multiplication are well defined,

$$\begin{aligned}\frac{||F_1||^2}{||F_0||^2} &= \frac{||F_0||^2 - \lambda_1^2}{||F_0||^2} = 1 - \frac{\lambda_1^2}{||F_0||^2}, \cdots, \frac{||F_N||^2}{||F_{N-1}||^2} \\ &= \frac{||F_{N-1}||^2 - \lambda_N^2}{||F_{N-1}||^2} = 1 - \frac{\lambda_N^2}{||F_{N-1}||^2}, \\ \frac{||F_N||^2}{||F_0||^2} &= \frac{||F_N||^2}{||F_{N-1}||^2}\frac{||F_{N-1}||^2}{||F_{N-2}||^2} \cdots \frac{||F_2||^2}{||F_1||^2}\frac{||F_1||^2}{||F_0||^2} = \prod_{\beta=1}^{N}\left(1 - \frac{\lambda_\beta^2}{||F_{\beta-1}||^2}\right).\end{aligned}$$

Denote $\alpha_\beta^2 = \frac{\lambda_\beta^2}{||F_{\beta-1}||^2}$, namely $\lambda_\beta = \alpha_\beta ||F_{\beta-1}||$. By the proved previously, $0 < \alpha_\beta \leq 1$. Let $\prod_{\alpha=1}^{n} \varphi_\beta^\alpha$ be eigenfunction associated with the greatest eigenvalue λ_β. Recall that

$$\lambda_\beta = \sup_{B^n} \int_\Omega F_{\beta-1}(x) \prod_\alpha \psi^\alpha d\Omega = \left\langle F_{\beta-1}, \prod \varphi_\beta^\alpha \right\rangle,$$

and for any $\alpha, 1 \geq \alpha > 0$, the inequality $(1-\alpha) \leq e^{-\alpha}$ holds, then we have

$$||F_N||^2 = ||F_0||^2 \prod_{\beta=1}^{N} (1 - \alpha_\beta^2) \leq ||F_0||^2 e^{-\sum_{\beta=1}^{N} \alpha_\beta^2},$$

or

$$||F_N|| \leq ||F_0|| e^{-\frac{1}{2}\sum_{\beta=1}^{N} \alpha_\beta^2} \tag{20}$$

Let $R_{N+1} = ||F_N||^2 = \sum_{\beta=N+1}^{\infty} \lambda_\beta^2$ be the sum of residual part of (19). It is easy to show the sum

$$\sum_{k=1}^{N} \alpha_k^2 = \sum_{k=1}^{N} \frac{\lambda_k^2}{R_k} \tag{21}$$

diverges as $N \to \infty$. In fact, let $m > 1$ be any integer. The following inequality always holds,

$$\frac{\lambda_{N+1}^2}{R_{N+1}} + \frac{\lambda_{N+2}^2}{R_{N+2}} + \cdots + \frac{\lambda_{N+m}^2}{R_{N+m}} > \frac{\lambda_{N+1}^2 + \cdots + \lambda_{N+m}^2}{R_{N+1}} = 1 - \frac{R_{N+m1}}{R_{N+1}}.$$

Due to the convergence of (19), for any N there exists a finite positive integer m such that $\frac{R_{N+m+1}}{R_{N+1}} < \frac{1}{2}$, or $\sum_{k=1}^{m} \frac{\lambda_{N+k}^2}{R_{N+k}} > \frac{1}{2}$. This observation indicates that the series (21) does not comply with the Cauchy necessary condition of convergence, thus (21) diverges to infinity as $N \to \infty$, and (20) tends to 0 exponentially in the sense of (20) as claimed in the Theorem. The unconditionality of convergence of (16) will be proved below.

In the above discussion we did not touch upon the properties of the set of eigenfunctions. We will see below, it is quite similar to the case of symmetrical integral operators, the set of all eigenfuctions generated by (15) constitutes an orthonormal system in $L_2(\Omega)$.

THEOREM 4. For arbitrarily given $F(x) \in L_2(\Omega), F \neq 0$, the set of all first eigenfunctions $\left\{\prod_\alpha \varphi_\beta^\alpha, \beta = 1, 2, \ldots,\right\}$ of the sequence of gradient operators $\Phi_\beta = (\Phi_{\beta 1}, \ldots, Phi_{\beta n})$, defined by (15), constitutes an orthonormal system as an ingredient part of some complete orthonormal basis of $L_2(\Omega)$.

By Theorem 2 and 3 we know that every eigenfunction of gradient operator Φ_N has unit norm, $||\prod_\alpha \varphi_\beta^\alpha|| = \prod_\alpha ||\varphi_\beta^\alpha|| = 1$, since $\varphi_\beta^\alpha \in S_\alpha$, the unit sphere of $L_2^{(\alpha)}(0,1)$ by construction. Recall that the eigenvalue λ_{N+1} and its eigenfunction $\prod_\alpha \varphi_{N+1}^\alpha$ are the solution of the problem

$$\sup_{\psi^\alpha \alpha} \int_\Omega F_N(x) \prod_\alpha \psi^\alpha d\Omega = \int_\Omega F_N(x) \prod_\alpha \varphi_{N+1}^\alpha d\Omega = \lambda_{N+1}. \qquad (22)$$

The definition of F_N implies that the following identity holds,

$$\left\langle F_\beta, \prod_\alpha \varphi_\beta^\alpha \right\rangle = 0, \forall \beta = 1, 2, ..., N, N+1.$$

Now, let $L(\varphi_N^\alpha)$ be the subspace spanned by φ_N^α in $L_2^{(\alpha)}(0,1)$. There exists unique orthogonal decomposition

$$\varphi_{N+1}^\alpha = a_\alpha \varphi_N^\alpha + b_\alpha \bar{\varphi}_{N+1}^\alpha, \bar{\varphi}_{N+1}^\alpha \in L^\perp(\bar{\varphi}_N^\alpha),$$

where $L^\perp$ is the orthogonal complement to $L(\varphi_N^\alpha)$, and a_α and b_α are the coefficients of normalization. Unfolding the product

$$\prod_\alpha (a_\alpha \varphi_N^\alpha + b_\alpha \bar{\varphi}_{N+1}^\alpha) = C_0 \prod \varphi_N^\alpha + C_1 \sum_{i=1}^n \bar{\varphi}_{N+1}^i \prod_{\alpha \neq i} \varphi_N^\alpha + \cdots + C_n \prod \bar{\varphi}_{N+1}^\alpha,$$

we see all terms on the right side, except the first, orthogonal to $\prod_\alpha \varphi_N^\alpha$. Put them into (22), due to the above identity, it becomes

$$\begin{aligned}\int_\Omega F_N \prod_\alpha \varphi_{N+1}^\alpha d\Omega &= \int_\Omega \left(F_{N-1} - \lambda_N \prod_\alpha \varphi_N^\alpha\right) \prod_\alpha \varphi_{N+1}^\alpha d\Omega \\ &= \int_\Omega F_{N-1}(x) \prod_\alpha \varphi_{N+1}^\alpha d\Omega \\ &= \sup_{\substack{\psi^\alpha \in B_\alpha \\ \psi^\alpha \perp \varphi_N}} \int_\Omega F_{N-1}(x) \prod_\alpha \psi^\alpha d\Omega = \lambda_{N+1}.\end{aligned}$$

Continue similar analysis with respect to $\prod_\alpha \varphi_\beta^\alpha, \beta = N-1, ..., 1$, one verifies that $\prod_\alpha \varphi_{N+1}^\alpha$ achieves supremum λ_{N+1} on the orthogonal complement of the linear subspace $L_N^{(\alpha)} = \{\varphi_1^\alpha, \ldots, \varphi_n^\alpha\}$ spanned by all preceding eigenfunctions. If denote its complement by $(L_N^{(\alpha)})^\perp$, we conclude

$$\lambda_{N+1} = \sup_{\psi^\alpha \in B_\alpha} \int_\Omega F_N(x) \prod \psi^\alpha d\Omega = \sup_{\substack{\psi^\alpha \in B_\alpha \\ \psi^\alpha \in (L_N^{(\alpha)})^\perp}} \int_\Omega F_0(x) \prod_\alpha \varphi_{N+1}^\alpha d\Omega. \tag{23}$$

This being true for all $N \in \mathbb{N}$ follows the set of eigenfunctions $\{\prod_\alpha \varphi_\beta^\alpha, \beta = 1, 2, ...\}$ constitutes a orthonormal system in $L_2(\Omega)$. Since for any orthonormal set S in a Hilbert space there is a complete orthonormal basis that contains S as its subset [13].

REMARK. It appears the remarkable maximum property of gradient operators expressed in (23), which is entirely analogous with compact self-adjoint operators in Hilbert spaces. Theorem 4 also shows there are as many different orthonormal bases as the cardinality of different elements in $L_2(\Omega)$.

THEOREM 5. If the dimension of underlying space $\mathbb{R}^n$ with $n = 2, 4$ and 6, the greatest eigenvalue of gradient operators defined above for arbitrarily given $F(x) \in L_2(\Omega), F(x) \neq 0$, have multiplicity no more than 1.

Suppose the contrary, if there exist two different eigenfunctions $\prod_\alpha \varphi_{N+1,1}^\alpha$ and $\prod_\alpha \varphi_{N+1,2}^\alpha$ corresponding to the same eigenvalue λ_{N+1}, which enable the following functional to achieve its supremum on B^n,

$$\begin{aligned}\lambda_{N+1} = \sup_{\psi^\alpha \in B_\alpha} f_N\left(\prod_\alpha \psi^\alpha\right) &= \sup_{\psi^\alpha \in B_\alpha} \int_\Omega F_N(x) \prod_\alpha \psi^\alpha \Omega \\ &= f_N\left(\prod_\alpha \varphi_{N+1,k}^\alpha\right), k = 1, 2.\end{aligned} \tag{24}$$

Denote $\varphi^\alpha_{N+1,1} = \varphi^\alpha_1$ and $\varphi^\alpha_{N+1,2} = \varphi^\alpha_2$ for brevity. Construct a new element $\varphi_3 = \prod_\alpha (t\varphi^\alpha_1 + (1-t)\varphi^\alpha_2), t \in [0,1]$, and substitute it into (24). We invoke the sufficient conditions for f to achieve its maximum at φ. If the dimension of the underlying space, $\mathbb{R}^n$, n is even, then the following equations must be satisfied [8,III],

$$D^k f\left(\prod_\alpha \varphi^\alpha\right) = 0, k = 1, 2, ..., n-1.$$

After unfolding the product of φ_3 , we get

$$Df = \sum_{i=1}^{n} f\left(\varphi^i_2 \prod_{\alpha\neq 1} \varphi^\alpha_1\right) = \sum_i \langle \varphi^i_2, \varphi^i_1 \rangle = 0,$$

$$D^2 f\left(\prod_\alpha \varphi^\alpha_1\right) = 2f\left(\sum_{i\neq j} h^i_1 h^j_2 \prod_{\hat{i}\hat{j}} \varphi^\alpha_1\right) - \sum_i \langle h^i_1, h^i_2 \rangle \lambda = 0$$

for arbitrary h^i_1 and h^i_2 taken from $L^{(i)}_2(0,1)$. Let $h^i_1 = h^i_2 = \varphi^i_2$. It yields

$$f_N\left(\sum_{i\neq j} \varphi^i_2 \varphi^i_2 \prod_{\hat{i}\hat{j}} \varphi^\alpha_1\right) = \frac{n}{2}\lambda_{N+1},$$

and

$$f_N(\varphi_3, t) = (t^n + t^{n-2}(1-t)^2\frac{n}{2} + t^2(1-t)^{n-2}\frac{n}{2} + (1-t)^n)\lambda_{N+1}.$$

It follows also from the assumption, $f_N(\varphi_3, 0) = f_N(\varphi_3, 1) = \lambda_{N+1}$, and f_N reaches minimum at $t = 1/2$. Let $t = 1/2$, it yields

$$f_N\left(\varphi_3, \frac{1}{2}\right) = f_N\left(\prod_a \frac{\varphi^a_1 + \varphi^a_2}{2}\right) = (2+n)2^{-n}\lambda_{N+1}.$$

Since $||\varphi^\alpha_1 + \varphi^\alpha_2|| = \sqrt{2}$, $\frac{\varphi^\alpha_1 + \varphi^\alpha_2}{\sqrt{2}} \in B_\alpha$, if $\rho(n) \geq 1$ then $(\rho(n))^{\frac{1}{n}} \geq 1$, so that $\prod_\alpha \frac{\varphi^\alpha_1 + \varphi^\alpha_2}{\sqrt{2}\rho^{\frac{1}{n}}} \in B^n$. A straightforward computation shows this is possible only for $n = 2, 4$ and 6. In these cases if $\varphi_1 = \prod_\alpha \varphi^\alpha_{N+1,1}$, and $\varphi_2 = \prod_\alpha \varphi^\alpha_{N+1,2}$ both render the supremum λ_{N+1} to the functional f_N, then the middle point φ_3 of segment joining φ^α_1 and φ^α_2, which is situated strictly in the interior of B^n, also provides the supremum λ_{N+1} to f_N. By the assumption, $\varphi_1 \neq \varphi_2$, it implies that there must be infinite amount of different elements in B^n at each of them f_N attains its supremum. This contradicts the trivial fact that the supremum can be achieved only on the boundary of B^n. Hence, we can conclude, for $n = 2, 4$, and 6, the multiplicity of the greatest eigenvalue of (15) is no more than 1.

The case $n = 2$ may cause particular theoretical interest[20]. Let $F(x, y)$ be defined on the unit rectangle B^2 of the plane and be square-integrable. By Theorem 2, it generates a gradient operator Φ, and (9) is reduced to

$$\Phi\psi = \int_0^1 F(x,y)\psi(y)dy = \lambda\varphi, \Phi^*\varphi = \int_0^1 F(x,y)\varphi(x)dx = \lambda\psi. \quad (25)$$

φ and ψ are eigenfunctions of self-adjoint operators

$$\Phi\Phi^*\varphi = \lambda^2\varphi, \Phi^*\Phi\psi = \lambda^2\psi.$$

Theorem 5 claims that for $n = 2$ the greatest eigenvalue of Φ have multiplicity no more than 1. Suppose the contrary. Let $\varphi_1(x)\psi_1(y)$ and $\varphi_2(x)\psi_2(y)$ provide the same supremum λ_1 on the boundary of B^2,

$$\begin{aligned} \lambda_1 &= \sup_{\varphi\psi\in B^2} \int_\Omega F(x,y)\varphi(x)\psi(y)dxdy \\ &= \int_\Omega F(x,y)\varphi_k(x)\psi_k(y)dxdy, k = 1, 2. \end{aligned} \quad (26)$$

By Theorem 4, $\varphi_1 \perp \varphi_2$ and $\psi_1 \perp \psi_2$. Let

$$\varphi_3\psi_3 = (t\varphi_1 + (1-t)\varphi_2)(t\psi_1 + (1-t)\psi_2), 0 \leq t \leq 1,$$

It is easy to check,

$$\int_\Omega F(x,y)\varphi_3(x)\psi_3(y)dxdy = (1 - 2t + 2t^2)\lambda_1, ||\varphi_3\psi_3|| = 1 - 2t + 2t^2,$$

thus

$$\int_\Omega F(x,y)\frac{\varphi_3(x)\psi_3(y)}{||\varphi_3\psi_3||}dxdy = \lambda_1, \forall t \in [0,1].$$

This means the functional reaches λ_1 in B^2 at any point of segment joining φ_1, φ_2 and ψ_1, ψ_2, which is situated in the interior of B^2. But the supremum of (26) can be attained only on the boundary of B^2. This contradiction implies λ_1 is not the supremum of the functional. One may notice, this is true if $F(x, y)$ replaced by $F_N(x, y) = F(x, y) - \sum_{\beta=1}^N \lambda_\beta\varphi_\beta\psi_\beta$ in (25). In particular, if $F(x, y) = F(y, x)$, it generates a self-adjoint operator, $\Phi = \Phi^*$, this is a special case of the fact described above.

From the geometrical point of view, it is known that the unit ball B of $L_2(\Omega)$ is strictly and uniformly convex[11,12]. It is believed that the $B \cap B^n$ possesses the same property. The equations (23) creates a supporting tangent hyperplane to B^n in $L_2(\Omega)$. A conjecture arises that the claim made in Theorem 5 would be true for any finite dimensional underlying spaces $\mathbb{R}^n$. But we have had direct proof only for n even and $n \leq 6$. So the general question remains still open.

It is worth to re-emphasize, Theorem 3 and Theorem 4 have shown that for any high-dimensional square-integrable function $F(x)$ there exists an optimal orthonormal system of its own, consisting of eigenfunctions of its gradient operator, in terms of which $F(x)$ can be expanded with shortest length and rapidest convergence. Since each element of the system is a product of n single-variable functions, this may be a reliable way for reduction of dimensionality and compact expression of information contained in $F(x)$ in one-dimensional spaces. The inequality (20) provides *a posteriori* accumulated error estimate. In the process of computing the remaining error can be precisely estimated after completion of each step of calculation, this is thus a difference from *a priori* error estimate.

We recall that $L_2(\Omega)$ and l_2, the space of square-summable sequences of reals, are isometrically isomorphic. Each element of $L_2(\Omega)$ has its spectral image in l_2 according to bases chosen in each spaces. If one identifies the square of norm of $F(x) \in L_2(\Omega)$ with the energy or information it carries, in terminology of physics, the outcome of Theorems presented in this paper is to assert that for any given F there exists an optimal basis in $L_2(\Omega)$ which furnishes the element with an image sharply concentrated on a few of spectrum-lines in l_2, if the latter is equipped with canonical basis. This may be in conspicuous contrast with a flat spread of spectral lines with respect to a casually chosen basis for spectral analysis as it happens in many cases of practices (see Figure1).

The problems we investigated in this paper are related to a topic posed and studied by Liapunov A. M. at the beginning of 20th century, he called it power series integral equations and imposed severe restriction on the given function. He required $F(x_1, x_2, ..., x_n)$ to be totally symmetric, that means the exchange of any two among n variables retains F unchanged[5,22]. The general properties of n-multilinear forms have been elucidated in [6, 7, 9]. Krasnoselsky M. A. proved that if F is strictly positive and totally symmetric, $F \in L_2(\Omega), 0 < m \leq F \leq M < \infty$, the

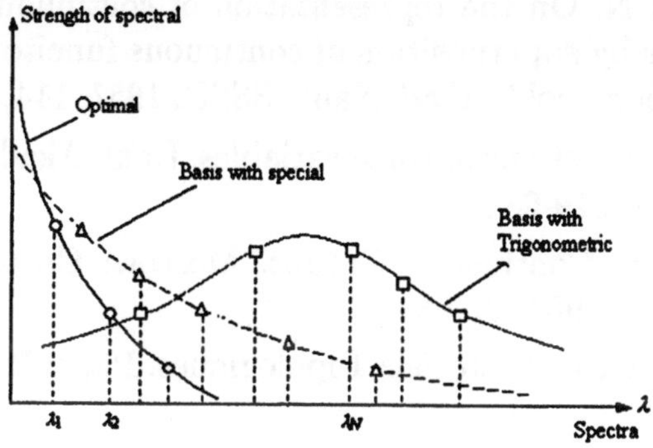

Figure 1. Distributions of spectra in different bases

following integral equation

$$\Phi\left(\prod_{i=1}^{n}\varphi(t_i)\right)$$

$$=\int_{\Omega}F(s,t_1,t_2,...,t_n)\varphi(t_1)\varphi(t_2)\cdots\varphi(t_n)dt_1dt_2\cdots dt_n=\lambda\varphi(s)$$

possesses at least one positive eigenvalue[22] . Wainberg M. M. had shown that for a totally symmetric $F \in L_2(\Omega)$ all components of the gradient operator Φ generated by F are compact, and the functional $\langle\Phi\left(\prod_{i=1}^{n}\varphi(t_i)\right),\varphi(s)\rangle$ is weak continuous respect to φ, it achieves its supremum value on the unit ball[5]. It is obvious, the results we obtained in this paper cover most cases studied by these earlier investigators.

One can observe that the results of this paper may find wide applications in computational mathematics and engineering sciences, particularly in the field of control theory and automation[20,23]. Take a typical example, if a hypersurface in $\mathbb{R}^n, x^n = F(x^1, x^2, ..., x^{n-1})$, is needed to be stored, the amount of data is measured as $N^{n-1} + N, N$ is the mean number of discrete samplings for each variable. If l terms are taken in (16) to represent F, the amount of data to be stored or processed will be reduced to nlN, a $1/N^{n-2}$ times less than previously needed. Engineering practice had shown, sometimes to take two to three terms of (16) would be precise enough to represent a given higher dimensional function by the sum of products of one-variable functions[23]. In this context, the proposed theory and the orthonormal system may serves as an optimal basis for a given function to be represented.

References

[1] Kolmogorov A N. On the representation of continuous functions of many variables by superposition of continuous funcitons of one variable and addition. Dokl. Akad. Nauk. SSSR, 1957, 114, No.5, 953-956

[2] Arnold V I. On functions of three variables. Dokl. Akad. Nauk. SSSR, 1957, 114, No.4, 679-681

[3] Pontryagin L S. Continuous Groups. Moscow: Gosudarstv. Izdat. Tehn.-Teor. Lit., 1954. 92-93

[4] Bourbaki N. Espaces Vectoriels Topologiques. Paris: Hermann & Cie, 1955. 307-311

[5] Wainberg M M. Some problems of differential calculus in linear spaces. YMN, 1952, 4(50), 88-102

[6] Ljusternik L A, Sobolev V I. Elements of Functional Analysis. Moscow: Nauka Press, 1965, 450-461, 489-492

[7] Lang S. Real and Functional Analysis. New York: Springer-Verlag New York Inc., 1990. 70-72

[8] Zeidler E. Nonlinear Functional Analysis and Its Applications. New York: Springer-Verlag New York Inc., 1990. II/A, 255-262; III, 603-605

[9] Guan Zhao-Zhi. Lectures on Functional Analysis. Beijing: High Education Press, 1958. 143-149, 274-293

[10] Kato T. Perturbation Theory for Linear Operators. New York: Springer-Verlag New York Inc., 1966. 275-276

[11] Beauzamy B. Introduction to Banach Spaces and their Geometry. Amsterdam-New York: North-Holland Publishing Co., 1982. 175-204

[12] Diestel J. Geometry of Banach Spaces-Selected Topics. Berlin-New York: Springer-Verlag, 1975. 29-94

[13] Yosida K. Functional Analysis. Berlin: Springer-Verlag, 1965. 86-88, 119-125

[14] Fihtengolts G M. Lectures on Differential and Integral Calculus, II. Moscow: Gosudarstv. Izdat. Tehn.-Teor. Lit., 1959. 292-294

[15] Sobolev S L. Equations of Mathematical Physics. 1954

[16] Gohberg I T, Krein M G. Introduction to Theory of Non-self- adjoint Linear Operators. Moscow: Nauka Press, 1965. 153-160

[17] Fan K. Maximum properties and inequalities for the eigenvalues of completely continuous operators. USA: Proc. Nat. Acad. Sci., 1951, 760-766

[18] Weyl H. Inequalities between the two kinds of eigenvalues of a linear transformation. USA: Proc. Nat. Acad. Sci., 1949, (35): 408-411

[19] Stinespring W. A sufficient condition for an integral operator to have a trace. J. Reine Angew. Math., 1958, No.3-4, 200-207

[20] Butkovski A G, Song J. On Construction of Functional Generator with Multiple Variables. Energetics and Automation, 1961 (2): 153-160

[21] Kostiutsenko A G, Krein S G, Sobolev V I. Linear Operators in Hilbert Spaces. In Book: Functional Analysis, NAYKA, 1964

[22] Krasnoselski M A. Topological Methods in Theory of Non- Linear Equations. Moscow: Gosudarstv. Izdat. Tehn.-Teor. Lit, 1956. 57-58

[23] Song J. Design of Time-Optimal Controller. Automatica and Telemechanics. 1958, 20(3): 273-228

[24] Song J. Optimal representation of multivariate functions or data in visualizable low-dimensional spaces, to appear in Chinese Science Bulletin, Vol 46, No. 16, 2001.

Chapter 2

MODELING OF COMPLEX SYSTEMS

K. J. Åström
Department of Automatic Control
Lund Institute of Technology, Lund University
Lund, Sweden

This paper is dedicated to Prof. Yu-Chi Ho to celebrate his outstanding contribution and to thank him for many inspiring discussions and a long standing friendship.

Abstract **The role of modeling is becoming increasingly important in design and operation of complex natural and man-made systems. Models are also appearing as components of control systems because of the increased use of model based control strategies such as Kalman filters and model predictive control. Mechanized tools are necessary to fully exploit modeling. The development of modeling is therefore closely associated with the development of computational tools for simulation. There has been a very dynamic development from the beginning in the 1920s. At that time, the technology was available only at a handful of university groups who had access to mechanical differential analyzers. Today modeling and simulation are available at low cost on the desk of everyone who needs it. This paper presents a perspective on modeling of complex systems leading up to the recent development of Modelica, which draws on object oriented methodology in computer science, differential algebraic equations in numerical mathematics and control theory.**

1. Introduction

Mathematical models of natural and man-made systems are extremely useful. They permit a very compact summary of knowledge. Newton's method is a prime example where the seemingly complicated motion of the planets could be captured by a simple equation which also could be used for prediction. The key idea formed the basis of reductionism which was the prevailing paradigm for research in natural science for centuries. Models have always played a key role in engineering. A large part of engineering education can be viewed as an education in modeling of different domains. Models are very effective for communication and education. Models are easier and safer to work with than the real systems. Models are indispensable when dealing with the emerging complex systems that confront control engineers. They are used extensively in design, they can assist in commissioning and operation of man-made systems. Models are a key for concurrent engineering. Many control systems have a process model built into the controller.

Modeling by itself is not sufficient for solving an engineering problem. It is also necessary to generate insight, numbers and curves from the model. Thus the use of models is strongly tied to the development of tools for working with models. This was expressed in the following way in [18]

> "Engineering can proceed no faster than the mathematical analysis on which it is based. Formal mathematics is frequently inadequate for numerous problems pressing for solution, and in the absence of radically new mathematics, a mechanical solution offers the most promising and powerful attack wherever a solution in graphical form is adequate for the purpose. This is usually the case in engineering problems."

This quote, which is still highly relevant, motivated Bush to build the first mechanical differential analyzer, see [17].

Modeling is a very rich field. It covers large parts of natural science and engineering. Statistics and design of experiments is an essential ingredient to determine parameters of modeling. Numerical mathematics is important for simulation of a model. Computer algebra also plays a role together with numerics for model transformation. Tools from computer science and software engineering are necessary to deal with the complexity of large systems.

The development and use of models have been very closely tied to tools to mechanize modeling, analysis and optimization. It has been revolutionized by the use of computers. Looking back at the development it is interesting to see that many of the old paradigms that were strongly influenced by the constraints of early mechanization were maintained when new tools were developed. It can thus be concluded that software for modeling has developed much slower than ideas and concepts. It is common that even "modern" software is strongly constrained by old ideas.

The paper is organized as follows. Different ways to deal with complexity are discussed in Section 2. Some modeling paradigms are reviewed in Section 3, with particular emphasis to the paradigms used for in modeling for control. Section 4 describes a paradigm shift in physical modeling and emerging modeling languages and software environments. An example of physical modeling that illustrates the needs for model reduction required in control is presented in Section 5.

2. How to Deal with Complexity?

There are many requirements on modeling of complex systems, see [6]. An essential feature is to be able both to have an overview and to look at all the details. A modeling task thus covers domains such as the real world, conceptual models, graphical and mathematical representations. It is not easy to cover these views for a complex system with many heterogeneous components. A number of different techniques have been developed over the years with inspiration from many different fields. Lately much of the inspiration have come from computer science, where it has been easy to experiment with many different representations, see [1]. Some techniques to deal with complexity are: abstraction and information hiding, layering, object orientation, multiple views, graphical and mathematical representations

2.1 Information Hiding

The idea of information hiding or encapsulation is a key idea in software engineering. It has, however, been used for a long time in many fields. The notion of block diagrams in control is an excellent example of information hiding as is illustrated in Figure 2.1. The block diagram hides all the technical details of a subsystem and focuses on the information transmission. The similarity of widely different control systems such as process control, autopilots and feedback amplifiers became clear because their block diagrams were very similar. This observation was one of the elements that contributed to the emergence of the discipline of automatic control in the 1950s. Very powerful tools for design, analysis and simulation were developed when the block diagrams were complemented with descriptions of the blocks in terms of transfer functions.

Figure 2.1, which is taken from one of the classic text of control, [50], illustrates the development of a block diagram. The details of widely different technical devices are hidden behind the pieces of cloth shown in the figure. The real block diagram is then obtained simply by showing the subsystems as square boxes and the interactions as lines with direction.

2.2 Layering

Layering is another principle that is useful when dealing with complex systems. The key idea is to divide a complex system into layers with

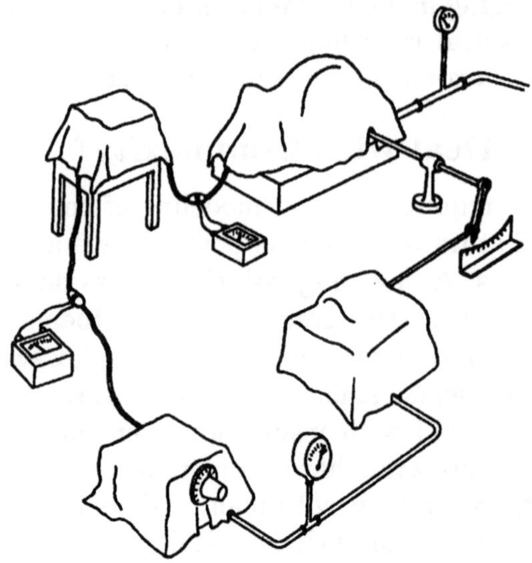

Figure 2.1. Block diagram as an illustration of information hiding. From [50].

well defined interfaces. Each layer can then be designed, and modified independently of the others as long as the interfaces remain the same. Layering has had a profound impact on software engineering and communication but it is also used in many other areas. In control we have for a long time been used to three types of standard models: functions, state equations and transfer functions. For example the blocks in a block diagram are typically represented either by functions, for static systems, and transfer functions or state equations for dynamic systems.

Another example of the use of layering is VLSI design. A significant break through occurred when it was discovered that circuits could be represented by simple graphics, where physics was hidden in a number of simple design rules that are easily represented graphically, see [46]. The system was augmented by powerful tools, which makes it possible to simulate the circuits. Standard representations of the final result of the design provided an effective layering between design and manufacturing which revolutionized VLSI design and made it available to universities and small groups world wide.

2.3 Object Orientation

Object orientation is a paradigm from computer science that gives a natural way to model real world, systems and abstract concepts. The basic idea is the notion of object which combines data structures and behavior in a single entity. It typically combines the ideas of identity, classification, polymorphism and inheritance, see [60], [34], [13]. Identity means that data is quantized into discrete, distinguishable entities called objects. Objects with the same data structure and behavior are grouped into a class. Each object is an instance of a class. Polymorphism means that the same operation may behave differently on different objects. Inheritance is a mechanism for sharing attributes and operations among classes based on a hierarchical relationship.

2.4 Multiple Views

When working with complex systems it is necessary to provide facilities for many different actors from many different backgrounds to see a system from many different view points. Examples of actors in a complex control system are designers, process engineers, operators, and managers. It is highly inefficient to develop models and systems for each category. A much better approach is to provide a base system and with user interfaces and browsers that permits many different view points. These ideas were developed by rzn in a project on knowledge based process control, see [5]. It would be highly desirable to have similar facilities in modeling systems where it is useful to provide interfaces for different views of model building and analysis, optimization, simulation, animation, and simulation with hardware in the loop. It is also necessary to keep in mind that there are many different users of models such as: model developers, experienced users and casual users.

Browsers that let a user navigate in the different views are highly desirable. Several browsers are now components of standard software. The Oberon system is a particularly neat example, see [57]. Anther interesting concept for a browser is the idea of information zooming developed in [25]. Very much work is, however, required to permit efficient use of multiple views and browsers for exploring models and results of simulation and optimization.

2.5 Software

Software tools for modeling, simulation, analysis, optimization, and control design are very important because it is impossible to work with complex systems without software tools. Unfortunately software for modeling is largely in the same poor state as other software. The practice of not publishing source code is a factor which strongly contributes to the poor quality. A good analogy is to ask the question: What would mathematics look like if theorems were published together with a man-

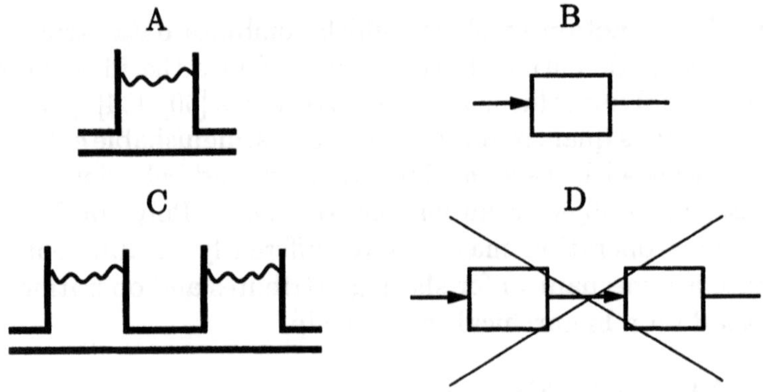

Figure 2.2. A simple hydraulic system with an inflow and a free outflow is shown in A. The block diagram representation of the system is shown in B. The system obtained by connecting two hydraulic systems is shown in C. This system cannot be represented by the series connection of the block diagrams in B.

ual on how to use them and proofs were hidden as company secrets? When modeling physical systems it is difficult to spell out all assumptions explicitly. The only way to find details about a model is to look at the source code of the system models. Even if all source code is not available I strongly believe that modeling software should be developed in such a way that all aspects of the model are transparent. Because of the complexity of modeling it seems unlikely that a single software package can serve all purposes. Hopefully software will develop so that different packages can easily be interconnected.

3. Modeling Paradigms

The modeling tools used today have grown out of specific modeling paradigms. They have often developed organically in an ad hoc fashion, see [8]. Only in very rare cases have there been efforts to step back and take a serious look at the fundamentals before indulging in extensive software development. In this section we will give a brief overview of some important modeling paradigms.

Block Diagrams

Block diagram modeling is the predominant paradigm in automatic control. It is a very powerful idea which contributed significantly to the emergence of the field of control because it made it possible to see the similarity of widely different systems. Block diagrams were originally simulated using analog computers were the basic operations are integration, addition, multiplication with a constant and generation of scalar functions. Such a diagram could be directly mapped to a an analog computer for simulation of the system. Much ingenuity went into developing representations of many different physical phenomena. When digital computer emerged a natural step was to let the digital computer mimic an analog computer. A large number of systems emerged rapidly. A major step forward was an initiative by the Simulation Council which led to a standard language for systems modeling called CSSL, see [64]. A particular implementation ACSL ([47]) became a de facto standard for computer modeling and simulation of control systems, see [58]. The limitation of representing all systems in terms of the basic analog computing elements still remained. The software Simnon, [22], which introduced continuous and discrete state space models as components, represented a major advance. Simnon also deviated from previous software by being much more interactive.

The block diagram environments have developed substantially and become very sophisticated. The ultimate block diagram environments used today are Simulink and VisSim, see [37]. The simulation environments are user friendly. Blocks can be picked up from a library and combined graphically. Aggregation and disaggregation are also made graphically. In Matlab there are also many tool-boxes for analysis of results, optimization and control design. It is interesting to observe that much of the analog computer technology is clearly noticeable in the modern tools.

A fundamental assumption of block diagrams is that the interaction is unidirectional from input to output. This means that it is not possible to represent the combination of the simple hydraulic systems in Figure 2.2 by combining the corresponding block diagrams. The reason is that the flow between the tanks are influenced by both tank levels. This fact has been known from the onset of using block diagrams. It is a severe restriction of block diagrams because it means that it is not possible to build libraries of components that can be connected in an arbitrary fashion.

A block is a causal description in the sense that the input is causing the output. When building model libraries it is highly desirable to have descriptions of subsystems that are not causal and let the causality be determined by the interconnections. A simple example is a resistor. The input is the voltage if the resistor is connected to a voltage source and a current if it is connected to a current source.

Another more subtle difficulty is that block diagram modeling inherently assumes that each block is represented as a state model. It may happen that states will vanish when two physical systems are combined. Two typical cases are illustrated in Figure 2.3. The first case shows a

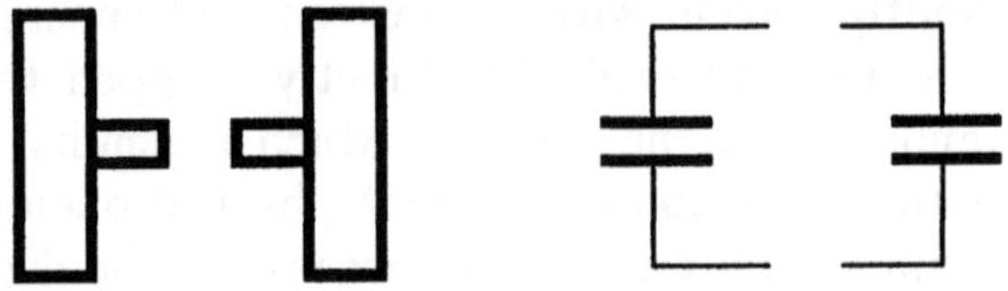

Figure 2.3. Illustrates two situations where states disappear when two systems are connected.

system consisting of two rotating masses, which has four states representing angles and angular velocities of the masses. When the masses are connected rigidly a system with only two states is obtained. The second system consists of two capacitors It has two states representing the voltages across the capacitors. When the capacitors are connected the voltages are the same and the system has only one state. States can thus disappear when systems are connected. These situations were encountered already in the time of analog computing were referred to as "algebraic loop problems" because the system obtained after the connection had a loop that only contained algebraic equations. Algebraic loops are also obtained in any situation when systems with a direct static coupling are connected in a closed loop. A typical example is a flow system where pressure dynamics is neglected.

Algebraic loops in analog computing were typically manifested as oscillations in the audio spectrum. The problem was remedied by inserting a small capacitor in the loop, representing a small capacitance in a closed loop consisting of resistors or an elastic connection of two masses. The resulting system then became a stiff differential equation which was of little concern in analog computing. The same trick does not work in digital computing because computing time may increase drastically. The word algebraic loop still survives in the modern block diagram languages. The only way around it when using block oriented modeling is to derive the model for the coupled system analytically. This is error prone and prohibits the use of model libraries.

One reason why block diagram modeling is still popular is that a common case where there is unidirectional interaction computer control, where the computer and the complete process are blocks. The difficulty of unidirectional coupling does however arise as soon as the process model is built up of separate blocks. We can thus conclude that block diagrams are not a suitable paradigm for serious physical modeling.

Circuit Theory

The difficulties caused by interconnection of subsystems, illustrated in Figure 2.2, has been dealt with very effectively in circuit theory, see [19]. A key observation is to introduce terminals instead of inputs and outputs. Pairs of terminals are called a port and associated with two variables: voltage and current. A model describes how the voltage and current at one port are related to the voltage and current at another port. In this way it is no difficulty in interconnecting the electrical system in Figure 2.2.

There are also two fundamental laws, Kirchoff's Voltage Law (KVL) and Kirchoff's Current Law (KCL) which governs the behavior of voltages and currents in interconnections. They express that all currents in a terminal sum to zero (KCL) and that the voltage drop around any closed circuit is zero (KVL). Effective model reduction methods such as Thevenin's and Norton's theorems were also developed. Particular attention was given to two-port systems which were classified as gyrators, transformers, amplifiers, transducers.

It was recognized at an early stage that the ideas from circuit theory could also be carried out to mechanical systems using analogies. There were also several attempts to create more general concepts, currents were thus generalized to flow variables and voltages to effort variables or across variables, see [63]. In this way it was realized that mass, spring and damper had electrical, fluid and thermal analogs.

Circuit theory has relatively few types of components. The components are quite simple (with the exception of complicated transistor models). A system may, however, have a very large number of components. It was thus necessary to develop techniques to deal with this. Efficient methods to do this were developed based on graph theory and matrix techniques. Another special feature is that in circuit theory the energy balance is the essential balance equation.

A very efficient simulation tool for electrical circuits called Spice was developed by Donald Pederson at Berkeley, see [48]. This program which was in the public domain had a major impact on modeling and simulation of circuits. Although Spice is a very effective way of dealing with electrical circuits it is complicated to model thermal phenomena using it.

Bond Graphs

The bond graphs, introduced in [54], can be considered as an abstraction of circuit theory. Bond graphs describe systems in terms of power flow. A bond is a an arrow that is associated with two variables called effort and flow. The effort, which is also called an across variable, is a generalization of voltage and the flow, which is also called a through variable, is a generalization of current. Bonds connect components and junctions. There are two types of junctions corresponding to Kirchoff's voltage and current laws. Bond graphs are particularly effective to model systems that can be modeled by one balance equations and it has had a large following, see [38], [59], [14], [29] and [16].

Process Control

Much of the modeling effort in process control started with static models representing the balance equations. Static modeling was later extended to cover dynamics see [62], [55], [52], [40], [44], [56], [10] and [49]. The unit processes, such as heat exchangers, reactors and columns, that are standard elements of process design gave a natural structure. Instrumentation systems are documented using so called P&I (Process and Instrumentation) diagrams. There is also an industrial standard IEC 1131 for control system, see [42]. These standards and practices form a natural base for modeling process systems. In process control it is also natural to combine modeling with optimization.

4. A Paradigm Shift

There is a wide variety of modeling tools for different specific physical domains e.g. Spice for electrical circuits, ADAMS for multi-body mechanical systems, and Aspen Plus and g-PROMS for chemical processes. These systems are very useful in their domains but it is difficult, impossible or cumbersome to combine them to model systems with components in different physical domains. The need for this type of modeling is becoming increasingly important because of the strong interest in areas such as mechatronics, micro-mechanical system and control. Before going into details we will discuss some basic requirements on modeling.

Requirements

Many different types of models are needed, from simple models for preliminary assessment and as components of the control system design to very detailed models validation. There are also many different types of users that have different requirements.

Modeling is a very time consuming task. To be efficient it is highly desirable that results can be reused. To do this it is necessary that submodels can be organized into libraries and that the descriptions of the models are transparent. It would be highly desirable to have representations of models that do not strongly depend on software of a particular vendor. Standardized descriptions of models are also useful to protect the substantial investment that goes into modeling.

4.1 Model Simplification

Model simplification or model reduction is an essential ingredient of modeling. Powerful model reduction techniques have been developed in control theory, see [31]. The methods are particularly useful because they are closely coupled to control design techniques, see [69]. They are, however, limited to linear systems.

System identification methods, combined with experiments, can also be used very effectively to give indication of model complexity, see [43]. When using system identification it is however important that the input signals match the intended use of the model.

Model reduction can also be made from a physical basis. The key question is to determine the resolution required to describe storage of mass, energy and momentum. A related question is to determine if balances should be described statically or dynamically. Distributed parameter models are extreme cases that permit a very high resolution of storage. Model simplification is closely related to the descriptions of uncertainty. For the purposes of control it is essential to have some way to express the uncertainty of a model. There is clearly a trade off between model complexity and model uncertainty.

Unfortunately there are no general methods for the key problems stated above for general systems. A user therefore has to resort to experimentation by making different simplifications and comparing the results. It is essential that a modeling environment can support such a procedure. Illustrations of model reduction are given in Section 5.

A Natural Approach

A simple and general approach to modeling that is commonly used in engineering is to split a system into subsystems, define interfaces, write the balance equations for the subsystems, add constitutive equations. This approach gives a description that is general and close to physics and convenient for building libraries. Splitting a system into subsystems was called tearing in [39].

A drawback with the method is that much manual work is required to assemble the subsystems into a model which is suitable for simulation or optimization. The work can however be automated. It is useful to perform consistency checks and to have assistance for model reduction.

Behavior of Subsystems

Having split a system into subsystems we need to describe the behavior of the subsystems. The subsystems can be modeled as lumped or distributed parameter systems. Here we will focus on lumped parameter systems. The paradigms of state models block diagrams are not well suited as was discussed in Section 3. Instead of representing a subsystem by a state model

$$\frac{dx}{dt} = f(x, u, t)$$
$$y = g(x, u) \tag{1}$$

it will be represented by a system of differential and algebraic equations (DAE),

$$F(y, \dot{y}, t) = 0 \tag{2}$$

where y is a vector and $\dot{y}$ the derivative of y. This may seem like a trivial extension but it has major implications. The problem illustrated in Figure 2.3 where state variables disappear can be described naturally using DAE. Connecting two inertias simply adds the constraint that the angles of the inertias are the same. It is also very easy to switch from dynamic to static variables simply by adding an equation that tells that a derivative is zero.

The state model (1) is a causal description because it has well defined inputs and outputs. The DAE model in (2) is not causal because variables are not assigned as inputs or outputs. This may appear like a contradiction because in the end all physical systems will be causal. Causality may, however, determined by the connections. For example a resistor that is connected to a voltage source will have voltage as input and current as output but the roles of input and output are reversed if it is connected to a current source.

The fact that differential algebraic equation is a natural representation for physical modeling has been observed in many fields. Early work on simulation of electrical circuits motivated the development of solvers for DAEs, see [30]. Differential algebraic equations was a key component in the development of the modeling language Dymola, see [23]. They were used in modeling of chemical processes in [53] and in power systems in [32] just to mention a few examples.

A state space model (1) can be regarded as a function for calculating the derivative of the state from the state and the input. A differential algebraic equation (2) can instead be viewed as a constraint between variables and their derivatives and it is therefore also called a constraint model. Many other names such as equation models, descriptor systems and behavior models have also been used.

In many cases it is also necessary to consider hybrid systems that combine continuous time behavior with switching. This has been investigated in [24], [3] and [10]. A nice side effect is that the hybrid features are also useful when describing tables for material properties.

Organizing the Work

A general straight forward approach to physical modeling can be described as follows. Decompose a system into subsystems. Describe the behavior of each elementary subsystem by the differential algebraic equation which captures storage of mass, momentum and energy. Add equations or tables that describe the material properties. Compose non-elementary subsystems and the total system by instantiating components with their physical parameters and connect them.

Even if the approach is simple in principle it may be complex because of the complexity of the subsystems and the large number of parameters involved. It is therefore necessary to organize the work so that it can be done systematically and so that the results can be reused. The approach can be used directly for a particular application. To make modeling cost effective it is highly desirable to add further organization so that the modeling can be reused. Language design and object oriented methods are natural tools to do this. For that reason this approach to modeling is also called object oriented modeling. Object orientation is also very useful for building model libraries. Combining libraries with graphical user interfaces gives very convenient environment for creating and manipulating models graphically. The inheritance mechanims are also very convenient to modify a model library. The effort required to model a system is reduced drastically when libraries are available because a model can then simply be assembled from components where components can be modified if they do not fit a particular case.

For complex systems there will be a large number of variables and equations. Many equations are trivial identities or other simple expressions. To obtain efficient code for analysis and simulation it is necessary to use extensive formula manipulation. Computer algebra is therefore also an important ingredient.

4.2 Software

The approach to physical modeling that has been outlined has many attractive properties. Because of the wide range of ingredients and the complexity of the task it is clear that good software is a prerequisite for applying it. it is interesting that such software is now appearing. Spice, [48], was a large Fortran system. The first effort to make a system for object oriented physical modeling was done in a PhD thesis, [23]. The system called Dymola was based on a formal definition of a modeling language. When this thesis was written the only object oriented software was Simula, see [12]. A complete translator was implemented and tested on a model of a power station, clearly showing the feasibility of the approach. More recent examples of systems for object oriented modeling are Omola, see [2], [45] and [3], gPROMS [10] and [49], NMF [61], ObjectMath [28], [68] U.L.M.-Allan [35], SIDOPS+ [15], Smile [36] ASCEND [56], DIVA [40], VHDL-AMS [9], [33].

4.3 Modelica

It is a major effort to develop software tools for physical modeling. To do this efficiently is is useful to develop suitable languages and model libraries. A formally defined modeling language gives a good layering and it also provides stability which is useful to protect the heavy investment in modeling. A good language should be parsimonious and have a strong expressive power. It should also lend itself to efficient implementation. Development of a suitable modeling language is by necessity an iterative procedure. Different language constructs must be tried on a very wide range of applications.

A collaborative effort to develop a language for physical modeling was initiated by Elmqvist in 1996. The idea was to obtain a non-proprietary modeling language for broad engineering domains. The language called Modelica was developed by a team of research groups and industries organized in the Modelica Association, see [26] and [65]. The contributors included developers of earlier languages such as Dymola, Omola, NMF, ObjectMath, Allan, SIDOPS+ and Smile. The team had over 30 participants representing more than 300 man-years of experience of modeling and language development.

The name Modelica is the trademark of the Modelica Association a non-profit organization. Modelica is not yet a formal standard. There are formal definitions of the language and its syntax. Commercial tools like Dymola from Dynasim (www.Dynasim.se) are able to translate the high-level, declarative Modelica to models to efficient numerical algorithms. The language is used extensively in the automotive and other industries, see [27]. Information about the language and the design rational is available on the web `http:///www.modelica.org`.

The Modelica language has many useful features such as units of variables, matrices and matrix equations, functions, hybrid modeling features and class parameters. A significant effort has also been devoted to development of model libraries. There are currently libraries for stan-

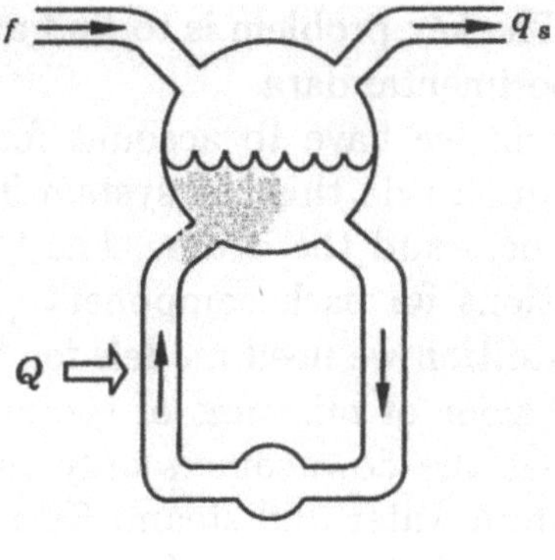

Figure 2.4. Schematic picture of steam generator.

dard control systems, multi-body systems, [4], electrical circuits, [20], power trains, [51], power systems, [41], hydraulic systems, [11], thermo-hydraulic systems [66], [21] and [67].

5. An Example

We will illustrate some of the ideas by modeling a steam generator, see [7]. The purpose of the modeling exercise is to develop nonlinear low-order models that can be used to design control strategies.

A schematic picture of a steam generator with a drum is shown in Figure 2.4. The heat, Q, supplied to the risers causes boiling. Feedwater, q_f, is supplied to the drum and steam, q_s, is taken from the drum to the super-heaters and the turbine. Gravity forces causes steam to rise and they generate a circulation in the risers-drum-downcomers loop. The presence of steam in the system causes the shrink-and-swell phenomenon which makes it difficult to control the water level in the drum. Consider, for example, the situation when the outlet steam is increased. The pressure in the drum then decreases, this causes steam bubbles to expand and the drum level increases initially. After some time the reduced amount of water causes the level to decrease. The net effect is an example of the inverse response effect caused by unstable zero dynamics.

A model for control design that also can be incorporated in a model based control system should preferably be simple. The complexity of a model is strongly influenced by how accurately the storage of mass, energy and momentum is represented, for example how many segments are needed to represent the risers and the drum. An extensive study of a steam generator is presented in [7]. System identification indicated that the essential dynamics can be described with a model of relatively low order, typically 4 to 6. This implies that a crude characterization of storage is sufficient. The key problem is to find a simple physical model that matches the experimental data.

To model the system we have to account for the storage of mass, energy and momentum. To do this the system is split into the subsystems: risers, downcomers and the drum. The basic balance equations then give three equations for each component (water and steam) and each subsystem. In addition we need models for the material properties that describe the behavior of mixtures of steam and water. It is reasonable to assume that the downcomers only contain water while the other subsystems contain water and steam. To make a first assessment of the model we assume that storage of mass, momentum and energy can be described by one variable in each subsystem. Counting the state variables we find that the system is described by $3 \times 2 \times 2 + 3 = 15$ balance equations. This number is significantly larger than indicated by the system identification.

An investigation of orders of magnitudes indicate that the momentum balance can be represented by a static equation because the dynamics is fast. The number of state variables are then reduced to 12, which is still too high. Detailed investigations of different types of models also indicate that for the particular modeling purpose the slip can also be neglected.

To proceed further we will consider a saturated mixture of water and steam in a given volume. Storage of mass and energy can be described by four variables e.g. mass and energy of steam and water. There is a very effective heat transfer in the system caused by evaporation and condensation, which introduces constraints between the variables. Let m and e be mass and energy and use the subscripts w for water and s for steam. The following equations are then obtained

$$\begin{aligned} m_s &= \alpha_v V \rho_s, \\ m_w &= (1-\alpha_v) V \rho_w, \\ e_s &= \alpha_v V \rho_s u_s = \alpha_v V(\rho_s h_s - pV), \\ e_w &= \alpha_v V \rho_w u_w = \alpha_v V(\rho_w h_w - pV), \end{aligned} \tag{3}$$

where u is internal energy, h enthalpy, ρ density, and p pressure. The variable α_v denotes the fraction of the volume that is occupied by the steam. It follows from these equations that the variables m_s and m_w are related, so are e_w and e_s. We can thus conclude that it is possible to characterize storage of mass and energy by two variables instead of four. There will be further reductions in the number of state variables when several subsystems that contain saturated steam are connected. Two variables are required for one system but only one additional variable is required for each additional system. No extra variable is required for volumes that contain only steam or water.

In the steam generator it is necessary to describe the distribution of steam in the risers. A detailed study is given in [7] where it is shown that a good approximation is to assume that the mass fraction of steam increases linearly along the risers. If the water entering the risers is saturated the steam distribution can be described by one variable. If the water entering the risers is sub-cooled an additional variable, that represents the position along the tube where boiling starts, is required. Here we will simply assume that the mixture entering the riser tubes is saturated.

Returning to the boiler consisting of the subsystems: drum, downcomers and risers we find that storage of mass momentum and energy can be described by four variables. No extra variables are required to describe storage in the downcomers and in the water above the level in the drum.

There are many possible choices of variables, for example m_w and p. It is convenient to use total balances for the mixture of steam and water. The metal masses that are in close thermal contact with the steam can also be included. When writing the equations it is convenient to choose the subsystems as follows: the total system, the risers and the drum. The balance equations can then be written as.

$$
\begin{aligned}
\frac{d}{dt}(m_{st} + m_{wt}) &= q_f - q_s, \\
\frac{d}{dt}(m_{st}u_{st} + m_{wt}u_{wt} + m_t C_p T_m) &= Q + q_f h_f - q_s h_s \\
\frac{d}{dt}(m_{sr} + m_{wr}) &= q_{dc} - q_r, \\
\frac{d}{dt}(m_{sr}u_{sr} + m_{wr}u_{sr} + m_{mr} C_p T_m) &= Q + q_{dc} h_w - (\alpha_r h_c + h_w) q_r, \qquad (4) \\
\frac{d}{dt}(m_{sd}) &= \alpha_r q_r - q_{sd} - q_{cd}, \\
\frac{d}{dt}(m_{sd}u_{sd}) &= \alpha_r q_r h_s - q_{sd} h_s - q_{cd} h_s \\
\frac{1}{2} k q_{dc}^2 &= \varrho_w A_{dc}(\varrho_w - \varrho_s) g \bar{\alpha}_v V_r.
\end{aligned}
$$

In this equations m denotes mass, V volume, m metal mass, ϱ specific density, u specific internal energy, h specific enthalpy, C_p specific heat, k friction coefficient, T temperature and q mass flow rate. The subscripts s, w, f and m refer to steam, water, feedwater, and metal respectively.

When double subscripts appear the second subscript t denotes total system, d drum and r risers.

The first, third and fifth equations are mass balances for the total system, the risers and the steam below the surface in the drum. The second, fourth and sixth equations are the corresponding energy balances. The last equation is the momentum balance for the risers-drum-downcomers loop.

Equation (4) looks formally like a six order system. Because of the assumption on the pressure of the system discussed above there are algebraic constraints which implies that Equation (4) is actually a differential algebraic equation with four state variables. There are many possible choices of state variables one possibility is: total water volume V_{wt}, pressure p, steam quality at the riser outlet α_r, and volume of steam under the liquid level in the drum V_{sd}. Straightforward but elaborate calculations show that equations can be written as

$$
\begin{aligned}
e_{11}\frac{dV_{wt}}{dt} + e_{12}\frac{dp}{dt} &= q_f - q_s \\
e_{21}\frac{dV_{wt}}{dt} + e_{22}\frac{dp}{dt} &= Q + q_f h_f - q_s h_s \\
e_{32}\frac{dp}{dt} + e_{33}\frac{d\alpha_r}{dt} &= Q - \alpha_r h_c q_{dc} \\
e_{42}\frac{dp}{dt} + e_{43}\frac{d\alpha_r}{dt} + e_{44}\frac{dV_{sd}}{dt} &= \frac{\rho_s}{T_d}(V_{sd}^0 - V_{sd}) + \frac{h_f - h_w}{h_c}q_f,
\end{aligned}
\tag{5}
$$

where $h_c = h_s - h_w$ and the coefficients e_{ij} are functions of h_s, h_w, ρ_s, ρ_w, $\partial\rho_s/\partial p$, $\partial\rho_w/\partial p$, $\partial h_s/\partial p$, $\partial h_w/\partial p$, t_s, and $\partial t_s/\partial p$ at the saturation pressure p which are obtained from steam tables. The details are given in [7].

Equation (5) has an interesting lower triangular structure where the state variables can be grouped as $(((V_{wt}, p), \alpha_r), V_{sd})$. The variables inside each parenthesis can be computed independently. The model can thus be regarded as a nesting of a second, a third, and a fourth order model. The second order model describes the dynamics drum pressure p and total volume of water in the system V_{wt}. The equations are global mass and energy balances. There is a very weak coupling between mass and energy balances caused by the condensation flow. The third order model also captures the steam dynamics in the risers and the fourth order model also describes the dynamics of steam below the water surface in the drum. The third equation is a combination of mass and energy balances for the risers, and the fourth equation is a mass balance for steam under the water level in the drum.

Linearizing Equation (5) shows that the system has a double pole at the origin and poles at $-h_c q_{dc}/e_{33}$ and $-1/T_d$. One pole at the origin is associated with water dynamics and the other with pressure dynamics. The pole associated with pressure dynamics is at the origin because the steam flow is chosen as a control variable. The pole moves into the left half plane when the drum is connected to the turbines. The poles $-h_c q_{dc}/e_{33}$ and $-1/T_d$ are associated with dynamics of steam in the risers and the drum.

An interesting feature of the model is that it requires only nine parameters: drum volume V_d, total volume of risers V_r, total volume of downcomers V_{dc}, drum area A_d at normal operating level, effective total metal mass m_t, friction coefficient in downcomers-risers loop k, residence time T_d of steam in drum, and an empirical parameter β in the equation for steam flow rate out of the drum. Most of these parameters can be obtained from construction data.

The calculations in [7] that lead to (5) can be considered as a manual index reduction of the differential algebraic equation (4). There are many alternatives to the chosen state variables and to the structuring of the system. Because of the extensive calculations required many possible alternatives were not investigated. An interesting feature of a modeling tool like Dymola is that these transformations can now be performed automatically. This simplifies the modeling drastically and also makes it easy to experiment with many different type of model simplifications and many different choices of state variables.

5.1 Comparison with experiments

Figure 2.5 shows responses in drum pressure and drum level for perturbations in steam flow rate at medium load. There is very good agreement between the model and the experiments. Notice in particular that drum level increases initially when steam flow rate is increased which is caused by the shrink and swell phenomena.

Figure 2.6 shows responses in drum pressure and drum level for perturbations in steam flow rate at full load. There is good agreement between the model and the experimental data. A comparison with Figure 2.5 shows that the shrink and swell effect is much smaller at full load. This effect is very important for the control design. It is interesting that the nonlinear model captures this quite well. These figures give the flavor of the comparisons of model with experiments, many more details are given in [7].

The example also illustrates the type of problems that we typically have to deal with in a modeling exercise and it also illustrates that

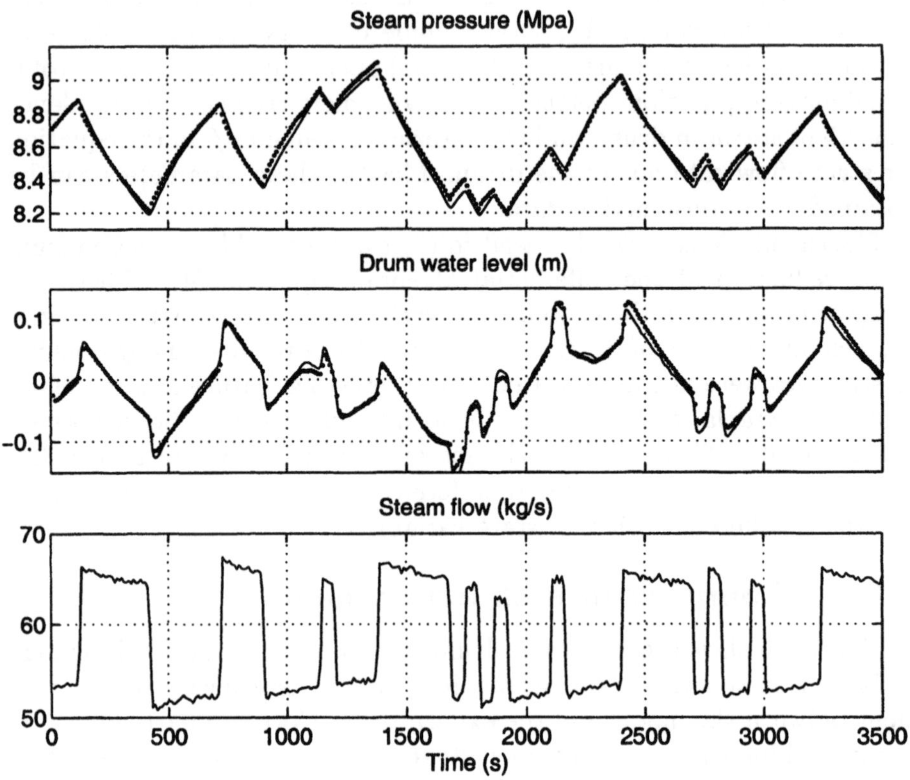

Figure 2.5. Comparison of model (solid line) and plant data (dots) for perturbations in steam flow rate at medium load, from [7].

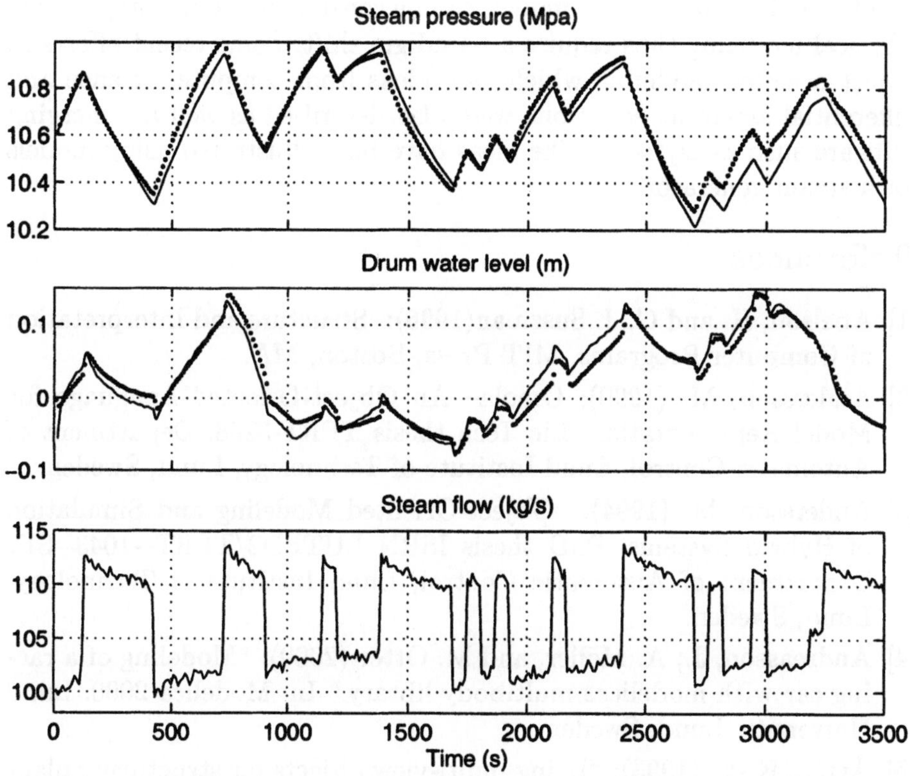

Figure 2.6. Comparison of model (solid line) and plant data (dots) for perturbations in steam flow rate at high load, from [7].

differential algebraic equations are a very convenient way to represent the behavior of dynamical systems.

6. Conclusions

In this paper we have given an overview of physical modeling from the perspective of control. The first observation is that block diagram modeling that is the predominant view of modeling in the control community is not suited for physical modeling because it is not possible to build model libraries where models can be constructed from components. Physical modeling thus require a paradigm shift. Some recent efforts in object oriented modeling which uses ideas from computer science and differential algebraic equations were also described as well as emerging software such as Dymola. The ideas have been illustrated using models for a steam generator.

References

[1] Abelson, H. and G. J. Sussman(1996): Structure and Interpretation of Computer Programs. MIT Press, Boston, MA.

[2] Andersson, M. (1990): Omola—An Object-Oriented Language for Model Representation. Lic Tech thesis TFRT-3208, Department of Automatic Control, Lund Institute of Technology, Lund, Sweden.

[3] Andersson, M. (1994): Object-Oriented Modeling and Simulation of Hybrid Systems. PhD thesis ISRN LUTFD2/TFRT--1043--SE, Department of Automatic Control, Lund Institute of Technology, Lund, Sweden.

[4] Andreasson, J., A. Möller, and M. Otter (2000): "Modeling of a racing car with modelicas multibody library." In Modelica 2000. Lund University, Lund, Sweden.

[5] Årzén, K.-E. (1993): "Using multi-view objects for structuring plant databases." Intelligent Systems Engineering, **2:3**, pp. 183–200.

[6] Åström, K. J., P. Albertos, M. Blanke, A. Isidori, W. Schaufelberger, and R. ẹ. Sanz (2001): Control of Complex Systems. Springer, London.

[7] Åström, K. J. and R. D. Bell (2000): "Drum boiler dynamics." Automatica, **36**, pp. 363–378.

[8] Åström, K. J., H. Elmqvist, and S. E. Mattsson (1998): "Evolution of continuous-time modeling and simulation." In Zobel and Moeller, Eds., Proceedings of the 12th European Simulation Multiconference, ESM'98, pp. 9–18. Society for Computer Simulation International, Manchester, UK.

[9] Barby, J. (1996): "The need for a unified modeling language and VHDL-A." In Proceedings of the 1996 IEEE International Symposium on Computer-Aided Control System Design, pp. 258–263. Dearborn, Mi, USA.

[10] Barton, P. and C. Pantelides (1994): "Modeling of combined discrete/continuous processes." AIChE J., **40**, pp. 966–979.

[11] Beater, P. (2000): "Modeling and digital simulation of hydraulic systems in desing and engineering education using modelica and hylib." In Modelica 2000. Lund University, Lund, Sweden.

[12] Birtwistle, G. M., O. J. Dahl, B. Myhrhaug, and K. Nygaard (1973): SIMULA BEGIN. Auerbach Publishers Inc.

[13] Booch, G. and J. Jacobson, I amd Rumbaugh (1998): The Unified Modeling Language User Guide (The Addison-Wesley Object Technology Series). Addison Wesley, Reading, MA.

[14] Breedveld, P. C. (1985): "Multibond graph elements in physical systems theory." Journal of the Franklin Institute, **319**, pp. 1–36.

[15] Breunese, A. P. and J. F. Broenink (1997): "Modeling mechatronic systems using the SIDOPS+ language." In Proceedings of ICBGM'97, 3rd International Conference on Bond Graph Modeling and Simulation, Simulation Series, Vol.29, No.1, pp. 301–306. The Society for Computer Simulation International.

[16] Broenink, J. F. (1999): "Object-oriented modeling with bond graphs and modelica." In International Conference on Bond Graph Modeling and Simulation, vol. IGBGM 99. San Francisco, CA.

[17] Bush, V. (1931): "The Differential Analyzer: A new machine for solving differential equations." Journal of the Franklin Institute, **212**, pp. 447–488.

[18] Bush, V. and R. D. Booth (1925): "Power system transients." AIEE Transactions, **44**, pp. 80–103.

[19] Chua, L., C. A. Desoer, and E. S. Kuh (1987): Linear and Nonlinear Circuits. McGraw Hill, New York, NY.

[20] Clauss, C., A. Schneider, T. Leitner, and P. Schwarz (2000): "Modelling of electrical circuits with modelica." In Modelica 2000. Lund University, Lund, Sweden.

[21] Eborn, J. and B. Nilsson (1996): "Simulation of a thermal power plant using an object-oriented model database." In IFAC'96, Preprints 13th World Congress of IFAC, vol. O, pp. 121–126. San Francisco, California.

[22] Elmqvist, H. (1973): "SIMNON—User's guide." Technical Report TFRT-3106. Department of Automatic Control, Lund Institute of Technology, Lund, Sweden.

[23] Elmqvist, H. (1978): A Structured Model Language for Large Continuous Systems. PhD thesis TFRT-1015, Department of Automatic Control, Lund Institute of Technology, Lund, Sweden.

[24] Elmqvist, H., F. Cellier, and M. Otter (1993): "Object-oriented modeling of hybrid systems." In Proceedings of European Simulation Symposium, ESS'93. The Society of Computer Simulation.

[25] Elmqvist, H. and S. E. Mattsson (1989): "A simulator for dynamical systems using graphics and equations for modelling." IEEE Control Systems Magazine, **9:1**, pp. 53–58.

[26] Elmqvist, H., S. E. Mattsson, and M. Otter (1999): "Modelica - a Language for Physical System Modeling, V isualization and Interaction." In Proceedings of Symposium on Computer-Aided Control System Desi gn, CACSD'99. IEEE, Hawaii. Plenary paper.

[27] Fritzon, P., Ed. (2000): Modelica 2000 - Workshop Proceedings. Lund University, Lund, Sweden.

[28] Fritzson, P., L. Viklund, D. Fritzson, and J. Herber (1995): "High-level mathematical modeling and programming." IEEE Software, **12:3**.

[29] Gawthrop, P. J. (1991): "Bond graphs: A representation for mechatronic systems," mechatronics." Mechatronics, **1**, pp. 127–156.

[30] Gear, C. W. (1971): "Simultaneously numerical solution of differential-algebraic equations." IEEE Transactions on Circuit Theory, **CT-18**, pp. 217–225.

[31] Glover, K. (1989): "A tutorial on model reduction." In Willems, Ed., From Data to Model, pp. 26–48. Springer, New York, NY.

[32] Hiskens, I. A. (2000): "Power system modeling for inverse systems." vol. November. Washington DC.

[33] IEEE (1997): "Standard VHDL Analog and Mixed-Signal Extensions." Technical Report IEEE 1076.1. IEEE.

[34] Jacobson, I., B. G., and R. J. (1998): Unified Software Development Process(Addison-Wesley Object Technology Series). Addison Wesley, Reading, MA.

[35] Jeandel, A., F. Boudaud, P. Ravier, and A. Buhsing (1996): "U.L.M: Un Langage de Modélisation, a modelling language." In Proceedings of the CESA'96 IMACS Multiconference. IMACS, Lille, France.

[36] Jochum, P. and M. Kloas (1994): "The dynamic simulation environment Smile." In Tsatsaronis, Ed., Second Biennial European Conference on System Design & Analysis, pp. 53–56. The American Society of Mechanical Engineers.

[37] Karayanakis, N. M. (1995): Advanced System Modelling and Simulation with Block Diagram Languages. CRC Press, New York, N. Y.

[38] Karnopp, D. C. and R. C. Rosenberg (1968): Analysis and simulation of multiport systems — The bond graph approach to physical system dynamics. MIT Press, Cambridge, MA, US.

[39] Kron, G. (1963): Diakoptics - The piecewise solution of large-scale systems. McDonald & Co, London.

[40] Kroner, A., P. Holl, W. Marquardt, and E. D. Gilles (1990): "Diva - an open architecture for dynamic simulation." Computers in Chemical Engineering, **14**, pp. 1289–1295.

[41] Larsson, M. (2000): "Objectstab - a modelica library for power system stability studies." In Modelica 2000. Lund University, Lund, Sweden.

[42] Lewis, R. (1998): Programming Industrial Control Systems Using IEC 1131-3. IEE, London.

[43] Ljung, L. (1987): System Identification—Theory for the User. Prentice Hall, Englewood Cliffs, New Jersey.

[44] Marquardt, W. (1991): "Dynamic process simulation - recent progress and future challenges." In Proc. Chemical Process Control IV. Cache, Elsevier, Amsterdam.

[45] Mattsson, S. E., M. Andersson, and K. J. Åström (1993): "Object-oriented modelling and simulation." In Linkens, Ed., CAD for Control Systems, chapter 2, pp. 31–69. Marcel Dekker, Inc, New York.

[46] Mead, C. and L. Conway (1980): Introduction to VLSI Systems. Addison Wesley, Reading, MA.

[47] Mitchell, E. E. L. and J. S. Gauthier (1976): "Advanced continuous simulation language (ACSL)." Simulation, pp. 72–78.

[48] Nagel, L. and D. O. Pederson (1973): "Simulation program with integrated circuit emphasis (SPICE)." Memorandum ERL-M382. Electronics Research Laboratory, College of Engineering, University of California, Berkeley, CA, USA.

[49] Oh, M. and C. Pantelides (1996): "A modelling and simulation language for combined lumped and distributed parameter systems." Computers and Chemical Engineering, **20**, pp. 611–633.

[50] Oppelt, W. (1964): Kleines Handbuch technischer Regelvorgänge. Verlag Chemie, Weinheim.

[51] Otter, M. (2000): "Package power train: A modelica library for modeling and simulation of vehicle power tranis." In Modelica 2000. Lund University, Lund, Sweden.

[52] Pantelides, C. C. (1988a): "SpeedUp — Recent advances in process simulation." Computers and Chemical Engineering, **12**, pp. 745–755.

[53] Pantelides, C. C. (1988b): "Speedup - recent advances in process simulation." Computers in Chemical Engineering, **12**, pp. 745–755.

[54] Paynter, H. M. (1961): Analysis and Design of Engineering Systems. MIT Press, Boston, MA.

[55] Perkins, J. D. and R. W. H. Sargent (1982): "SPEEDUP: A computer program for steady-state and dynamic simulation and design of chemical processes." In Mah and Reklatis, Eds., Selected Topics in Computer-Aided Process Design and Analysis, AIChE Symposium Series 78.

[56] Piela, P. C., K. M. Westerberg, and A. W. Westerberg (1991): "Ascend an object oriented computer environment for modeling and analysis - the modeling language." Computers in Chemical Engineering, **15**, pp. 53–72.

[57] Reiser, M. and W. N. (1992): Programming in Oberon - Steps Beyond Pascal and Modula. Addison-Wesley, Reading, MA.

[58] Rimvall, M. and F. Cellier (1986): "Evolution and perspectives of simulation languages following the CSSL standard." Modeling, Identification and Control, **6**, pp. 181–199.

[59] Rosenberg, R. C. and D. C. Karnopp (1983): Introduction to Physical System Dynamics. McGraw-Hill, New York, NY.

[60] Rumbaugh, J., M. Blaha, W. Premerlani, F. Eddy, and W. L. (Contributor) (1991): Object-Oriented Modeling and Design. Prentice Hall, Englewood Cliffs, NJ.

[61] Sahlin, P., A. Bring, and E.F.Sowell (1996): "The Neutral Model Format for building simulation, Version 3.02." Technical Report. Department of Building Sciences, The Royal Institute of Technology, Stockholm, Sweden.

[62] Sargent, R. W. H. and A. W. Westerberg (1964): "SPEED-UP in chemical engineering design." Trans. Inst. Chem. Eng. (London), **42**, pp. 190–197.

[63] Shearer, J. L. and B. T. Kulakowski (1990): Dynamic Modeling and Control of Engineering Systems. Macmillan, New York, NY.

[64] Strauss (ed.), J. C. (1967): "The SCi continuous system simulation language (CSSL)." Simulation, **9**, pp. 281–303.

[65] Tiller, M. M. (2001): Introduction to Physical Modeling with Modelica. Kluwer, Dordrecht, The Netherlands.

[66] Tummescheit, H. and J. Eborn (1998): "Design of a thermo-hydraulic model library in Modelica." In Zobel and Moeller, Eds., Proc. of the 12th European Simulation Multiconference, ESM'98, pp. 132–136. SCS, Manchester, UK.

[67] Tummescheit, H., J. Eborn, and J. F. Wagner (2000): "Development of a modelica base library for modeling of thermo-hydraulic systems." In Modelica 2000. Lund University, Lund, Sweden.

[68] Viklund, L. and P. Fritzson (1995): "ObjectMath - An object-oriented language and environment for symbolic and numerical processing in scientific computing." Scientific Programming, **4**, pp. 229–250.

[69] Vinnicombe, G. (2001): Uncertainty and Feedback: $\mathcal{H}_\infty$ loop-shaping and the ν-gap metric. Imperial College Press, London.

Chapter 3

POWER CONTROL IN AD HOC WIRELESS NETWORKS: AN ARCHITECTURAL SOLUTION FOR A MELEE OF MULTIPLE AGENTS, COST CRITERIA, AND INFORMATION PATTERNS* †

Swetha Narayanaswamy, Vikas Kawadia, R. S. Sreenivas
and
P. R. Kumar
Dept. of Electrical and Computer Engineering, and
Coordinated Science Laboratory
University of Illinois
1308 West Main Street
Urbana, IL 61801
USA

Dedicated to Yu Chi Ho

Abstract **Wireless networking is a fertile source of problematical situations with multiple agents, multiple cost criteria, and multiple information patterns. It offers an arena for imaginative solutions for a range of problems.**

One example is the power control problem. Choosing the optimal transmit power level of a node in a system as complex as a wireless ad hoc network is an interesting optimization problem. There are several objectives, including maximizing network capacity and minimizing battery usage. Also, the chosen power level affects the functioning of

*This material is based upon work partially supported by the U.S. Army Research Office under Contract DAAD19-00-1-0466, the Office of Naval Research under Contract N00014-99-1-0696, and DARPA under Contract N00014-01-1-0576. Any opinions, findings, and conclusions are those of the authors and do not necessarily reflect the views of the above agencies.
†Please address all correspondence to the last author at University of Illinois, Coordinated Science Laboratory, 1308 West Main Street, Urbana, Illinois 61801, USA. Email: prkumar@uiuc.edu. Web: black.csl.uiuc.edu/~prkumar.

various other network protocols as well. Not only are there theoretical issues, but there are also many issues about harmoniously integrating any power control scheme in the existing network hierarchy.

The COMPOW (for common power) protocol we detail simultaneously increases the network's traffic carrying capacity, provides power aware routes, and reduces the contention for the shared medium. The nodes of the network can run this algorithm in a distributed and asynchronous manner. The software has been implemented in the Linux kernel. To our knowledge it is the first power control protocol to be implemented on a real wireless ad hoc network.

1. Introduction

The field of ad hoc networks is a fertile source of problems in multi-criteria optimization of dynamical systems with multiple agents and information patterns. We will present a solution to one such problem, power control, which we have developed and implemented over networks of laptops.

An ad hoc network consists of several possibly mobile nodes, each with a wireless transceiver. Packets are transported from their sources to their destinations by a sequence of hops where they are relayed from one to node to another, i.e., in a multi-hop manner, see Figure 3.1.

The key property of the wireless radio medium is that it is a shared medium. By reducing their power and timing their transmissions, the nodes can cooperate in sharing the channel. This illustrates the essential multi-agent nature of the problem.

The nodes may be mobile and not fully aware of each others' positions or the terrain, which introduces further problems of obstacles, shadowing, multipath effects, etc. The same wireless medium which they need to share is also the one over which they have to communicate. Such communication therefore requires cooperation which in turn requires communication. Thus the problem is replete with nodal, local, regional and other information patterns.

There are also numerous cost criteria of interest. Nodes may wish to maximize the throughput capacity of the network or to minimize their power consumption. Also, each node may have its own individual cost criterion.

This gives rise to many fertile problems, see [1]. Examples include the media access control problem [2, 3, 4], routing [5, 6, 7, 8, 9, 10], transport [11, 12], and power control (see [13] for references). Protocols for all these problems have to operate in a distributed asynchronous manner since the information patterns of the nodes are different. Further, they need

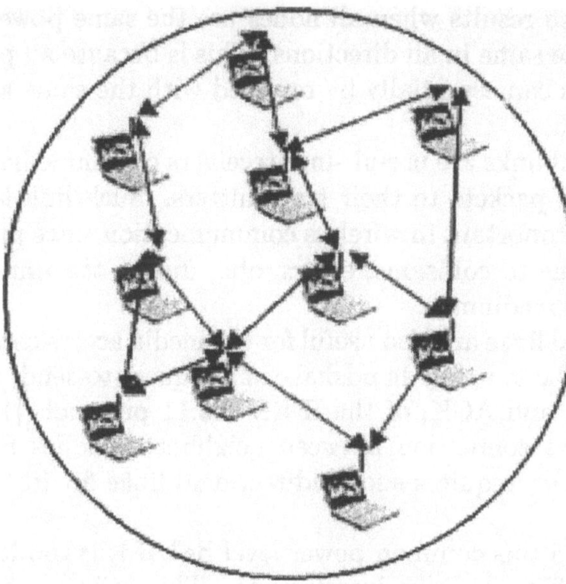

Figure 3.1. An ad hoc network.

to adapt to node locations, even node existence, node mobility, traffic patterns, etc. These protocols will also have to be robust to several types of errors in packet reception due to the unreliable nature of the wireless medium where packets are lost, have errors, etc.

We will present a solution to one problem in operating ad hoc networks, the power control problem: How should nodes choose the power levels of their transmissions? The protocol we outline, called COMPOW (for common power) has been implemented in the Linux kernel, and tested on networks of laptops. To our knowledge, it is the first scheme for power control that has been implemented and tested in a real wireless network testbed. For full details we refer the reader to [13].

2. The common power solution

Our solution is for all nodes to use the smallest common power level such that the entire network is connected. We begin by providing the reasoning behind this approach and showing how it satisfies several objectives.

First, we note that several problems in operating networks are considerably simplified when all links are bidirectional. Such bidirectionality is achieved if all nodes have a common range for their transmission, since if node A can hear node B, then node B can also hear node A. Bidirectionality also results when all nodes use the same power, even if the range is not the same in all directions. This is because all paths followed by radio waves can essentially be reversed with the same attenuation in either direction.

Bidirectional links are useful since receivers can immediately acknowledge receipt of packets to their transmitters. Such link level acknowledgments are important in wireless communication since packets are frequently lost due to collisions, errors, etc., due to the unreliable nature of the wireless medium.

Bidirectional links are also useful for the media access control problem. For example, the four way handshake of a request-to-send, clear-to-send, DATA packet, and ACK, of the IEEE 802.11 protocol ([14, 15]) which seeks to resolve contention between neighboring nodes for the shared wireless medium, requires such bidirectional links for its proper operation.

What should this common power level be? If it is too large, then the nodes may suffer excessive interference. This reduces the capacity of wireless networks, as shown in [16]. However if it is too small, then the network becomes disconnected, see [17]. Thus we propose the choice of the smallest common power level at which the network is connected.

One may ask if this is an optimal strategy from the viewpoint of maximizing the traffic carrying capacity of the network. The answer is that it is nearly so. It is shown in [16] that, under a Protocol Model, if all nodes choose such a low common power level, then the per node throughput for a network of n randomly located nodes with random destinations is $O(\frac{1}{\sqrt{n \log n}})$ bits/sec. This differs from $O(\frac{1}{\sqrt{n}})$, the upper bound on the optimal achievable per node throughput, by only the factor $\frac{1}{\sqrt{\log n}}$, which is relatively small.

A low common power level also reduces the contention for the wireless medium. If there are n nodes in a domain of A square meters, then there are about $\pi r^2 n/A$ nodes within the range r of a transmission. If the average distance from a source to a destination is L meters, then each source destination pair requires about L/r hops. Thus, if every source wishes to transport of W bits per second to its destination, then the effective relaying burden on every node is WL/r bits per second. Thus the $\pi r^2 n/A$ nodes in a geographical range proximity have to collectively transmit $(WL/r)(\pi r^2 n/A) = cr$ bits per second, where $c = \pi n WL/A$. This contention is clearly minimized when r is minimized.

Next we note that a small range also reduces power consumption. Suppose that the attenuation of transmitted power P with distance ρ is cP/ρ^α, where $\alpha \geq 2$. If the minimum power for receiving a clear signal is γ, the required transmit power is $P = \gamma\rho^\alpha/c$. Thus we can define the power cost of a link of length ρ as ρ^α, ignoring the constant γ/c. Hence the power cost of a route from a source to a destination which consists of h hops of link lengths $\rho_1, \rho_2, \ldots, \rho_h$ is $\sum_{i=1}^{h} \rho_i^\alpha$. We say that such a route is a power optimum route if it has the minimum power cost over all routes from that source to that destination. Let us say that a graph G, with vertices at the node locations, is a minimum power graph if it contains only edges which lie along power optimal routes, and the graph contains at least one power optimal route from every source to every destination. (The graph G need not be unique since there may be more than one power optimal route between a source and a destination).

Theorem 1 *For $\alpha = 2$ there is a minimum power graph which is planar, with the nodes at their actual locations, and straight lines edges that do not cross. For $\alpha > 2$, the minimum power graph can be chosen as a subgraph of that for $\alpha = 2$.*

Figure 3.2 below (inspired by a graph in [18]) shows the minimum power graphs for 500 randomly located nodes. On the left is the minimum power graph for $\alpha = 2$, and on the right for $\alpha = 4$. As can be seen, due to the non-crossing property with straight-line edges, these graphs only contain edges to nearby nodes. Thus, choosing a low common power level is also commensurate with low power routing.

3. The algorithm

Thus we have arrived at the following formulation of the power control problem: Determine the smallest common power level at which the network is connected.

Now we describe how an individual node can determine this lowest common power level at which the network is connected. Let $\{P_{min}, \ldots, P_{max}\}$ denote the discrete set of power levels that nodes may operate at. [1] Fix attention on a particular node x. Suppose that all the nodes in the graph operate at the same common power level P. Let $N(P)$ be the number of nodes to which x is connected, directly or indirectly through other nodes, that is, through multi-hop forwarding. Note that $N(P_{max})$ is the maximum number of nodes that x can possibly be connected to. Hence, if for a power level $P < P_{max}$, $N(P) = N(P_{max})$, then the nodes will not sacrifice any connectivity by operating at level P rather than

[1]One should note that power levels are typically discrete. For example, the CISCO 340 Series Aironet Cards allow four different power levels: 1mW, 5mW, 15mW, and 30mW, while the CISCO 350 series Aironet Cards allow six different power levels 1mW, 5mW, 20mW, 30mW, 50mW, 100mW.

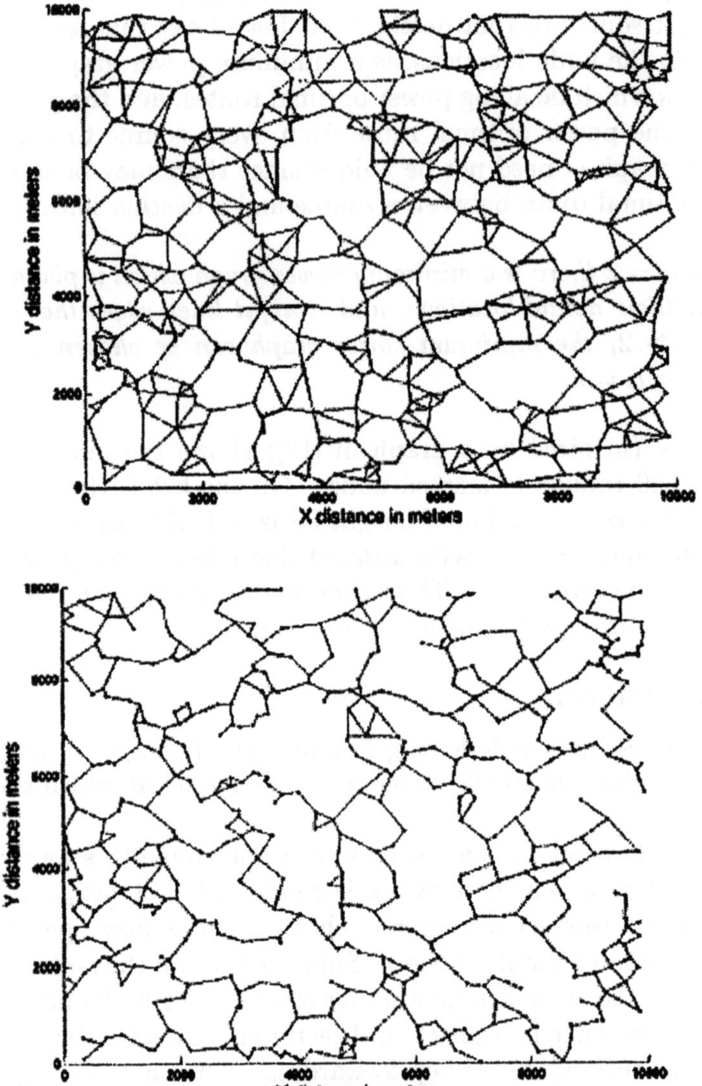

Figure 3.2. A network with 500 nodes randomly located in a square. The graph on the left shows the edges lying minimum power routes for $\frac{c}{\rho^2}$ path loss. The graph on the right is for $\frac{c}{\rho^4}$. It is clearly a sub-graph of that on the left.

P_{max}. This leads to the following choice for the power level $P(t)$ that all nodes can operate at:

$$P(t) = P_i \text{ if } N(P_i) = N(P_{max}) \text{ and } N(P_j) < N(P_{max}) \text{ for all } P_j < P_i.$$

One issue that arises is: How can a node x determine $N(P)$, the number of nodes that it is connected to when all nodes operate at the common power level P? This can be done by taking help from the routing layer. For table driven routing protocols, $N(P)$ is just the number of entries in the routing table. In this paper we will only consider such table driven routing protocols.

However, to determine $N(P)$, all the nodes will have to cooperate and transmit at the common power level P. How to do that in a modular way and to harmoniously integrate it with the existing IP stack is an architectural issue which we address in the next section.

4. The architecture of the COMPOW power control protocol

The question now arises as to how one should implement the above minimum common power for connectivity solution. We need to pay attention to the architecture of the answer. Indeed we will see that by carefully designing the architecture, the algorithm itself becomes trivial.

It should be noted that the success of the Internet is due to its architecture. The careful choice of hierarchical layers, and the notion of peer-to-peer protocols has allowed the plug-and-play feature, which has allowed the Internet to proliferate.

The first issue we must attend to is: In which layer should power control reside? The problem with power control is that it impacts on several layers. For example the power level affects the quality of the links, and hence it impacts on the physical layer. At the same time, higher power allows more routes, and so it impacts on the network layer. Finally, since high power can cause interference and thus congestion, it impacts on the transport layer.

We argue that power control should reside in the network layer. To see this, note that our strategy requires the smallest power level at which all the nodes are connected. The connectivity of a network can only be decided at the network layer, not below it. If one is employing a routing protocol which utilizes routing tables, then the number of nodes that one is connected to can be determined from the routing table by counting the number of entries in it. At the same time, the routing table is itself determined by the set of links which are made available, which in turn depends on the power level employed. Thus, the power level to be chosen depends on the routing table, which in turn depends on the power level. Hence we propose a joint solution for the power control and routing problems in the network layer.

Now we address the issue arising from the need for asynchronous and distributed operation at the nodes. How can a node determine the connectivity of the network at a power level when other nodes may be employing different power levels? We propose an architectural solution which cleanly resolves this problem. We will employ *parallel modularity* at the network layer. We will simply run several routing algorithms in parallel at the network later, one for each power level of interest.[2] Each routing table is implemented by a routing daemon.

This architecture has many desirable properties. As noted above, the routing tables are constructed in a completely distributed way by each routing daemon. The functioning of the route daemons is thus not affected by transition periods when some of the nodes in the network have not converged to the common power level and are using different power levels. The route daemons will compute the correct reachable set at their own power levels and thus eventually the whole network will converge to the common power level. Last but not the least, we can reuse implementations of the routing protocols and thus save development time. The implementation is also modular in the sense that any table driven routing protocol can be plugged in and the scheme would work just as well - the plug and play feature at work. We have chosen the DSDV routing algorithm [6] for simplicity.

By examining its routing tables at the various power levels, each node can determine for itself the lowest common power level at which the maximum networkwide connectivity is attained. This is the power level at which DATA packets originating at the node, or being relayed by the node, are transmitted.

The routing control packets spawned by a routing daemon are simply transmitted at the power level corresponding to that routing table. Routing tables at the same power level, located at neighboring nodes, exchange information by exchanging control packets which are transmitted over UDP. This is done by exploiting the port demultiplexing property of UDP. Each routing demon is simply assigned a port number.

Figure 3.3 shows the architecture of the COMPOW power control protocol.

[2]The number of power levels, for e.g., in the CISCO 340 and 350 Series is only four and six, respectively, and is not large. So the number of routing daemons is feasible.

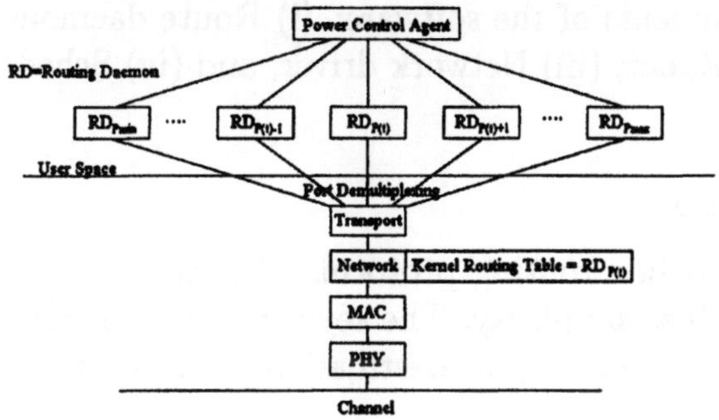

Figure 3.3. The architecture of the COMPOW power control protocol which exploits the port demultiplexing feature of UDP.

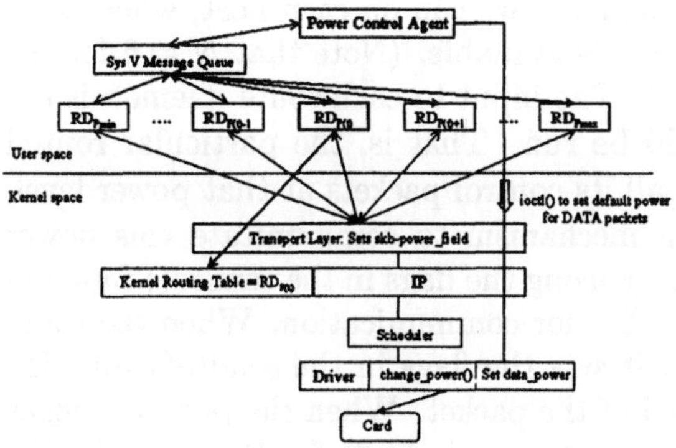

Figure 3.4. The software implementation of COMPOW in the Linux kernel.

5. The software implementation of COMPOW

We run a routing daemon at each power level. These routing daemons are run in parallel at the network layer and construct their routing tables independently by sending out sporadic control packets. Each daemon sends its own control packets at the power level for which it was designated to construct a routing table. To coordinate all these routing daemons, we also run a power control agent in user space. The agent gets input from the routing daemons and computes the common network power level. This scheme is illustrated in Figure 3. The interaction between the agents and the route daemons is described below in further detail.

We implemented COMPOW in the Linux 2.2.16-22 kernel, and it should be easy to port it to any later kernel. The implementation is illustrated in Figure 3.4.

There are four main components of the software: (i) Route daemons (also called "routed's"), (ii) Agent, (iii) Network driver, and (iv) Scheduler.

5.1 Route daemons

Route daemons run any ad hoc routing protocol. We chose to use DSDV [6] for specificity as well as simplicity. The key property of DSDV is that it avoids loops in routes by having the destinations issue sequence numbers.

DSDV was implemented by modifying the netkit-routed package which implements the RIP protocol. Sequence numbers were added to all the routing updates, which allows rejecting stale information and avoiding loops.

N of these DSDV routing daemons are run on each host, where N is the number of discrete power levels available. (Note that $N = 6$ for the CISCO Aironet series of cards). The input to each route daemon is the power level at which it should be run. That is, the particular routed (route daemon) will send out all its control packets at that power level.

The route daemon needs a mechanism to communicate this power level to the kernel. We do this by using the flags in the sendto() function call, since route daemons use UDP for communication. When the route daemon sends down a packet, it sets the flags in the sendto() function call, to indicate the power level of the packet. When the packet reaches the kernel, the sk_buff (packet) structure is built for this packet. We added a new field, power_field, to the sk_buff structure, the need for which is justified in the discussion of the Network driver below. The skb→power_field is set to the power level of the packet based on the flags that have been set by the route daemon. Thereafter, the skb traverses IP and reaches the network device driver.

Peer to peer communication between routing daemons running at the same power level but on different nodes is established by using UDP's port demultiplexing service. That is, a port is reserved for a routing daemon running at each power level on each node.

Also note that at any time there is only one *master* routing daemon, i.e., the route daemon at the current power level $P(t)$, that is allowed to modify the kernel routing table. The other routing daemons keep building their routing tables but communicate it only to the power control agent and *not* to the kernel. The master routing daemon communicates all its updates to both the power control agent and the kernel. The designation of the particular routing daemon to serve as the *master* routing daemon is decided by the power control agent (called "agent" below), whose functionality we describe next.

5.2 Agent

The agent is a program running in user space on each node. It controls all the routing daemons. The route daemons send an update to the agent whenever the number of entries in their routing table changes. Every update triggers a re-computation of the current power level $P(t)$ by the agent. The agent decides the new power level as the smallest power level at which the maximum number of entries in the routing table is attained, as explained earlier.

If the newly computed power level is different from the current power level, then the agent commands the route daemon at the current power level to stop writing to the kernel IP routing table, by sending a STOP command. When the route daemon gets the STOP command, it updates the entries in the kernel routing table. Then the agent authorizes the route daemon running at this new power level to write to the kernel, by issuing a START command. Note that the START and STOP commands do not kill or start a route daemon; they are only permissions to start or stop modifying the kernel's IP routing table.

The agent and the route daemons communicate using System V message queues. We chose message queues for this inter-process communication because messages in the message queue have a type field which can be used to differentiate messages. Thus, multiple processes can communicate using the same message queue. To achieve the same goal using FIFO or pipes we would need $2N$ of them, where N is the number of power levels. There is a thread running in each route daemon that continuously monitors the message queue for messages (commands) from the agent. Similarly the agent monitors the message queue for updates from the routing daemons. Using message queues also makes the startup process asynchronous. That is, the order of starting up of the route daemons and the agent does not affect the algorithm.

5.3 Network driver

The network driver (in our case: airo.c 1.6 2001/02/08) for the Aironet cards had to be changed in two ways to support changing of power level.

First, for control packets which are spawned by the route daemon, the power level is decided by the route daemons, but actually can be set only by the device driver. Thus the power level information should travel through the network stack hierarchy along with the packet. For this, a new field called power_field was added to the packet data structure (sk_buff). One should note that route daemons communicate the power level to the IP stack using the flags in the sendto() function call and the kernel passes this information to the driver by setting the value in skb→power_field to the transmit power level of the packet. Thus there are two steps in the process of transmitting control packets at a certain power level: route daemon to IP, and IP to driver. The driver reads the transmit power level for the packet from the packet's skb→power_field and changes the power value in the card's config structure: config.txPower. Note that this changing of txPower has to be preceded and followed by MAC_DISABLE and enable_MAC primitives respectively, for enabling and disabling the card.

For non-route daemon spawned packets, which we call DATA packets, the skb→powerfield is 0 by default, which indicates that the current power level decided by the power control agent should be used for transmitting these packets. To support this mechanism, a second change had to be incorporated in the driver, as follows. When the agent decides the current default power level $P(t)$, it communicates this to the driver by sending down an ioctl (SIOCDATAPOWER) to the driver, with $P(t)$ as a parameter. The driver will store the new data power level (data_power) in its private structure and use it to set the value of txPower on the card when sending out DATA packets.

We have tried to make as few and as little changes in the device driver as possible so as to keep our protocol portable. We believe that any card which is designed to support any power control protocol will have to provide at least the above mentioned mechanisms to change transmit power from the user space.

5.4 Scheduler

The Scheduler component is situated between the IP layer and the network driver. It implements a scheduling policy to reduce the latency of power switchovers.

Changing the transmit power level on the card is a costly activity in terms of the time taken for the currently available off-the-shelf equipment that we are aware of. The transmit power in our equipment is a configuration parameter on the card, and the card consequently needs to be disabled before it can be changed. Thus, changing the transmit power involves three steps: Disable the card, write the new power in the card's config structure, and then finally enable the card. Measurements in the driver show that this time to change the power level is about 6ms. However, when the same latency is estimated from measurements at the network layer, by monitoring the round trip time of ping packets, it turns out to be an order of magnitude higher. We are still investigating the cause of this high latency but the bottom line is that switching power levels too often is not desirable. Clearly, it would be desirable to have hardware and supporting software with reduced latency in switching power levels.

Thus arises the issue of optimizing to reduce latency of switching from one power level to another. To optimize this, we *serve* packets in a round robin clearing (or exhaustive service) fashion to eliminate needless switches of power level.

The scheduler could also be implemented in the device driver, but that would make the protocol less portable, as every driver would have to support our scheduling scheme. By implementing it just above the device driver in the dev queues, we maintain portability across different devices. We have implemented one – round robin with exhaustive clearing. (In the future, we plan to implement many scheduling schemes and give the user space agent the choice to select the best one). We intercept a packet just before it is placed in the queue, and insert it right after packets of its power level in the queue. If packets of its power level are not already present in the queue, then the packet is appended to the end of the queue.

Now we describe the typical operation of the scheme. The agent starts up and starts monitoring the message queue. Asynchronously, the route daemons also start up and start building their routing tables independently. When a route is added or deleted in one of the routing tables, the respective routed (route daemon) writes its identity (the power level at which it is running) and the new number of entries to the message queue. The agent re-computes the new power level $P(t)$ and executes the following three steps: (i) It sends a STOP command to the current "master routing daemon," i.e., the routing daemon at the current power level. (ii) It sends a START command to the new "master routing daemon." (iii) It sends down an ioctl (SIOCDATAPOWER) to the driver indicating that $P(t)$ should be the new data power level.

The Network driver transmits a packet at the appropriate power level by reading the skb→power_field and the data_power communicated by the agent through an ioctl(). After a while, the algorithm converges to a common power level which is used by all nodes. The functioning of the route daemons is not affected by transition periods when some of the nodes in the network have not converged to the common power level and are using different power levels. The route daemons will compute the correct reachable set at their own power levels and thus eventually the whole network will converge to the common power level. If any of the the routing tables changes, then a re-computation of the common power level is triggered, and the network settles to the new common power level. The algorithm runs continuously, and the only overhead is the extra control packets generated by the multiple routing daemons.

Experimentation was done on our testbed of Compaq Presario laptops, each running Redhat Linux 7.0 and equipped with a Cisco Aironet 350 wireless card, and proper functioning of the entire software was established by experimentation.

6. Concluding remarks

The next phase of our experiment is to optimize the protocol for power switching latency as well as other improvements in maintaining routing tables. We also plan to conduct a performance analysis vis a vis mobility. We further plan to investigate the use of on-demand routing protocols, such as DSR [7] or AODV [8], for example.

References

[1] P. R. Kumar, "New Technological Vistas for Systems and Control: The Example of Wireless Networks," *IEEE Control Systems Magazine*, Special Section on Networks and Control, pp. 24–37, vol. 21, no. 1, Feb. 2001.

[2] V. Bharghavan, A. Demers, S. Shenkar, and L. Zhang, "MACAW: a media access protocol for wireless LANs," in *Proceedings of the SIGCOMM '94 Conference on Communications Architectures, Protocols and Applications*, pp. 212–225, August 1994.

[3] Z.J. Haas and J. Deng, "Dual busy tone multiple access (DBTMA): A medium access control for multihop networks," *IEEE Wireless Communications and Networking Conference 1999 (WCNC'99)*,New Orleans, LA, September 21–24, 1999.

[4] R. Rozovsky and P. R. Kumar, "SEEDEX: A MAC protocol for ad hoc networks." University of Illinois, May 2001.

[5] S. Ramanathan and M. Steenstrup, "A survey of routing techniques for mobile communication networks," *Mobile Networks and Applications*, vol. 1, no. 2, pp. 89–104, 1996.

[6] C. E. Perkins and P. Bhagwat, "Highly dynamic destination-sequenced distance-vector routing (DSDV) for mobile computers," *Computer Communications Review*, vol. 24, no. 4, pp. 234–244, 1994. (ACM SIGCOMM '94).

[7] D. B. Johnson and D. A. Maltz, "Dynamic source routing in ad hoc wireless networks," in *Mobile Computing* (T. Imielinski and H. Korth, eds.), Kluwer Academic Publishers, 1996.

[8] Charles E. Perkins, Elizabeth M. Royer, Samir R. Das, "draft-ietf-manet-aodv-03.txt," Mobile Ad Hoc Networking Working Group, http://explore.kwangwoon.ac.kr/ ssbae/theory/internet-drafts/draft-ietf-manet-ao dv-03.txt, 25th June 1999.

[9] V. D. Park and M. S. Corson, "A highly adaptive distributed routing algorithm for mobile wireless networks," in *Proceedings of INFOCOM'97*, (Kobe, Japan), April 7-11, 1997.

[10] P. Gupta and P. R. Kumar, "A system and traffic dependent adaptive routing algorithm for ad hoc networks," in *Proceedings of the 36th IEEE Conference on Decision and Control*, (San Diego, CA), pp. 2375–2380, Dec. 1997.

[11] Van Jacobson, "Congestion avoidance and control," *ACM Computer Communication Review*, pp. 314–329, August 1988.

[12] Van Jacobson and M. J. Karels, "Congestion avoidance and control," in *SIGCOMM 1988*, August 1988.

[13] S. Narayanaswamy, V. Kawadia, R. S. Sreenivas, and P. R. Kumar, "The COMPOW protocol for power control in ad hoc networks: Theory, architecture, algorithm, implementation, and experimentation." University of Illinois, May 2001.

[14] IEEE Protocol 802.11, "Draft standard for wireless LAN: Medium access control (MAC) and physical layer (PHY) specifications." IEEE, July 1996.

[15] IEEE Computer Society LAN MAN Standards Committee, "Wireless LAN medium access control (MAC) and physical layer (PHY) specifications." IEEE Standard 802.11-1997. The Institute of Electrical and Electronics Engineers, New York, NY, 1997.

[16] P. Gupta and P. Kumar, "The capacity of wireless networks," *IEEE Transactions on Information Theory*, vol. IT-46, pp. 388–404, March 2000.

[17] P. Gupta and P. R. Kumar, "Critical power for asymptotic connectivity in wireless networks," in *Stochastic Analysis, Control, Optimization and Applications: A Volume in Honor of W.H. Fleming* (W. M. McEneany, G. Yin, and Q. Zhang, eds.), pp. 547–566, Boston, MA: Birkhauser, March 1998.

[18] T. N. Shepard, "A channel access scheme for large dense packet radio networks," in *Proceedings of ACM SIGCOMM*, August 1996.

Chapter 4

SOME EXAMPLES OF OPTIMAL CONTROL

A. E. Bryson

Tribute to Larry Ho

As a long-time friend and colleague of Larry's, it is a great pleasure to be here to help celebrate his 67th birthday and retirement from teaching. He has been and still is an outstanding contributor to our field of automatic control and systems theory. However, that is only part of the reason why we are all here. The other part is that he is a wonderful gentleman; he is always considerate, kind, tolerant, and patient in his dealings with students and colleagues - in fact with everyone. Larry, we admire and love you.

Introduction

I would like to show you some of the unusual and, I believe, instructive applications of optimal control that I have collected over the last 30 years. Professional software packages like MATLAB have made it possible for amateurs like me to calculate and display many of these examples, so I shall show you computer-animated movies of controlling plants like bicycles, unicycles, and trucks with trailers.

Lateral Stabilization and Turning of a Bicycle

A dynamic model of the lateral motions of a bicycle was given in [9], 1948, who, in turn, refer to Bouasse [1], 1910. In roll it is an inverted pendulum $\ddot{\phi} = a(\phi/c - \delta - cu)$ where $\dot{\delta} = u$ (see [2], pp. 237-238 and Figure 4.1). In yaw the rolling-without-slipping constraints on the wheels yield the simple kinematic model $\dot{\psi} = \delta$. Thus a rate-of-turn of the front wheel $u = \dot{\delta}$ produces a yaw angular acceleration $\ddot{\psi}$. The rider stabilizes in roll by using the D'Alembert force associated with $\ddot{\psi}$. If the rider senses a fall to the right, he initiates a rate-of-turn of the front

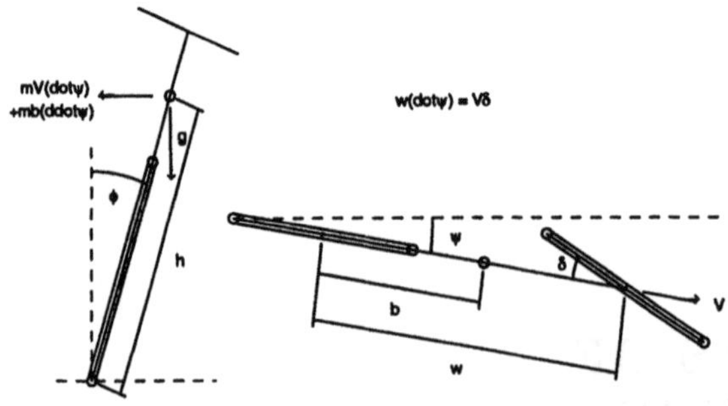

Figure 4.1. Nomenclature for a Bicycle Robot

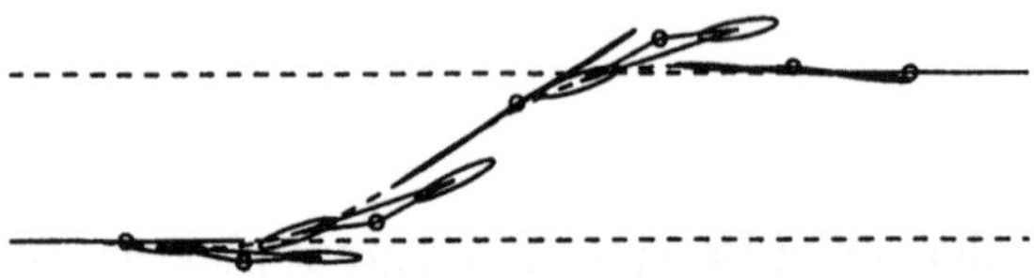

Figure 4.2. Stroboscopic Movie of Bicycle Making an S-Turn

wheel to the right, which produces a D'Alembert force to the left which opposes the gravitational fall-down torque. Thus, almost imperceptible rates-of-change of the steering angle of the front wheel stabilize the bike in roll. The idea of a turn-rate into the direction of the fall may be the reason some people have difficulty learning to ride a bike. Surprisingly, the gyroscopic torques associated with the rotating wheels are insignificant at normal cruising speeds but become significant at higher speeds, particularly for motorcycles.

In a steady turn the centrifugal force associated with the turn rate $\dot{\psi}$ just balances the fall-down gravitational torque (see Figure 4.1). However, to initiate a turn to the left, the rider must first turn slightly to the right to produce the D'Alembert force that tips the bike to the left. Figure 4.2 is a stroboscopic movie of a bike moving at constant velocity making an S-turn to the left. Note the front wheel goes off the track to the right before it moves left. Try riding a bike along a line in the street, and start a turn to the left; you will easily observe that the front wheel first moves to the right.

The transfer function from the control to the yaw angle (u to ψ) has a RHP pole and zero that cancel each other. The roll TF has the same RHP pole but no cancelling zero. However, when controlling both roll and yaw, there is an uncancelled 'compromise zero' in the right half plane of the symmetric root locus, making the system non-minimum phase. This is consistent with the controlled behaviour described above.

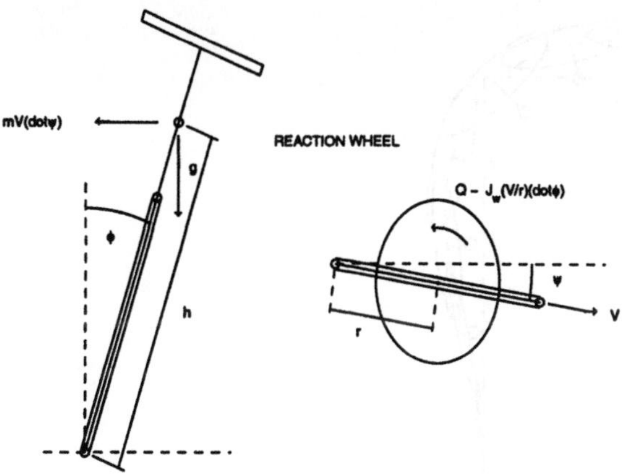

Figure 4.3. Nomenclature for a Unicycle Robot

The terminal control law used here has time-varying LQ gains which produces substantial control at the beginning AND at the end to meet the specified final conditions, so that there is no settling time. The quadratic performance index has weights only on squared-deviations from the desired terminal state and on the integral-square of the control u.

Lateral Stabilization and Turning of a Unicycle

Unicycle lateral stabilization is more diificult than bike stabilization (see Figure 4.3). There is no $mb\ddot{\psi}$ force so the rider must use the centrifugal force $mV\dot{\psi}$. If he senses a fall to the right, he rotates his shoulders to the left. This produces a rightward yaw torque on the unicycle that produces a yaw angular acceleration $\ddot{\psi}$ that produces a yaw angulr velocity $dot\psi$ that produces a centrifugal force $mV\dot{\psi}$ that opposes the gravity fall-down torque $mgh\phi$ (a truly "cascaded" control). As with a bicycle, the rider turns into the direction of fall. Yaw motion is coupled to roll motion by the gyroscopic yaw torque $J_w\Omega p$ associated with changing the direction of the unicycle wheel angular momentum $J_w\Omega$ ($\Omega = V/r$)) with the roll rate $p = \dot{\phi}$.

In a robot unicycle a reaction wheel with a nominally vertical axis produces a control yaw torque Q by means of an electric motor (this simulates the action of the rider's shoulders). This was suggested by Arnold Schoonwinkel in his Ph. D. thesis ([7], 1987). He built a working unicycle robot which he demonstrated on a Stanford tennis court.

Figure 4.4 is a stroboscopic movie of a unicycle traveling at constant velocity making a 90 degree right turn. It first turns slightly to the

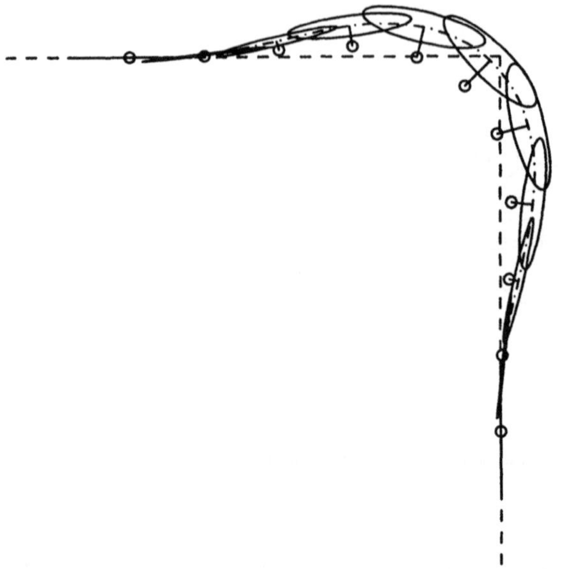

Figure 4.4. Stroboscopic Movie of a Unicycle Making an S-Turn

left before turning right. This is consistent with the fact that the yaw transfer function has a right half-plane zero. The terminal control law uses time-varying LQ gains which produces substantial control at the beginning AND at the end to meet the specified final conditions, so that there is no settling time.

Backing Up a Truck with a Trailer

Backing up a truck with a trailer is difficult because the system is unstable (see [2], pp. 397-399). The states are θ_1, θ_2 = angles of the (truck, trailer) with respect to an inertial reference line, and x, y= inertial coordinates of a point on the trailer (see Figure 4.5). The control is the front wheel steering angle δ which is bounded.

Backing up from an arbitrary position and orientation to a specified position and orientation takes, in addition, some good path planning. Figure 4.6 and Figure 4.7 are stroboscopic movies showing minimum distance paths from two different orientations and positions in a parking lot to a specific point and orientation at a loading dock. This problem was suggested and solved by Prof. Bernard Widrow and his student Derrick Nguyen ([4]) by training a neural network. The bang-bang-zero-bang solution shown here was obtained using parameter optimization (three switch times and the final time). The calculus of variations solution with a constraint on δ is 'bang-bang-singular-bang' where the singular

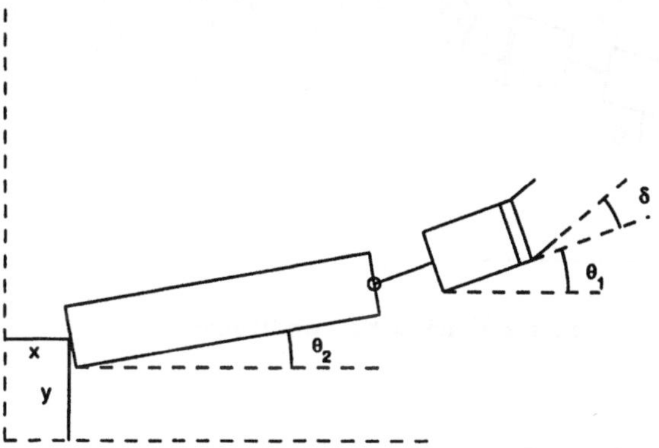

Figure 4.5. **Nomenclature for a Truck with a Trailer**

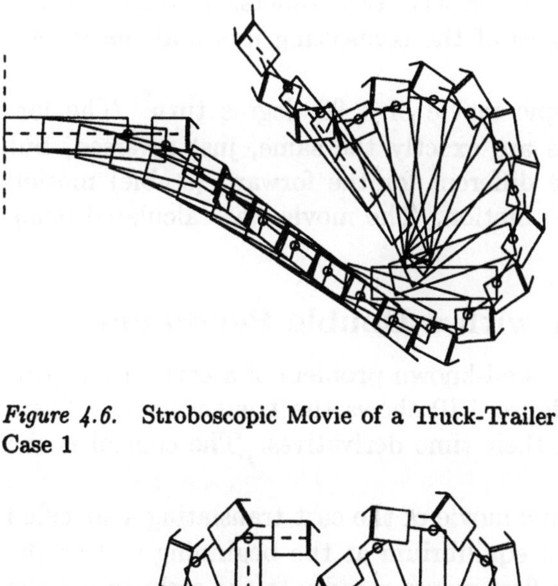

Figure 4.6. **Stroboscopic Movie of a Truck-Trailer Backing Up to a Loading Dock; Case 1**

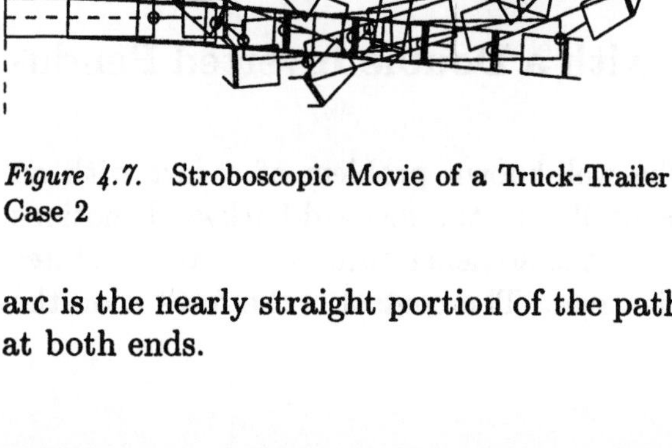

Figure 4.7. **Stroboscopic Movie of a Truck-Trailer Backing Up to a Loading Dock; Case 2**

arc is the nearly straight portion of the path with 'chattering' junctions at both ends.

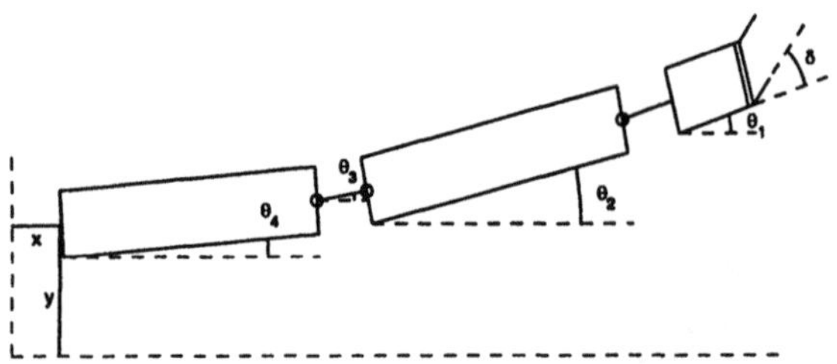

Figure 4.8. Nomenclature for a Truck with Two Trailers

Backing Up a Truck with Two Trailers

Backing up a truck with two trailers is even more difficult. Figure 4.8 shows the nomenclature for a truck with two trailers. Two more states are added, θ_3, θ_4 = the angles of the connecting link and the second trailer.

Figure 4.9 is a stroboscopic movie of a 90 degree turn. The forward and backward motions are exactly the same, just reversed, but the feedback gains are quite different for the forward (stable) motion and the backward (unstable) motion. The movie was calculated using time-varying LQ gains.

Translation of a Cart with a Double Pendulum

This is an extension of the well-known problem of a cart with a pendulum (overhead crane). Figure 4.10 shows the nomenclature. There are six states y, θ_1, θ_2 and their time derivatives. The control is the force u on the cart.

Figure 4.11 is a stroboscopic movie of the cart translating a specified distance with the pendula in equilibrium at the beginning and at the end of the motion. The cart first moves rapidly to the right to get the pendula swinging, then moves slowly back, then slowly forward, and finally rapidly to the right to come over the top of the pendula in the equilibrium condition at the new position.

Translation of a Cart with a Double Inverted Pendulum

This is an extension of the well-known problem of a cart with an inverted pendulum (which is similar to the forward-backward motions of a unicycle). Figure 4.12 shows the nomenclature. There are six states y, θ_1, θ_2 and their time derivatives. The control is the force u on the cart.

Figure 4.9. **Stroboscopic Movie of a Truck with Two Trailers Making a 90 Degree Turn**

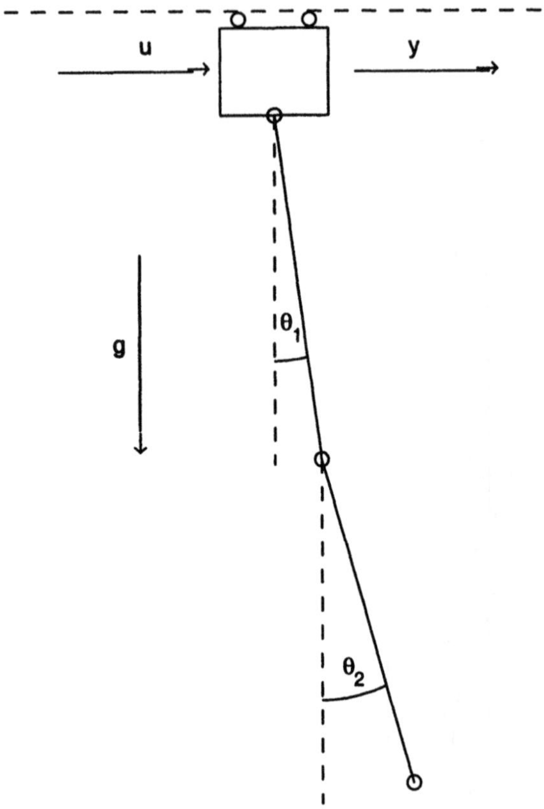

Figure 4.10. Nomenclature for a Cart with a Double Pendulum

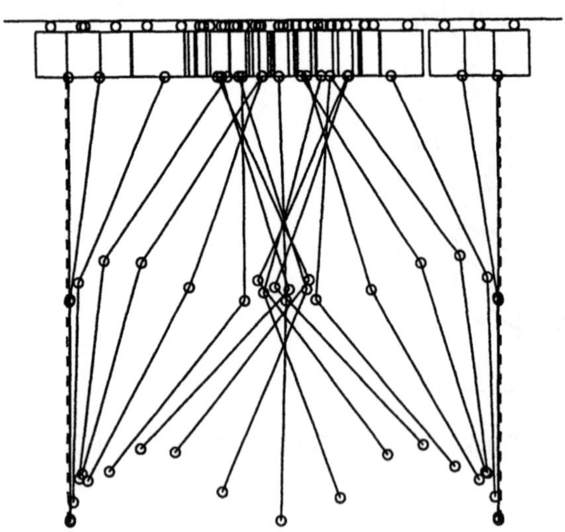

Figure 4.11. Stroboscopic Movie of the Translation of a Cart with a Double Pendulum

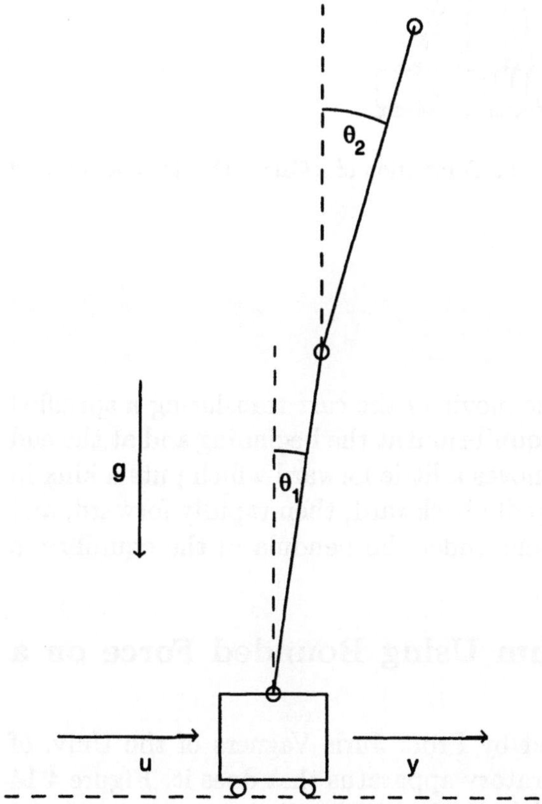

Figure 4.12. **Nomenclature for a Cart with a Double Inverted Pendulum**

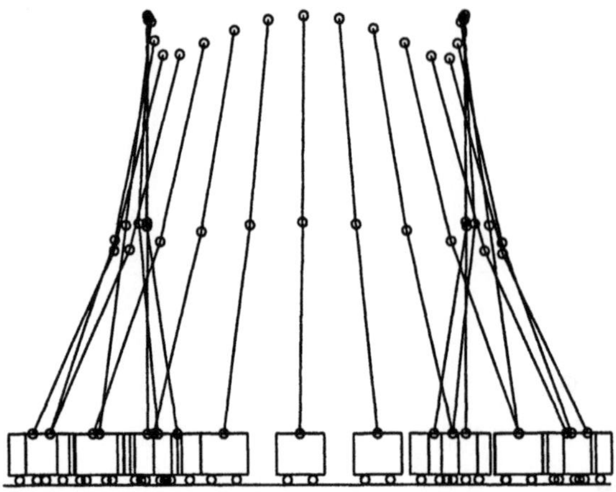

Figure 4.13. Stroboscopic Movie of the Translation of a Cart with a Double Inverted Pendulum

Figure 4.13 is a stroboscopic movie of the cart translating a specified distance with the pendula in equilibrium at the beginning and at the end of the motion. The cart first moves a little forward which puts a kink in the inverted pendula, then rapidly backward, then rapidly forward, and finally slowly backward to come under the pendula in the equilibrium condition.

Erection of a Pendulum Using Bounded Force on a Cart

This problem was suggested by Prof. Juris Vagners of the Univ. of Washington, who built a laboratory apparatus that does it. Figure 4.14 shows the nomenclature (see [2], pp. 374-375). The states are y, θ and their time derivatives; the control is again the force on the cart u which, in this example, is bounded.

Figure 4.15 is a stroboscopic movie of the erection ('swing-up'). The bound on the cart force equals the weight of the cart which is equal to the weight of the pendulum bob. The pendulum starts from rest in the down position and ends at rest stabilized in the inverted position with no net motion of the cart. The solution was obtained by COV and nonlinear progamming (NLP), the first to get an approximate solution (find the number of switches), and the latter to refine the switch times. The pendulum is swung up and the cart then moves underneath it.

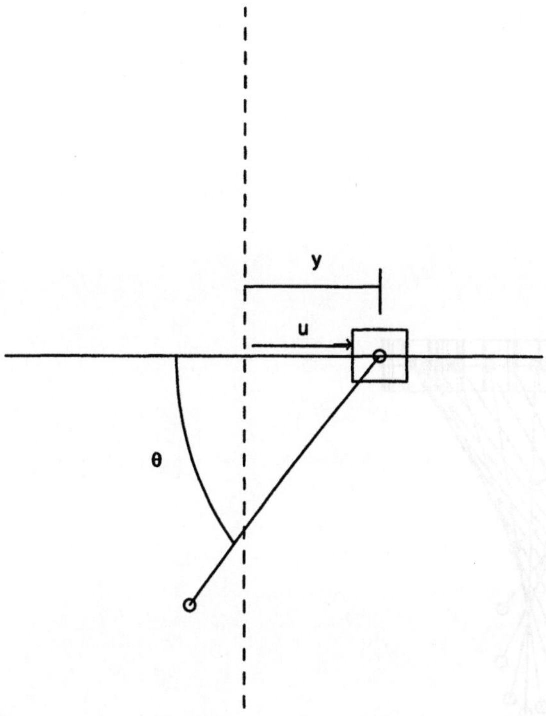

Figure 4.14. **Nomenclature for Erection of a Pendulum Using Bounded Force on a Cart**

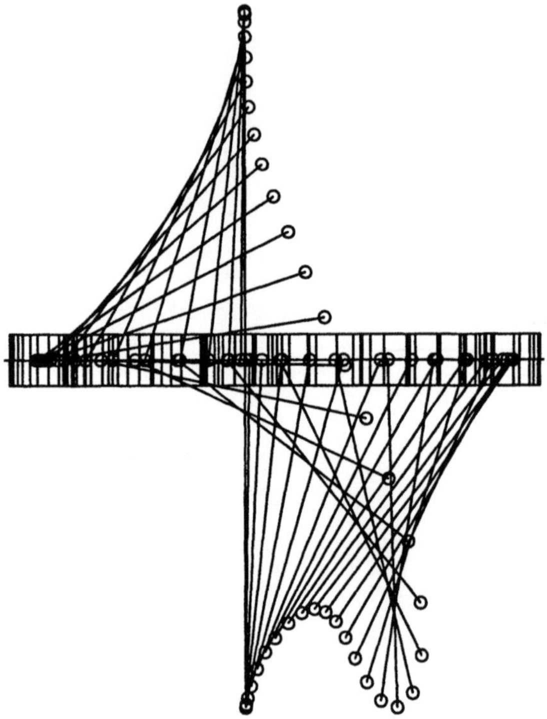

Figure 4.15. Stroboscopic Movie of the Minimum Time Erection of a Pendulum Using Bounded Force on a Cart

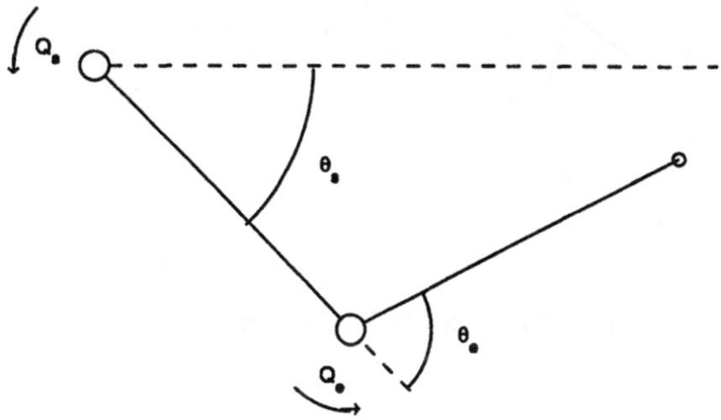

Figure 4.16. Nomenclature for Two Link Robot Arm

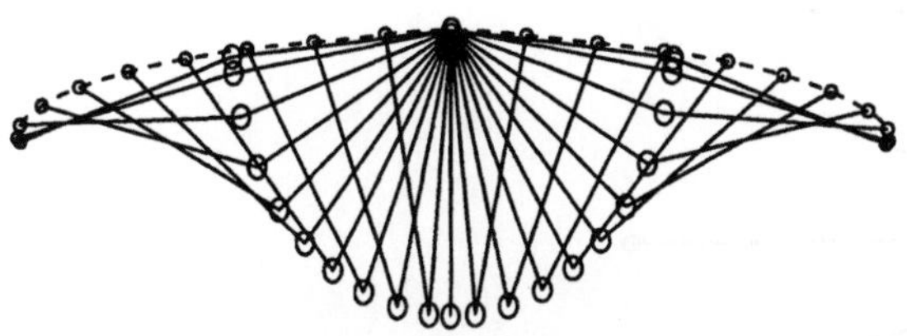

Figure 4.17. Stroboscopic Movie of a Two Link Robot Making a Minimum Time Pick and Place Motion

Pick and Place Motion with a Two-Link Robot Arm

This problem was inspired by the work of Prof. Robert Cannon and his student William Ballhaus in the Aerospace Robotics Lab at Stanford (1994). The solution was computed by Rhonda Slattery as part of her Stanford Ph. D. thesis ([8]) and also appears in [2], pp. 375-377. There are bounded torquers at the shoulder and the elbow (see Figure 4.16). The EOM are nonlinear and the four states are θ_s, θ_e and their time derivatives. The two controls are Q_s, Q_e = the torques at the shoulder and at the elbow.

Figure 4.17 is a stroboscopic movie of of the minimum time path to move an object (using a gripper at the tip) from a given position to another specified position, in this case nearly the full reach of the arm. The torques at the elbow and the shoulder are bounded, so the solution is bang-bang (in some cases there are singular arcs, but not in the case show below). The orientation of the arm with respect to the initial and final locations is also determined optimally. The solution was obtained using COV and the switch times were refined using NLP. The arm moves in such a way as to minimize its moment of inertia. It is similar to the motion a human would use when moving an object at shoulder level.

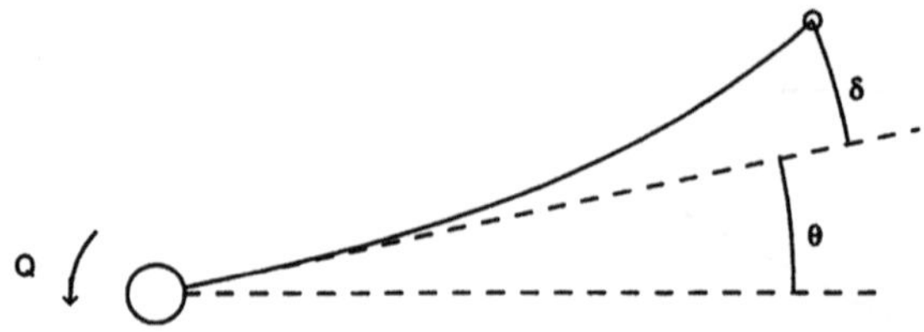

Figure 4.18. Nomenclature for a Flexible Robot Arm

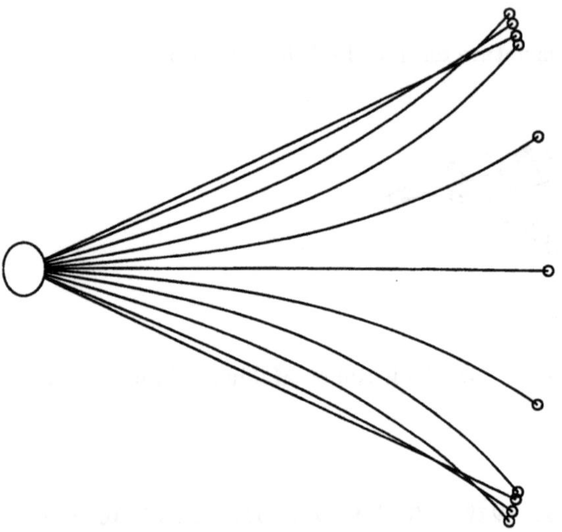

Figure 4.19. Stroboscopic Movie of a Minimum Effort Pick and Place Motion of a Flexible Robot Arm

Pick and Place Motion with a Flexible Robot Arm

This problem was inspired by the work of Prof. Robert Cannon and his student Eric Schmitz ([6]). There is a torquer at the shoulder, and the arm is modeled as a cantilever beam using the first bending mode and the rigid-body mode (see Figure 4.18). The four states are θ, δ and their time derivatives, and the control is the shoulder torque.

Figure 4.19 is a stroboscopic movie of a pick and place motion with minimum integral-square control, beginning and ending with the beam in equilibrium. The tip moves in the wrong direction at first, overshoots the final position, and then comes back. This solution was also obtained using time-varying LQ logic so that, again, there is substantial shoulder torque at the end of the motion as well as at the begining, producing zero settling time.

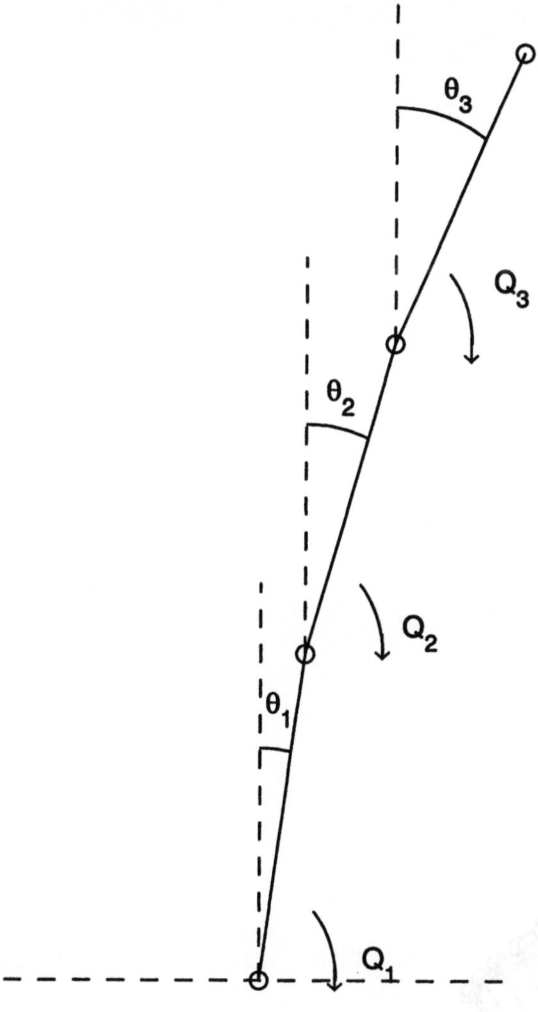

Figure 4.20. Nomenclature for Stick-Person Robot

Stick-Person Robot Doing Knee Bends

Figure 4.20 shows a stick-person robot modled as a triple inverted pendulum with torquers at the ankle, the knee, and the hip. This model was inspired by the Ph. D. thesis of Christopher Runge [5]. The six states are the three stick angles and their time derivatives; the controls are the three torques.

Figure 4.21 is a stroboscopic movie of the robot doing 'knee bends'. The motion was calculated using time-varying feedforward and feedback to the three torquers using an LQ follower code [3].

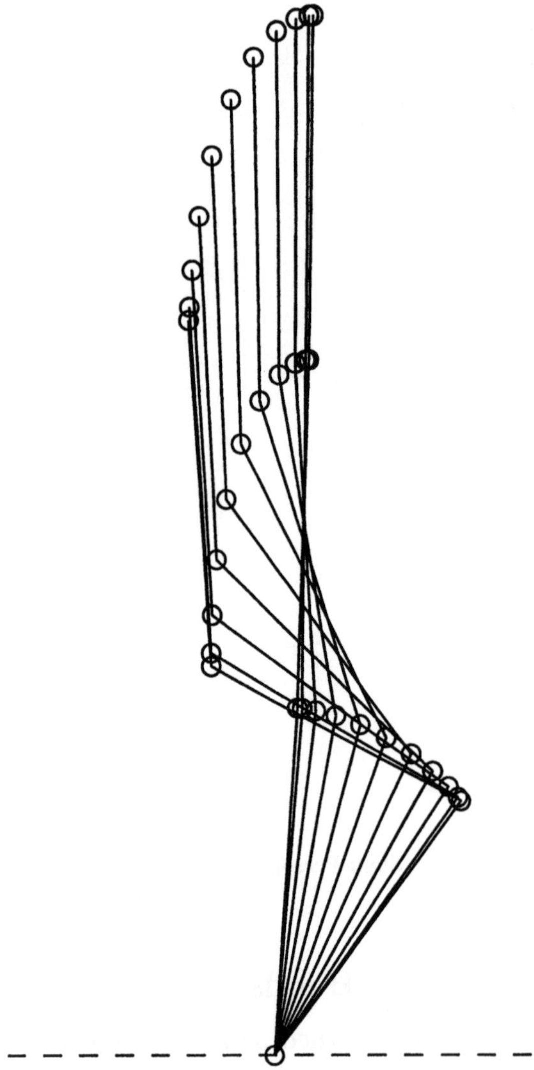

Figure 4.21. Stroboscopic Movie of a Stick-Person Robot Doing Knee Bends

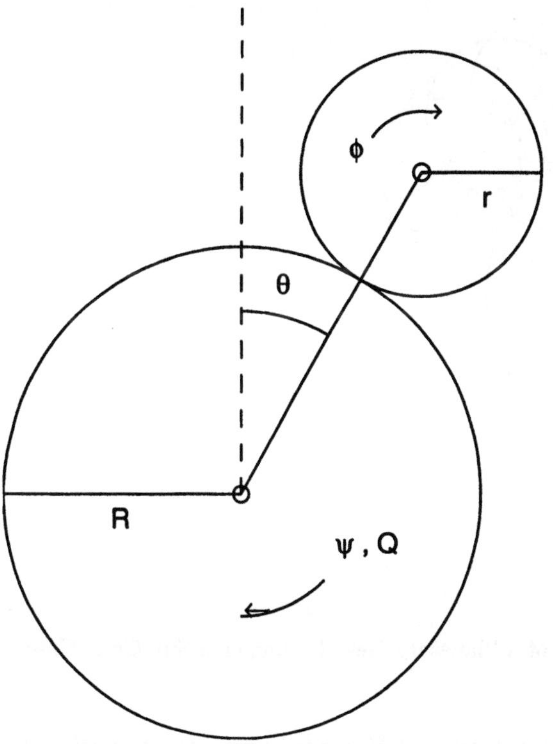

Figure 4.22. Nomenclature for Planetary Gear Rolling on a Sun Gear

Planetary Gear Rolling on a Sun Gear

This problem was suggested by Prof. Steven Rock's lab experiment at Stanford of balancing a ping-pong ball on top of a rotating disk. The ball rolls without slipping in a groove on the disk. When his students first tried it with a conventional constant-gain SISO control law, they got a surprise; starting with the ball a little off the top, the ball was brought to the top of the disk and balanced there, BUT the disk (and the ball) were left spinning at a constant rate. Constant spin is, of course, an equilbium solution as well as no spin (a similar thing happens with a cart stabilizing an inverted pendulum). The set up here has been changed to a planetary gear (PG) being rotated by a fixed sun gear (SG) (see Figure 4.22). A link connects the center of the planetary gear to the fixed center of the sun gear which allows the PG to rotate freely around the SG; thus it would be unstable at the top just like the ball on a disk. The advantage of this set up is that the PG will not fall off the SG; the system is linear (unlike the erection of the pendulum by force on the cart).

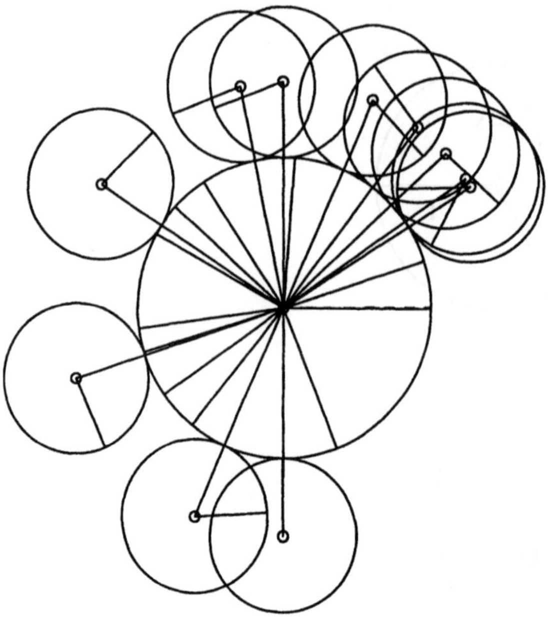

Figure 4.23. Stroboscopic Movie of a Planetary Gear Rolling on a Sun Gear; Case 1

Figure 4.23 is a stroboscopic movie of the minimum effort motion to take the PG from straight down to straight up, ending with zero angular velocity of both gears. The PG overshoots the top then rolls back.

Figure 4.24 is a stroboscopic movie of the minimum effort motion, adding the final condition that a specified point on the SG must end at the top. This produces a substantial overshoot, a substantial undershoot, and a roll back to the top. A similar motion occurs with a cart translating with an inverted pendulum when the final position of the cart is specified. The motions in the last two figures used time-varying LQ gains which produce substantial control effort at the end as well as at the beginning to meet the specified terminal conditions (no settling time).

Summary

At the symposium we showed simulation movies (in color) of eleven optimal control problems with terminal boundary conditions. They are shown here as 'stroboscopic' black-and-white movies. Three of the examples are nonlinear with bounded controls (exhibiting bang-bang behaviour) and the rest are linear-quadratic (LQ) problems. The LQ examples have time-varying (TV) feedback gains and TV feedforward controls; they were calculated using a Riccati equation code from [3]. Many

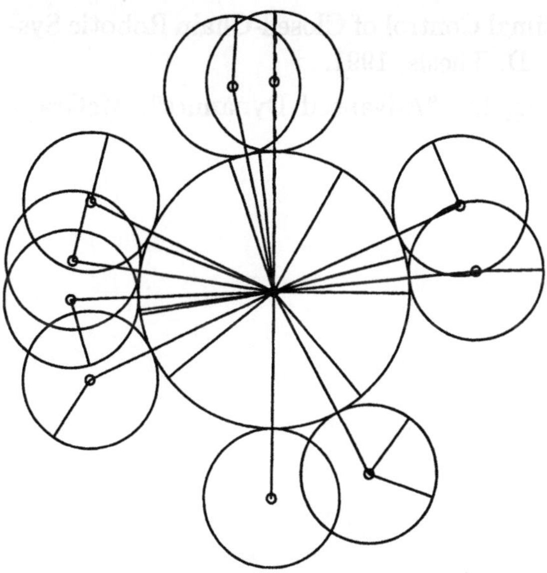

Figure 4.24. Stroboscopic Movie of a Planetary Gear Rolling on a Sun Gear; Case 2

of the LQ problems are non-minimum phase, exhibiting controls that start in the 'wrong' direction.

References

[1] Bouasse, Henri, "Cours de Mecanique Rationelle et Experimentale", Delagrave, Paris, 1910.

[2] Bryson, A. E., "Dynamic Optimization", Addison-Wesley-Longman, Menlo Park, CA, 1999.

[3] Bryson, A. E., "Applied Linear Optimal Control", Cambridge Univ. Press, New York, 2002.

[4] Nguyen, D. K., "Application of Neural Networks in Adaptive Control", Stanford Univ. Ph. D. Thesis, 1991.

[5] Runge, C. F., Zajac, F. E., Allum, J. H. J., Risher, D. W., Bryson, A. E., and Honegger, F., "Estimating Joint Torques from Kinesiological Data Using Optimal Linear System Theory", *IEEE Trans. on Biomedical Engineering*, Vol. 42, 12, 1995.

[6] Schmitz, E., "Experiments on the End-Point Position Control of a Very Flexible One-Link Manipulator", Stanford Univ. Ph. D. Thesis, 1985.

[7] Schoonwinkel, Arnoldus, "Design and Test of a Computer Stabilized Unicycle", Stanford Univ. Ph. D. Thesis, 1987.

[8] Slattery, Rhonda A., "Optimal Control of Closed-Chain Robotic Systems", Stanford Univ. Ph. D. Thesis, 1991.

[9] Timoshenko, S., and Young, D., "Advanced Dynamics", McGraw-Hill. NY, 1948.

Chapter 5

ADAPTIVE CONTROL USING MULTIPLE MODELS: A METHODOLOGY

Kumpati S. Narendra, Osvaldo A. Driollet and Koshy George
Center for Systems Science
Department of Electrical Engineering
Yale University, New Haven, CT 06520-8267, USA

Abstract A procedure based on multiple models, switching, and tuning for adaptively controlling dynamical systems in time-varying environments was introduced in the early 1990s. During the past decade, this procedure has evolved into a general methodology for adaptive control. Some of the recent advances in this new area are discussed in the paper.

1. Introduction

Models play an important role in general, and automatic control in particular. A model is merely an abstraction of reality, and depending upon the simplifying assumptions made by the designer, different models of the same plant can be obtained. The designer of control systems will naturally choose to include those characteristics in the model which she/he deems to be important for control purposes. Linear deterministic and stochastic models, and nonlinear models have been treated in the control literature.

The area of adaptive control was born in the 1950s when engineers attempted to design systems that would monitor their own performances and adjust their controllers to improve them. Even though their interest centered around adaptation in changing environments, from the beginning, for the sake of mathematical tractability, they confined their attention to the control of linear time-invariant systems with unknown constant parameters. The accepted philosophy among adaptive control theorists was that if an adaptive system was fast and accurate when plant parameters are constant, it would also prove satisfactory when the parameters varied with time, provided the latter occurred on a relatively slower time scale. During the 1970s adaptive control made great strides and the proof of stability of ideal adaptive systems was given by many authors in 1980 [1, 2, 3]. Following this, in the 1980s, the field of robust adaptive control came into existence to deal with the problems that arise when ideal assumptions made for deriving elegant mathematical solutions are no longer valid under real world conditions. These include external disturbances, unmodeled dynamics, and time variations in plant parameters, and a wide spectrum of methods were developed to deal with them [4].

As systems became more complex and controllers were called upon to cope with sudden and large variations in plant parameters, the limitations of classical adaptive control became increasingly evident. Adaptive controllers performed very well when initial parameter errors were small, but resulted in large transients when the parameter errors were large. This in turn made them unsatisfactory in many practical applications. It was also realized that such behavior was intrinsic to the class of adaptive systems being investigated.

The use of multiple models is not new in control theory. Multiple Kalman filters were proposed in the 1960s and 1970s by Magill [5], Lainiotis [6] and Athans et al. [7] to improve the accuracy of the state estimate in control problems. Moose and co-workers [8] used these ideas in maneuverable target tracking while Kaufman and colleagues [9] applied it to medical control. Fault detection and control in aircraft was proposed by Maybeck and Pogoda in [10] and in the following years Maybeck has used the concept extensively in controlling aircraft systems. Bar-Shalom and co-workers have adapted the above concepts in air traffic control [11]. In all the above cases, no switching is involved, and only a linear combination of the control determined by the different models is used to control the system.

In the late 1980s switching assumed importance in the context of stabilization of adaptive systems. The new trend was started in 1986 by Mårtensson [12] to relax the conditions under which unknown linear systems can be adaptively stabilized. This was soon followed by both direct and indirect switching methods appearing in the literature. However, it was soon realized that direct switching schemes had little practical utility. In indirect switching schemes, multiple models were used both to determine when and to which controller one should switch at every instant [13, 14, 15] and were found to be attractive for practical applications. Robust set point control using multiple models was studied in [16] and [17]. In recent years there has also been a great deal of research activity in extending the multiple model approach to the modeling and control of nonlinear systems. In a recent book [18], articles dealing with multiple model approaches based on classical control theory, statistical methods, and fuzzy architectures have been collected together. While the book contains numerous interesting heuristic ideas, the question of stability is not addressed in it.

The idea of using both switching and tuning with multiple models to improve performance in rapidly time-varying environments was first proposed in [19] and proved to be stable in [20], when the plant is linear and time-invariant with unknown parameters. The possibility of extending the concepts to nonlinear systems using neural networks was also suggested in [21]. Since considerable computational resources are currently available, the control strategies that would be optimal for different situations can, in theory, be computed and stored by the controller. When plant parameters vary rapidly, only a procedure for determining the appropriate strategy would be needed, i.e., a rational way of choosing one of the stored strategies to control the plant. As described in the following sections, the predictive abilities of the different models were compared using performance indices based on estimation errors, and the controller corresponding to the minimum was deemed the most appropriate for the given situation.

The approach described above, which started as a specific method for adaptively controlling a plant, has gradually evolved over the past years into a general methodology. It has brought about numerous changes in the attitudes of adaptive control theorists, two of which are worth noting. The first is concerned with the very nature of adaptation. When a finite number of fixed models are used to decide which controller is to be chosen at any instant, the control parameters change discontinuously and by finite amounts (switching). This choice of control parameters is also an adaptive process. Thus the multiple model approach freed the control theorist from the traditional view that all adaptation has to incremental. It soon came to be realized that discontinuous jumps in control parameters may be needed to cope with sudden changes in system characteristics, while conventional adaptive control (tuning) based on incremental changes could be used to improve performance at a slower rate.

The second fact that is worth noting is that any adaptive approach that has proved successful with a single model in a specific region in parameter space can be extended in a relatively straightforward manner to a union of such regions in parameter space, with the use of multiple models. The latter merely expands the region of efficacy of conventional adaptive controllers. It is this fact that has been mainly responsible for the evolution of the multiple model approach into a methodology.

Numerous adaptive/robust control methods have been developed in the past for stochastic systems, and specific classes of nonlinear systems using a single model. In this paper, it is shown that all these methods can be extended to the multiple model case.

2. Adaptive Control — Continuous-Time Deterministic Systems

The multiple model, switching, and tuning (MMST) approach for adaptively controlling a linear time-invariant continuous-time dynamical system was first proposed in [19]. Most of the basic concepts related to this approach, which are currently being used in other contexts, were introduced in this paper. While detailed mathematical derivations cannot be included here due to space limitations, the principal ideas contained in [19] are described briefly in this section.

Statement of the Problem:. A single-input single-output plant Σ_p is described by the equation

$$\Sigma_p : \qquad y_p = W_p(s)u \tag{1}$$

where u and y are respectively the input and the output, and $W_p(s)$ is a transfer function with zeros in the left-half plane (minimum phase). The relative degree of $W_p(s)$ is known but its parameters are assumed to be unknown. A reference model Σ_m is described by the following equation

$$\Sigma_m : \qquad y_m = W_m(s)r \tag{2}$$

where r is a bounded reference input, $W_m(s)$ is a stable minimum phase transfer function with the same relative degree as the plant. The standard Model Reference Adaptive Control (MRAC) problem is to determine a bounded control input u such that $y_p(t)$ tracks the desired output $y_m(t)$ asymptotically with zero error, i.e.,

$$\lim_{t\longrightarrow\infty} |y_p(t) - y_m(t)| \triangleq \lim_{t\longrightarrow\infty} |e_c(t)| = 0$$

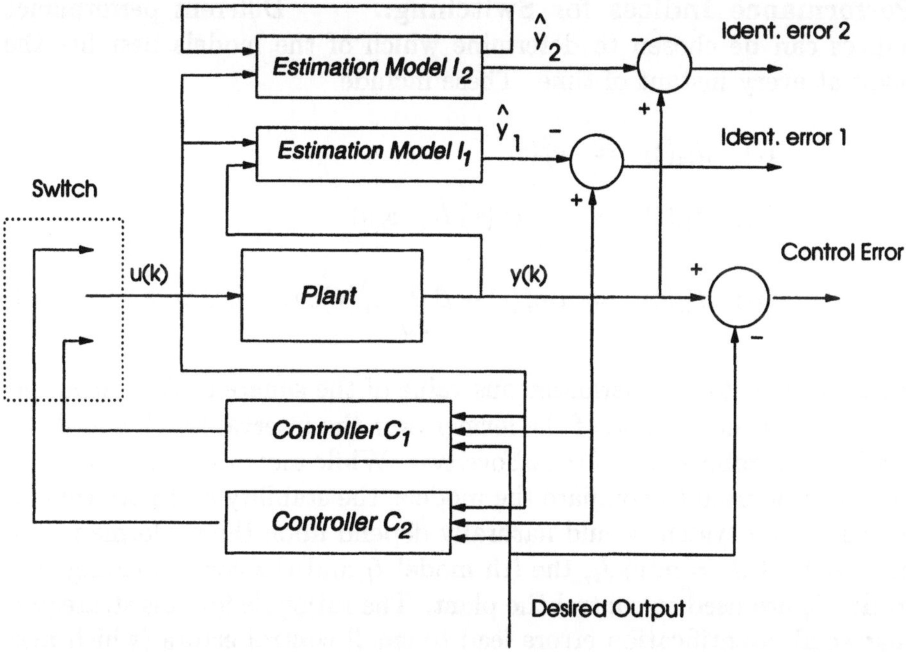

Figure 5.1. **Multiple model, switching and tuning scheme.**

The Adaptive Control System (MMST):. **Numerous (equivalent) solutions to the above problem are known. If indirect control is used, the parameters of Σ_p are estimated using an identification model, and at every instant the control input to the plant is chosen assuming that the estimated plant parameters are the same as the true plant parameters. This is generally referred to as the certainty equivalence principle and we shall refer to u as the certainty equivalence control. As stated earlier, the solution to this problem was given in 1980 by several authors.**

In the MMST approach the single identification model is replaced by M identification models I_i ($i = 1, 2, \ldots, M$) in parallel. The structure of the adaptive control system is shown in Fig. 1. The M estimates of the output are denoted by $y_i(t)$, and the corresponding output estimate errors are $e_i(t) = y_i(t) - y_p(t)$. Corresponding to each identifier I_i, there exists a controller C_i which generates a certainty equivalence control u_i based on I_i. The question then is to decide which of the control inputs $u_i(t)$ is to be used at every instant t. This is based on performance indices $J_i(e_i(t))$ as described below.

Performance Indices for Switching:. Different performance indices can be chosen to determine which of the models best fits the plant at every instant of time. These include

$$\begin{aligned}
(i)\ J_{1i}(t) &= e_i^2(t), \\
(ii)\ J_{2i}(t) &= \int_0^t e_i^2(\tau)d\tau, \text{ and} \\
(iii)\ J_{3i}(t) &= \alpha e_i^2(t) + \beta \int_0^t e_i^2(\tau)d\tau, \quad \alpha, \beta > 0. \qquad (3)
\end{aligned}$$

$J_{1i}(t)$ represents the instantaneous value of the square of the output error, $J_{2i}(t)$ is the integral of the former over the interval $[0,\ t]$, and $J_{3i}(t)$ is a linear combination of the above two. While any one of the above indices can be used to compare the models, the stability and performance of the overall system would naturally depend upon the performance index used. If $J_l = \min_i J_i$, the lth model I_l and the corresponding controller C_l are used to control the plant. The rationale for this strategy is that small identification errors lead to small control errors (which may not always be true). Since J_{1i} leads to rapid switching, and J_{2i} to infrequent switching between controllers but larger control errors, J_{3i} was chosen in [20] as the performance index. Further, to avoid switching at arbitrarily high frequencies, an interval $T_{\text{min}} > 0$ is allowed to elapse between switchings.

Structure of the Models (Fixed and Adaptive):. Numerous questions arise while choosing the models I_i. One of the more important questions is whether I_i should have constant parameters or adaptive parameters. After considerable simulation and theoretical studies, it was decided to use a combination of the two, i.e., some models were chosen to have fixed coefficients while others were adjusted adaptively. The reasoning behind such a choice goes as follows:

1 To construct models suitable for adaptive control, the plant has to be parameterized in such a fashion that its output can be expressed as $y = p^{*T}\bar{\omega}$, p^*, $\bar{\omega} \in \mathbb{R}^{2n}$ where the unknown parameter vector p^* of the plant occurs linearly, and $\bar{\omega}$ is a known vector of time functions.

2 The models I_i are consequently chosen as $\hat{y}_i = \hat{p}_i\bar{\omega}$. The model I_i is said to be fixed if $\hat{p}_i$ is a constant, and adaptive if $\hat{p}_i$ is varied adaptively. Computation of $\hat{p}_i$ in the former case is trivial while the computational overhead in the latter case is considerably greater.

3 By judiciously choosing constant vectors $\hat{p}_i$ in parameter space (i.e., fixed models) based on the past performance of the plant, we assure that the plant parameter vector p^* is sufficiently close to one of them.

4 Two additional adaptive models are also included to improve the performance. One is a free running adaptive model, while the second is initialized at exactly the same point in parameter space where one of the fixed models has the minimum performance index. In [20], it was shown that switching between $M - 2$ fixed models ($M > 2$) and two adaptive models using the performance index J_{3i}, and tuning the last two was theoretically stable and gave the best results in simulation studies.

As stated earlier, this is the same basic procedure used in Sections 4 and 5 for linear stochastic control and nonlinear adaptive control.

3. Adaptive Control — Discrete-Time

In the previous section, adaptive control of a linear continuous-time deterministic dynamical system using multiple models, switching and tuning was described. In this section we discuss the extensions of the same methods to discrete-time systems. Three main reasons can be given [22] for considering discrete-time systems. Most complex systems are controlled by computers, which are discrete in nature, and this is an obvious reason for describing them. The second and significantly more important one is the fact that the presence of random noise can be dealt with more easily in the case of discrete-time systems. Finally, our ultimate aim is to apply the proposed methodology to nonlinear systems using artificial neural networks, and this, in turn, requires the use of discrete-time models.

The proof of stability in both deterministic and stochastic cases of multiple model-based discrete-time adaptive control has been treated in detail in [22]. Hence, we merely present in this section the highlights of the proof in the deterministic case and indicate the principal theoretical difficulties that are encountered in the stochastic case. Following this, recent modifications in the approach to improve the performance of the adaptive system [23] are presented, and simulation results are included to demonstrate the improvements in performance that can be achieved.

Parameter Estimation:. Consider the discrete time equation

$$y(k) = \phi^T(k-1)\theta_0 \qquad k \in N \tag{4}$$

where k denotes the time instant, $\phi(k-1) \in \mathbb{R}^n$ is the regression vector at time k, $\theta_0 \in \mathbb{R}^n$ is an unknown constant parameter vector that has to be estimated, and $y(k)$ is a scalar output at time k. The aim of the estimation procedure is to estimate θ_0 recursively using all available data. The Recursive Least Squares (RLS) algorithm, which is both fast and robust and has stood the test of time, is one of the most commonly used. It has the following form:

$$\hat{\theta}(k) = \hat{\theta}(k-1) + P(k-1)\phi(k)e(k) \tag{5}$$

where $P(k)$ is a positive definite matrix computed on-line as

$$P(k-1) = P(k-2) - \frac{P(k-2)\phi(k-1)\phi^T(k-1)P(k-2)}{1+\phi^T(k-1)P(k-2)\phi(k-1)} \tag{6}$$

and $P(-1)$ is any positive definite matrix P_{-1}. The consequences of using the RLS algorithm are summarized in Lemma 1 and are relevant to all the developments in this section.

Lemma 1 *If a system is described by equation (4) and the RLS algorithm given by (5) and (6) is used, we have the following results:*

$$\text{(a)} \quad ||\hat{\theta}(k) - \theta_0||^2 \leq K_1 ||\hat{\theta}(0) - \theta_0||^2 \tag{7}$$

$$\text{(b)} \quad \lim_{N\to\infty} \sum_{k=1}^{N} \frac{e(k)^2}{1+\phi^T(k-1)\phi(k-1)} < \infty \tag{8}$$

$$\text{(c)} \quad \lim_{N\to\infty} \sum_{k=1}^{N} \frac{e(k)}{[1+\phi^T(k-1)\phi(k-1)]^{\frac{1}{2}}} = 0 \tag{9}$$

$$\text{(d)} \quad \lim_{k\to\infty} ||\hat{\theta}(k) - \hat{\theta}(k-N)|| = 0 \quad \textit{for any finite } N. \tag{10}$$

For a detailed proof of the above lemma the reader is referred to [24]. From inequalities (7)–(10) it follows that the parameter error vector is bounded, the changes in the parameter estimates over a finite interval tend to zero, and most important of all, even if the regression vector grows in an unbounded fashion, the prediction error $e(k)$ can grow only at a slower rate. As seen in the following sections, Lemma 1 is central to the proof of stability both when a single model is used and when multiple models are used.

Statement of the Adaptive Control Problem (Deterministic Case):. The system Σ that has to be controlled is described by the equation

$$\Sigma: \qquad y(k+d) = \sum_{i=0}^{n_A-1} a_i y(k-i) + \sum_{j=0}^{n_B-1} b_j u(k-j) \tag{11}$$

where the parameters $a_i (i = 1, 2, \ldots, n_A)$ and $b_j (j = 1, 2, \ldots, n_B)$ are unknown. The objective of the deterministic control problem is to determine a bounded control input $u(k)$ such that the output $y(k)$ of Σ asymptotically tracks a specified arbitrary bounded reference output $y^*(k)$. It is assumed that the controller of Σ knows the value of $y^*(k+d)$ at time k.

An indirect adaptive control approach is used to determine the control input $u(k)$. The predictive model of Σ is described by

$$\hat{y}(k+d) = \sum_{i=1}^{n-1} \hat{a}_i(k) y(k-i) + \sum_{j=0}^{m-1} \hat{b}_j(k) u(k-j) \tag{12}$$

where $\hat{a}_i(k)$ and $\hat{b}_j(k)$ are the parameter estimates of a_i and b_j respectively at time k.

Adaptive Control Using a Single Model:. If the model given by eqn. (12) is used to determine the control input, the parameters $\hat{a}_i(k)$ and $\hat{b}_j(k)$ are estimated at every instant k as described earlier. The control input $u(k)$ at that instant is then computed using the certainty equivalence principle by which the estimated values are assumed to be the true values of the plant parameters. In eqn. (12) all the terms other than $u(k)$ are known, and hence, $u(k)$ can be computed so that $\hat{y}(k+d)$ is the same as the desired value.

Proof of Stability:. If Σ is assumed to be minimum phase, it has been shown [22] that all its signals are bounded and that the output error tend to zero asymptotically. The proof can be succinctly stated in three steps:

1 $e_c(k) = e(k) + \phi^T(k-d)[\hat{\theta}(k-1) - \hat{\theta}(k-d)]$, where $e_c(k)$ is the control error and $e(k)$ is the estimation error.

2 From Lemma 1 and the equation in 1 relating $e_c(k)$ and $e(k)$ we have

$$\lim_{k\to\infty} \frac{e_c(k)^2}{[1+\phi^T(k-1)\phi(k-1)]^{\frac{1}{2}}} = 0 \tag{13}$$

or $e_c(k)$ (even if it is unbounded) can grow only more slowly than the regression vector.

3 By the assumption that Σ is minimum phase we have $|u(k-d)| \leq m_1 + m_2 \max_{1\leq\tau\leq k} |y(\tau)|$, $m_1, m_2 > 0$ which in turn implies that $||\phi(k-d)|| \leq m_3 + m_4 \max_{1\leq\tau\leq k} |y(\tau)|$, $m_3, m_4 > 0$.

From 1–3 it therefore follows that

$$||\phi(t-d)|| \leq c_1 + c_2 \max_{1\leq\tau\leq k} e_c(\tau) \quad m_1, m_2 > 0.$$

Since $e_c(k)$ grows slower than $y(k)$ according to (13), and at the same rate as $y(k)$ according to (14) (since $y(k) = e_c(k) + y^*(k)$, where $y^*(k)$ is bounded), we have a contradiction. Hence $y(k)$ is bounded. This implies that the input and the regression vector are bounded, and hence, $e_c(k)$ and $e(k)$ tend to zero asymptotically.

Proof of Stability with Multiple Models:. In the deterministic case, the proof given above for stability using a single model can be extended in a straightforward manner to the case of multiple models, when the parameters of the models are adjusted adaptively. In place of a single model, we have M models with parameter estimates $\hat{a}_i^l$ and $\hat{b}_j^l$ corresponding to the lth model. The parameters of all the models are estimated using the same algorithm, except that the initial conditions are different. Assuming that a model is chosen arbitrarily at any instant, and the control input is determined using that model, it can be shown that the control error of the plant will remain bounded and tend to zero asymptotically. An important consideration in proving stability is that the regression vectors of all the models are identical.

Comment 1 For the continuous-time problem, as stated in Section 2, it was first recommended in [19] that $M-2$ fixed models, one free running adaptive model and a reinitialized adaptive model should be used to obtain satisfactory performance with reasonable computation overhead. The same procedure is also found to be the best compromise in the discrete-time case.

Comment 2 Since switching between models is based on the performance criterion J_{3i} in equation (3), which in turn depends on output identification errors $e_i(k)$, all the methods proposed converge (generically) to one of the two adaptive models. This is because the outputs of the fixed models do not (generically) tend to zero for time-varying reference inputs.

4. Adaptive Control Problem — The Stochastic Case

The discussions in the previous subsections provide the background for considering the stochastic adaptive control problem using multiple models. The stochastic adaptive control problem using a single identification model was solved in [25], and recently the results were extended to the multiple model case in [22]. Without going through all the details of the proof of stability, we merely indicate the principal differences between the deterministic and stochastic cases, and discuss briefly the theoretical questions that arise in the proof of convergence when multiple models are used.

When an additional input $w(k)$ (besides the control input $u(k)$) is present, the extended ARMA model

$$\Sigma_s : \qquad A(q^{-1})y(k) = q^{-d}B(q^{-1})u(k) + C(q^{-1})w(k), \qquad (14)$$

is a natural description of the linear system, where A and C are monic polynomials and B is a stable polynomial, of degrees n_A, n_C and n_B respectively in the backward shift operator q^{-1}. It is generally assumed that $C(q^{-1})$ has roots strictly inside the unit circle. $w(k)$ is assumed to be a white noise sequence with zero conditional mean and finite variance so that
(i) $E[w(k)|k-1] = 0$
(ii) $E[w(k)^2|k-1] = \sigma^2$ and
(iii) $\lim_{N\to\infty} \sup \frac{1}{N}\sum_{k=1}^{N} w^2(k) < \infty \quad a.s.$

The basic idea behind estimation methods based on whitening of the prediction error is to construct a predictor that will produce an estimation error that will become white asymptotically. If the parameters of the plant are known, and the ARMAX plant representation (15) is used, the output of a predictor model $\hat{y}(k+1)$ that will give a white prediction error will have the form

$$\hat{y}(k+1) = -A^*(q^{-1})y(k) + B^*(q^{-1})u(k-d) + C^*(q^{-1})w(k) \qquad (15)$$

since the prediction error $y(k+1) - \hat{y}(k+1)$ is $w(k+1)$. However, Eq. (16) merely provides motivation for the choice of the estimator structure, since $w(k)$ in (16) cannot be measured.

When the plant parameters are unknown they are estimated on-line, and $A^*(q^{-1})$, $B^*(q^{-1})$ and $C^*(q^{-1})$ are replaced by $\hat{A}^*(q^{-1})$, $\hat{B}^*(q^{-1})$

and $\hat{C}^*(q^{-1})$ respectively. Further, in view of our objective that the prediction error should become asymptotically white, $w(k)$ in Eq. (16) is replaced by the a posteriori prediction error $e(k)$ (defined in what follows in Eq. (23)). In such a case, the model in the estimation procedure has the form

$$\hat{y}^0(k+1) = -\hat{A}^*(q^{-1})y(k) + \hat{B}^*(q^{-1})u(k-d) + \hat{C}^*(q^{-1})e(k) \tag{16}$$

or alternately

$$\hat{y}^0(k+1) = \hat{\theta}^T(k)\phi(k) \tag{17}$$

where

$$\hat{\theta}^T(k) = [\hat{a}_1(k), \ldots, \hat{a}_{n_A}(k), \hat{b}_1(k), \ldots, \hat{b}_{n_B}(k), \hat{c}_1(k), \ldots, \hat{c}_{n_C}(k)] \tag{18}$$

is the estimate of the plant parameter vector, and

$$\begin{aligned}\phi(k)^T = [&-y(k), \ldots, -y(k-n_A+1), u(k-d), \ldots, u(k-d-n_B+1), \\ &e(k), \ldots, e(k-n_C+1)]\end{aligned} \tag{19}$$

is the regression vector . $\hat{y}^0(k+1)$ in Eqs. (17) and (18) is called the a priori estimate of $y(k+1)$ since it is the estimate obtained by using the parameter estimate at time k. At time $k+1$, once the parameter estimate $\hat{\theta}(k+1)$ has been determined, we have an a posteriori estimate of the output denoted by $\hat{y}(k+1)$, which is defined by

$$\hat{y}(k+1) = \hat{\theta}^T(k+1)\phi(k) \tag{20}$$

Using $\hat{y}^0(k+1)$ and $\hat{y}(k+1)$ we define the a priori and a posteriori estimation errors, $e^0(k+1)$ and $e(k+1)$ as

$$e^0(k+1) = y(k+1) - \hat{y}^0(k+1) \tag{21}$$
$$e(k+1) = y(k+1) - \hat{y}(k+1) \tag{22}$$

It is also seen that the regression vector $\phi(k)$ given by Eq. (20) has for its elements $y(k), u(k-d+1), e(k)$ and their past values. While the former signals are accessible, $e(k)$ as defined by Eq. (23), depends upon the estimation model used.

Given system (14) with coefficients of the polynomials A, B and C unknown, the objective is to determine a minimum-variance control input such that

$$\text{(i)}\ \lim_{N\to\infty} \sup \frac{1}{N}\sum_{k=1}^{N} u^2(k) < \infty \qquad a.s.$$

$$\text{(ii)}\ \lim_{N\to\infty} \sup \frac{1}{N}\sum_{k=1}^{N} y^2(k) < \infty \qquad a.s.$$

$$\text{(iii)}\ \lim_{N\to\infty} \frac{1}{N}\sum_{k=1}^{N} E[y(k) - y^*(k)|k-1]^2 = \sigma^2\ a.s. \tag{23}$$

Comment 3 Even in the non-adaptive case where the parameters of the system Σ_s are known, if the performance criterion $E[y(k+d)-y^*(k+d)]^2$ is to be minimized, it can be shown that the best that the controller can do is to minimize $E[y^0(k+d) - y^*(k+d)]^2$ where $y^0(k+d)$ is the optimal d-step ahead prediction of $y(k)$. This implies that a suitable parameterization must be found such that $y^0(k+d)$ can be expressed in terms of the quantities that are defined at or before k. If such a representation can be expressed as

$$y^0(k+d) = \phi_0^T(k)\theta_0$$

where θ_0 is the unknown parameter vector, a procedure similar to that in the deterministic case can be used. In [24] it was shown that this leads to the solution of the minimum variance control problem.

4.1 Analytical Difficulties in the Multiple Model Case

The difficulties in the adaptive control problem using multiple models arise from the fact that the regression vector $\phi_0(k)$ contains in addition to the past values of the output $y(k)$ and the input $u(k)$, the past estimates $y^0(k), y^0(k-1), \ldots$. Hence, during adaptation, when the system switches between the different controllers, the regression vector is no longer the same. The principal contribution of [22] was the demonstration that all the regression vectors are equivalent in a certain sense, and that the overall system continues to behave as in the single model case.

4.2 Stochastic Adaptive Control Using Different Estimation Models

In all the control problems — both deterministic and stochastic — considered thus far, the structure of the M models used to identify the plant are identical. It is only the initial values of the adaptive parameters that are different. In this section we describe recent extensions [23] of the stochastic control problem described earlier which include different estimation models. The need for using structurally different multiple estimation models arises from the fact that it is not known a priori which of them is best suited for the adaptive control problem. Once again, we shall not dwell on the details of the proof but merely focus on the key issues involved in the adaptive procedure.

Stochastic Parameter Estimation Methods:. Identification of linear systems in the presence of noise is a well investigated field. Numerous estimation procedures have been proposed and their convergence properties have been investigated in detail. These include the Extended Least Squares (ELS), the Output Error with Extended Prediction Model (OEEPM), the Recursive Maximum Likelihood (RML), and the Recursive Generalized Least Squares (RGLS) methods. If two or more of these are used to estimate the unknown parameters of the plant, and the choice of the controller is based on the estimation procedure which yields the minimum identification error according to an index of performance, our interest is in deciding whether the overall system will be stable, and whether the improvement in performance will justify the additional effort involved. These questions are addressed in [23] and the results can be summarized as follows:

- If (for example) the ELS and OEEPM models are used to identify Σ_s and the corresponding controllers are chosen randomly, the overall system converges to the minimum variance control. This is because both ELS and OEEPM converge to minimum variance control when used separately. In addition, both procedures result in equivalent sets of inequalities (similar to (7)–(10)) related to parameter and output errors.

- When other estimation procedures are used, switching to the controllers corresponding to them must be carried out only when the respective conditions for convergence are satisfied. In some cases, even when the plant output can be shown to be bounded, the outputs of some of the estimators may become unbounded and may have to be reinitialized.

- Simulations studies have clearly demonstrated that significant improvement in performance can be achieved using multiple estimation models.

Simulation:. The plant to be controlled is described by the difference equation

$$y(k+1) = 0.97y(k) + 0.264y(k-1) + 0.66u(k) + 0.528u(k-1) + 0.5w(k) + 0.4w(k-1) + w(k+1).$$

The parameters of the plant are unknown and $w(k)$ is a white noise sequence with $E[w(k)|k-1] = 0$ and $E[w^2(k)|k-1] = \sigma^2$. The variance of the noise, σ^2, is a parameter that is varied in the simulation study. The objective is to asymptotically track a desired output $r(k) = y^*(k+1)$ with minimum error. The desired output was chosen to be $y^*(k+1) = 0.5\sin(\frac{2\pi k}{50}) + 0.5\sin(\frac{2\pi k}{150})$.

Three free running estimation models were used in the adaptive procedure i.e. ELS, OEEPM and RML models. The amplitude of the noise was increased successively from zero (no noise) over the interval $0 \le k \le 100$, to two different noise sequences over the intervals $101 \le k \le 300$ ($\sigma^2 = 0.003$), and $k > 300$ ($\sigma^2 = 0.0141$) respectively. The RML model was initialized with the estimates of the ELS model after 150 steps. It is seen in Figure 2 that the system switches from the ELS model to the RML model and finally to the OEEPM model.

5. Nonlinear Adaptive Control

In the past ten years linear adaptive control methods have been extended to nonlinear control problems in which the unknown parameters appear linearly. The difficulties encountered with the transient responses when parameter errors are large, described in Sections 1–4 in the linear case, are further exacerbated in the nonlinear control problem, and hence the need for using multiple models is even greater in such cases.

Trajectories of nonlinear dynamical systems can exhibit a finite escape time. Hence, if such systems are to be controlled adaptively, the control action has to be taken over a relatively short interval of time. Even though the global stability of nonlinear systems has been established using suitable Lyapunov functions, simulation studies reveal that the transient responses of such systems have large amplitudes and sometimes even grow without bound, if the initial value of the parameter error is large. This may be attributed to the round-off errors caused by the computer's finite precision arithmetic. In view of the above considerations, if adaptive control is to be used in practical nonlinear systems, it appears to be mandatory that the initial parameter errors be sufficiently small, and this in turn calls for an approach similar to that proposed in

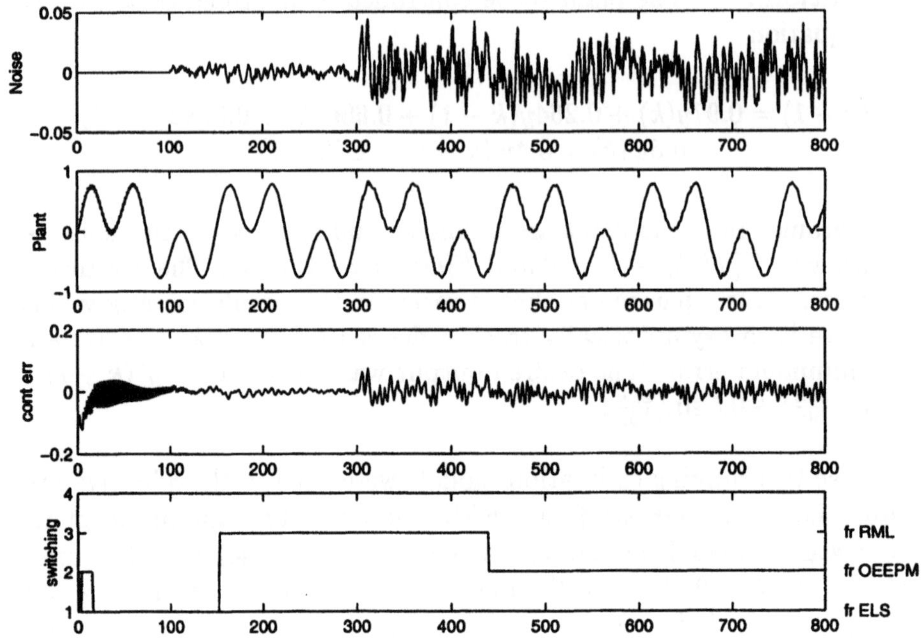

Figure 5.2. Switching according to $J_i = \frac{1}{N}\sum_{k=1}^{N} e_i^2(k)$.

this paper.

In this section, we consider the adaptive control of continuous-time nonlinear dynamical systems in which the unknown parameters occur linearly. In all cases, the entire state vector of the plant is assumed to be accessible, and the system has a triangular structure as described in [26, 27]. We first discuss the simpler case where the input can directly compensate for the nonlinear term (matched case) and later consider the more complex case where such compensation is not possible. The stability proofs given in the literature for the problems considered are based on direct adaptive control and do not involve estimation models. The main contribution of this section is consequently the development of an indirect procedure (which is essential for the multiple model approach) and establishing its global stability.

Case 1:. A plant Σ_{nl} is described by the differential equation

$$\Sigma_{nl}: \quad \begin{aligned} \dot{x}_{(1)} &= x_{(2)} \\ \dot{x}_{(2)} &= x_{(3)} \\ &\vdots \\ \dot{x}_{(n-1)} &= x_{(n)} \\ \dot{x}_{(n)} &= f^T(x_{(1)}, \cdots, x_{(n)})\theta + u \end{aligned}$$

Equivalently,

$$\Sigma_{nl}: \quad \dot{x} = Ax + b\left(f^T(x)\theta + u\right) \tag{24}$$

where $\theta \in \mathbb{R}^p$ is the unknown parameter vector, $x = (x_{(1)}\ x_{(2)}\ \cdots\ x_{(n)})^T$ is the state vector which can be measured, $f(x)$ is a known continuous vector field and $b = \begin{pmatrix} 0 & \cdots & 0 & 1 \end{pmatrix}^T \in \mathbb{R}^n$. The unknown parameters are assumed to belong to a compact set S. The objective is to ensure that the state $x(t)$ of the above system tracks the state $x_m(t)$ of the following reference model:

$$\dot{x}_m = A_m x_m + br \tag{25}$$

where A_m is stable matrix in companion form and $r(t)$ is a piecewise continuous bounded function.

The Methodology:. As in the linear deterministic and stochastic adaptive control problems discussed in earlier sections, M identification models are chosen with different values of the unknown parameter and the one which minimizes the performance function J_{2i} or J_{3i} described in Section 2 is chosen at any instant to control the system.

Control Using a Single Model:. From the structure of the matrices A_m and A, it is obvious that a constant vector k exists such that $A_m = A + bk^T$. If the input $u(t)$ is chosen as $v(t) + k^T x(t)$, equation (24) reduces to the form

$$\dot{x} = A_m x + b\left(f^T(x)\theta + v\right) \tag{26}$$

To estimate the unknown parameter θ, an estimation model Σ_{nle} is defined as

$$\Sigma_{nle}: \quad \dot{\hat{x}} = A_m \hat{x} + b\left(f^T(x)\hat{\theta} + v\right) \tag{27}$$

If $\hat{x}(t) - x(t) \triangleq e(t)$, and $\hat{\theta}(t) - \theta \triangleq \tilde{\theta}(t)$, we obtain the error equation

$$\dot{e}(t) = A_m e(t) + bf^T(x)\tilde{\theta}(t) \tag{28}$$

Using standard arguments from linear adaptive control theory, the adaptive law for adjusting $\hat{\theta}(t)$ can be chosen as

$$\dot{\hat{\theta}} = \dot{\tilde{\theta}} = -e^T(t)Pbf(x)$$

so that $V(e,\tilde{\theta}) = e^T Pe + \tilde{\theta}^T\tilde{\theta}$ is a Lyapunov function with $\dot{V} = -\tilde{x}^T Q\tilde{x} \leq 0$ where $P = P^T > 0$ is the symmetric positive definite matrix solution of the Lyapunov equation,

$$A_m^T P + PA_m = -Q, \qquad Q = Q^T > 0$$

Hence it directly follows [28] that $e \in \mathcal{L}^2 \cap \mathcal{L}^\infty$ and $\tilde{\theta} \in \mathcal{L}^\infty$. If the input $v(t)$ in eqn. (26) is chosen to be $v(t) = r(t) - f^T(x)\hat{\theta}(t)$, then the control error $e_c = x_m - x$ satisfies $\dot{e}_c = A_m e_c + bf^T(x)\tilde{\theta}(t)$ which is identical to (28) except for initial conditions. Thus, $\dot{e} - \dot{e}_c = A_m(e - e_c)\ \forall\, t \in [0,\infty)$ and represents stable dynamics, so that $e_c \in \mathcal{L}^\infty$. From equation (29) it follows that $\dot{e} \in \mathcal{L}^\infty$, and $\lim_{t\longrightarrow\infty} e(t) = 0$.

Control Using Multiple Models:. When multiple identification models $(I_1, \ldots, I_M)$ are used to estimate the parameter θ, they are described by the differential equations

$$\dot{x}_i = A_m x_i + b\left(f^T(x)\theta_i + v\right) \qquad 1 \leq i \leq M \tag{29}$$

where θ_i is the estimate of θ given by the ith model. The procedure for the choice of the model to be used for control purposes follows along the same lines as before, and is based on a performance criterion J_i as described in Section 2.

Theorem 1 *If the plant is described by eqn. (24), the identification models by eqn. (29), all adaptive parameters are adjusted according to the law*

$$\dot{\theta}_i = -e_i^T Pbf(x) \tag{30}$$

and switching between models (and hence controllers) is based on a performance criterion J_{3i} described earlier, all the signals in the system remain bounded, and the plant state $x(t)$ will track the desired output $x_m(t)$ asymptotically with zero error in the sense

$$\lim_{t\longrightarrow\infty} \|x(t) - x_m(t)\| = 0$$

Proof: When all models are adaptive, the error equations have the form

$$\begin{aligned} \dot{e}_i &= A_m e_i + bf^T(x)\tilde{\theta}_i \\ \dot{\tilde{\theta}}_i &= -e_i^T Pbf(x) \end{aligned} \tag{31}$$

where $\tilde{\theta}_i = \theta_i - \theta$ is the parameter error of I_i. If $V_i(e_i, \tilde{\theta}_i) = e_i^T P e_i + \tilde{\theta}_i^T \tilde{\theta}_i$ is a candidate Lyapunov function for I_i, it readily follows that along the trajectories of the system, $\dot{V}_i = -e_i^T Q e_i \leq 0$, where $P = P^T > 0$ is the positive definite solution to the Lyapunov equation $A_m^T P + P A_m + Q = 0$, $Q = Q^T > 0$. Hence, $e_i \in \mathcal{L}^2 \cap \mathcal{L}^\infty$ and $\tilde{\theta}_i \in \mathcal{L}^\infty$ for $i = 1, \ldots, M$. At any instant let the control law $v(t) = -f^T(x)\hat{\theta} + r(t)$ be used, where $\hat{\theta} = \theta_j$ for some j. If $\hat{\theta} = \theta_j$ over an interval, the identification error dynamics and the control error dynamics of I_j are identical since $\dot{e}_j = A_m e_i + b f^T(x)\tilde{\theta}_j$ and $\dot{e}_c = A_m e_c + b f^T(x)\tilde{\theta}_j$; i.e,

$$\dot{e}_j - \dot{e}_c = A_m(e_j - e_c) \tag{32}$$

Since $e_j(t)$ is bounded for $t \geq 0$ and $j = 1, \ldots, M$, and eqn. (32) is satisfied at every instant for some j, it follows that $e_c(t)$ is bounded. This in turn implies that the plant state $x(t) \in \mathcal{L}^\infty$. From eqn. (31) it then follows that all the identification errors $e_i(t) \longrightarrow 0$ asymptotically. Hence, from eqn. (32), we have

$$\lim_{t \longrightarrow \infty} e_c(t) = 0$$

Comment 4 If all the identification models are adaptive, it follows that exact asymptotic tracking is achieved for arbitrary switching between controllers C_i. Also, stability and performance are decoupled. When fixed models are present, stability can no longer be guaranteed for arbitrary switching between models, and the controller which minimizes a performance index is chosen. In the generic case when there are fixed and adaptive models, the proof of stability is based on the fact that the identification error of the adaptive model grows at a smaller rate when compared to that of the fixed model. Therefore, there exists a finite time T such that the system switches to the controller corresponding to the free running adaptive model.

Simulation 1:. We illustrate the drawbacks of using a single model, and the advantages of using multiple models with the following third order example with three unknown parameters:

$$\begin{aligned}
\dot{x}_{(1)} &= x_{(2)} \\
\dot{x}_{(2)} &= x_{(3)} \\
\dot{x}_{(3)} &= \theta_{(1)} x_{(2)}^2 + \theta_{(2)} x_{(1)} x_{(2)}^2 + \theta_{(3)} x_{(3)}^3 + u
\end{aligned}$$

The parameters are known to lie in the compact set $\mathcal{S} = [-2, 2] \times [-3, 3] \times [-4, 4]$, and our objective is to ensure that the state $x(t)$ of this plant

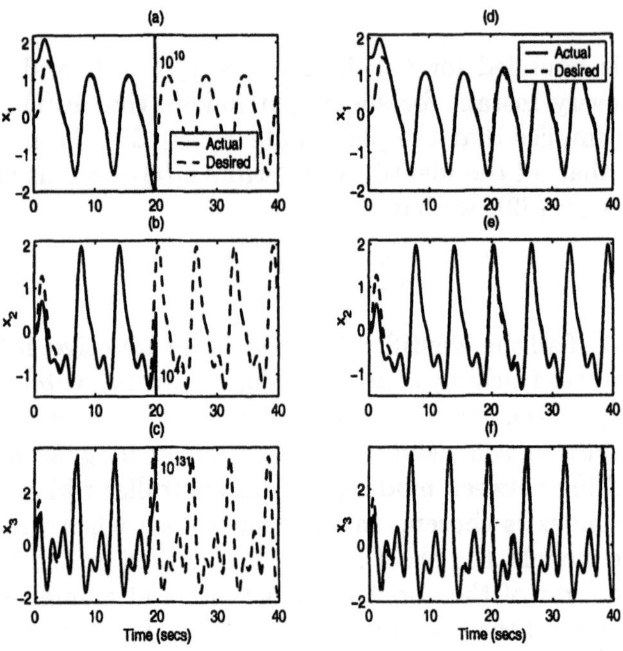

Figure 5.3. The states of a third order nonlinear system when adaptively controlled using one model (Parts (a)–(c)) and using multiple models (Parts (d)–(f)).

tracks the state $x_m(t)$ of the following reference model:

$$\begin{aligned}\dot{x}_{m,(1)} &= x_{m,(2)}\\ \dot{x}_{m,(2)} &= x_{m,(3)}\\ \dot{x}_{m,(3)} &= -k_1 x_{m,(1)} - k_2 x_{m,(2)} - k_3 x_{m,(3)} + r\end{aligned}$$

where $k = \begin{pmatrix} k_1 & k_2 & k_3 \end{pmatrix}^T = \begin{pmatrix} -3 & -5 & -4 \end{pmatrix}^T$ and $r(t) = 5\sin t + 4\sin 2t + 4\sin 3t$. For the purposes of the following simulation experiments, the true plant parameters are assumed to have the following values:

$$\theta = \begin{cases} \begin{pmatrix} 2.0 & -3.0 & 4.0 \end{pmatrix}^T & 0 \le t \le 15 \\ \begin{pmatrix} 0.0 & -0.8 & 1.8 \end{pmatrix}^T & 15 < t < \infty \end{cases}$$

Fig. 5.3(a)–5.3(f) show the response of the adaptive system with $x(0) = \begin{pmatrix} 1.5 & 0 & 0 \end{pmatrix}^T$ and $\hat{\theta}(0) = \begin{pmatrix} 1.75 & -2.75 & 3.75 \end{pmatrix}^T$. While the system with one model is able to respond well and track the states of the reference model, we observe from Fig. 5.3(a)–5.3(c), that all the states of the adaptive system become unbounded at $t = 20$, as a result of the sudden change in the parameters. Using multiple models, not only are all the state variables bounded as shown in Fig. 5.3(d)–5.3(f), but the tracking performance also remains satisfactory despite the changes in the parameters. It was impossible to achieve equivalent performance using a single adaptive model.

Case 2:. In case 1, the plant input $u(t)$ was chosen to compensate for the nonlinear term $f^T(x)\theta$ if the parameter θ is known. Hence, the only modification that was needed was to use an input $-f^T(x)\hat{\theta} + v(t)$, and adjust $\hat{\theta}(t)$ adaptively. In the present case we consider a plant where such direct cancellation is not possible. For brevity, we consider the following second order plant:

$$\begin{aligned}\dot{x}_{(1)} &= x_{(2)} + \theta x_{(1)}^2\\ \dot{x}_{(2)} &= u\end{aligned}$$

The direct adaptive control of the above plant using back stepping has been studied earlier, and global stability has been established using a quadratic Lyapunov function. However, as before, for such a controller to perform well, the initial parameter estimation errors must be sufficiently small. Here, we modify the existing backstepping technique in order to apply the multiple model methodology as shown in Fig. 5.4. Global stability can be shown by standard arguments using the Lyapunov function $V(e_1, e_2, e_i, \tilde{\theta}_1, \tilde{\theta}_2) = \frac{1}{2}\left(e_1^2 + e_2^2 + e_i^2 + \tilde{\theta}_1^2 + \tilde{\theta}_2^2\right)$. This modified direct control scheme is then extended to multiple model scheme discussed in this paper in order to make the single model controller effective in a larger parameter space. Fig. 5.5 shows the response of the multiple model scheme applied to the above second order plant where the true value of $\theta = -5.5$. It is known a priori that $\theta \in [-10, 10]$.

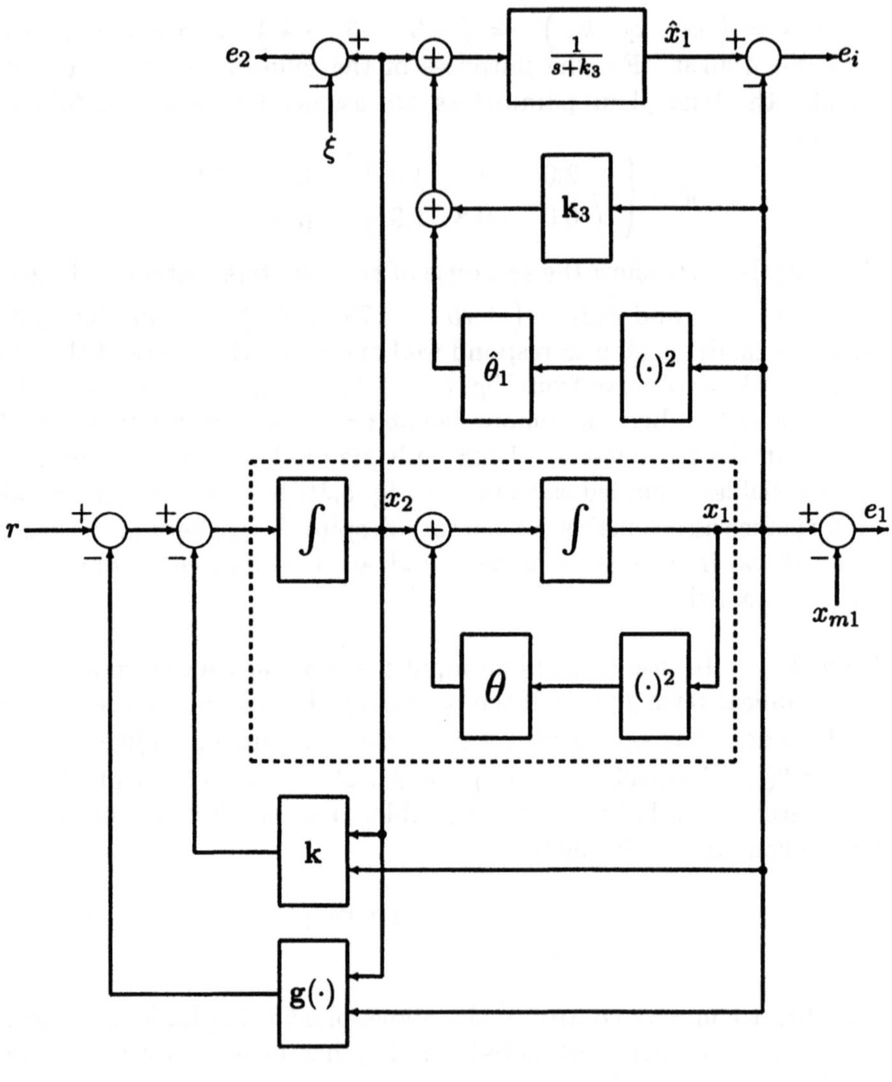

Figure 5.4. Modified direct adaptive control for a benchmark example.

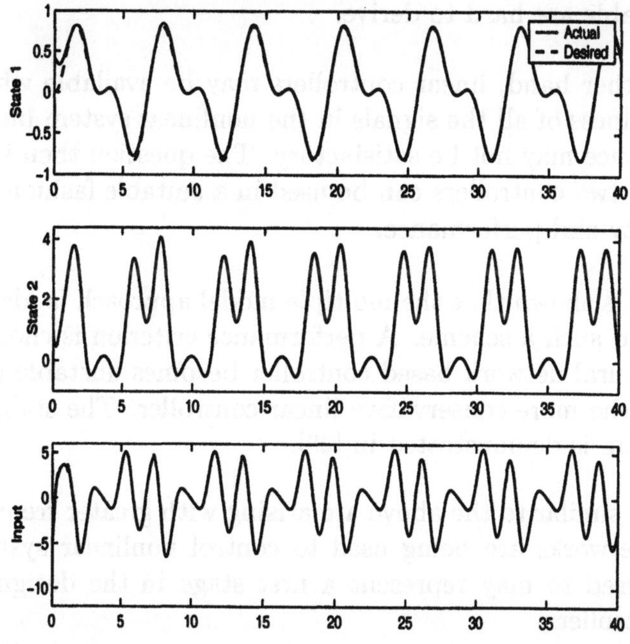

Figure 5.5. The performance of a second order nonlinear system when adaptively controlled using multiple models.

6. Adaptive Control Using Linear and Nonlinear Models

As stated in the introduction, models of dynamical systems can take many forms depending upon the objectives of the designer. One of the intriguing questions that the first author has been addressing over the past few years is concerned with the possibility of using multiple models to realize their individual advantages. In the context of the problems that are of interest in this paper, the question can be posed as follows: Is it possible to use two models so that both stability and performance can be realized simultaneously?

Work reported recently [29] demonstrates that in some special cases it may be possible to prove rigorously that an affirmative answer can be given to the above question.

It has been empirically demonstrated that the use of neural networks based controllers can result in improved performance of complex nonlinear systems. However, obtaining conditions under which such systems would be stable are hard to derive.

On the other hand, linear controllers may be available which assure the boundedness of all the signals in the nonlinear system but the overall performance may not be satisfactory. The question then is to decide whether the two controllers can be used in a suitable fashion to achieve both stability and performance.

In [29], it is shown that the multiple model approach is ideally suited to implement such a scheme. A performance criterion is chosen so that when the neural network based controller becomes unstable the system switches to the more conservative linear controller. The stability of the overall system is demonstrated in [29].

Problems similar to the above are arising with greater frequency even as neural networks are being used to control nonlinear systems. The results referred to may represent a first stage in the design of a new class of controllers.

7. Applications

The methodology described in this paper is finding a wide following in industry in control contexts where large changes in system dynamics are anticipated, and adaptive control using a single model is found to be inadequate. A few typical examples are described below.

In aircraft systems, fast and accurate flight control reconfiguration is of paramount importance for increasing aircraft survivability in the presence of subsystem failures and structural damage. The MMST approach appears to be eminently suited to cope with such situations. For example, the case when one of the effectors undergoes float, lock-in-place, or hard-over failure has been investigates in detail in [30]. Computer simulations of a linearized model of the F/A - 18A aircraft during carrier landing maneuvers using five models demonstrated that excellent performance under a critical effector failure could be achieved.

In [32], the MMST approach has been applied to a flexible transmission system. Flexible systems with low damping factor, are very difficult to control, particularly in the presence of large load variations. A fixed high performance controller at one load may lead to instability to another load. Fixed parameter robust controllers, on the other hand, designed for the different anticipated load variations, could not assure high performance for all loadings. Adaptive control using a single model of the plant resulted in large transients as expected, for sudden changes in load. The MMST approach, however, together with a closed loop output error parameter estimation technique resulted in improved stability and performance.

Nonlinearities, product transitions, constrains and delays are some of the difficulties encountered in industrial process control. In [31] the MMST approach is applied to control a MIMO (multi-input multi-output) nonlinear stable two-phase chemical reactor. The authors conclude that the response obtained was faster and more accurate than using a multiple PI controller strategy. In particular, the authors found the approach very attractive when the process was known to transition to unknown operating states.

8. Comments and Conclusions

During the past decade the multiple-model based switching and tuning (MMST) approach has developed into a general methodology for adaptively controlling dynamical systems even in the presence of rapidly varying environments. The use of multiple models enlarges the region in parameter space in which adaptation becomes very effective. Through switching, models that are close to the plant in parameter space are chosen to determine the control input to be applied to the plant. Hence, switching avoids catastrophic failures when large parametric changes occur in the system. Tuning, on the other hand, corresponds to traditional adaptive control and improves the performance of a stabilized plant at a more leisurely rate.

The methodology also brings out the importance of adaptive control methods based on a single model. The stability of MMST controllers is closely related to the stability of such systems. In this paper, the methodology is applied to linear stochastic systems, and nonlinear systems which have been investigated extensively in the past using a single model. In all cases, stability is proved and simulation results indicate substantial improvements in performance. It is therefore not surprising that MMST is increasingly being used in many industrial contexts.

All the methods considered thus far increase the effectiveness of adaptive control methods in larger regions in parameter space. However, in the future, the most exciting research in this rapidly developing area, may also be in problems briefly referred to in Section 6, where efforts are made to combine the advantages of different models. While the results obtained thus far apply to situations that are quite restrictive, they nevertheless suggest the possibility of achieving superior performance in complex environments.

References

[1] K. S. Narendra, Y.-H. Lin, and L. S. Valavani. Stable adaptive controller design — Part II: proof of stability. *IEEE Transactions on Automatic Control*, 25(3):440–448, June 1980.

[2] A. S. Morse. Global stability of parameter adaptive control systems. *IEEE Transactions on Automatic Control*, 25(3):433–439, June 1980.

[3] G. C. Goodwin, P. J. Ramadge, and P. E. Caines. Discrete-time multivariable adaptive control. *IEEE Transactions on Automatic Control*, 25(3):449–456, June 1980.

[4] P. A. Ioannou and J. Sun. *Robust Adaptive Control.* Prentice-Hall, Upper Saddle River, NJ, 1996.

[5] D. T. Magill. Optimal adaptive estimation of sampled stochastic processes. *IEEE Transactions on Automatic Control*, 10:434–439, 1965.

[6] D. G. Lainiotis. Partitioning: A unifying framework for adaptive systems — I: Estimation; II: Control. *Proceedings of the IEEE*, 64:1126–1143 and 1182–1197, August 1976.

[7] M. Athans, D. Castañon, K.-P. Dunn, C. S. Greene, W. H. Lee, N. R. Sandel, Jr, and A. S. Willsky. The stochastic control of the F-8C aircraft using a multiple model adaptive control (MMAC) method — Part I: Equilibrium flight. *IEEE Transactions on Automatic Control*, 22(5):768–780, October 1977.

[8] R. L. Moose, H. F. Van Landingham, and D. H. McCabe. Modeling and estimation for tracking maneuvering targets. *IEEE Transactions on Aerospace Electronic Systems*, 15:448–456, May 1979.

[9] C. Yu, R. J. Roy, H. Kaufman, and B. W. Bequette. Multiple-model adaptive predictive control of mean arterial pressure and cardiac output. *IEEE Transactions on Biomedical Engineering*, 39(8):765–778, August 1992.

[10] P. S. Maybeck and D. L. Pogoda. Multiple model adaptive controller for the STOL F-15 with sensor/actuator failures. In *Proceedings of the 28th IEEE Conference on Decision and Control*, pages 1566–1572, Tampa, Florida, USA, December 1989.

[11] X. R. Li and Y. Bar-Shalom. Design of an interacting multiple model algorithm for air traffic control tracking. *IEEE Transactions on Control Systems Technology*, 1(3):186–194, September 1993.

[12] B. Mårtensson. *Adaptive Stabilization*. PhD thesis, Lund Institute of Technology, Lund, Sweden, 1986.

[13] R. H. Middleton, G. C. Goodwin, D. J. Hill, and D. Q. Mayne. Design issues in adaptive control. *IEEE Transactions on Automatic Control*, 33(1):50–58, January 1988.

[14] A. S. Morse, D. Q. Mayne, and G. C. Goodwin. Applications of hysteresis switching in parameter adaptive control. *IEEE Transactions on Automatic Control*, 37(9):1343–1354, September 1992.

[15] S. R. Weller and G. C. Goodwin. Hysteresis switching adaptive control of linear multivariable systems. *IEEE Transactions on Automatic Control*, 39(7):1360–1375, July 1994.

[16] A. S. Morse. Supervisory control of families of linear set-point controllers — Part 1: Exact matching. *IEEE Transactions on Automatic Control*, 41(10):1413–1431, October 1996.

[17] A. S. Morse. Supervisory control of families of linear set-point controllers — Part 2: Robustness. *IEEE Transactions on Automatic Control*, 42(11):1500–1515, November 1997.

[18] R. Murray-Smith and T. A. Johansen. *Multiple Model Approaches to Modeling and Control*. Taylor and Francis, London, UK, 1997.

[19] K. S. Narendra and J. Balakrishnan. Performance improvement in adaptive control systems using multiple models and switching. In *Proceedings of the Seventh Yale Workshop on Adaptive Learning Systems*, pages 27–33, Center for Systems Science, Yale University, New Haven, USA, May 1992.

[20] K. S. Narendra and J. Balakrishnan. Adaptive control using multiple models. *IEEE Transaction on Automatic Control*, 42(2):171–187, February 1997.

[21] K. S. Narendra, J. Balakrishnan, and M. K. Ciliz. Adaptation and learning using multiple models, switching and tuning. *IEEE Control System Magazine*, 15(3):37–51, June 1995.

[22] K. S. Narendra and C. Xiang. Adaptive control of discrete-time systems using multiple models. *IEEE Transactions on Automatic Control*, 45(9):1669–1686, September 2000.

[23] K. S. Narendra and O. A. Driollet. Stochastic adaptive control using multiple models for improved performance in the presence of random disturbances. *International Journal of Adaptive Control and Signal Processing*, 15(3):287–317, May 2001.

[24] G. C. Goodwin and K. Sin. *Adaptive Filtering, Prediction and Control.* Prentice Hall, Englewood Cliffs, NJ, 1984.

[25] G. C. Goodwin, P. J. Ramadge, and P. E. Caines. Discrete-time stochastic adaptive control. *SIAM Journal Control and Optimization*, 19:829–853, 1981.

[26] M. Kristić, I. Kanellakopoulis, and P. Kokotović. *Nonlinear and Adaptive Control Design.* John Wiley & Sons, Inc., New York, USA, 1995.

[27] D. Seto, A. M. Annaswamy, and J. Baillieul. Adaptive control of nonlinear systems with a triangular structure. *IEEE Transactions on Automatic Control*, 39(7):1411–1428, July 1994.

[28] K. S. Narendra and A. M. Annaswamy. *Stable Adaptive Systems.* Prentice-Hall, Englewood Cliffs, New Jersey, USA, 1989.

[29] L. Chen and K. S. Narendra. Nonlinear adaptive control using neural networks and multiple models. *Automatica*, 37(8), August 2001. To appear.

[30] J. D. Bošković and R. Mehra. Stable multiple model adaptive flight control for accommodation of a large class of control effector failures. In *Proceedings of the American Control Conference*, pages 1920–1924, San Diego, California, USA, June 1999.

[31] R. Gundala, K. A. Hoo, and M. J. Piovoso. Multiple model adaptive control design for a multiple-input multiple-output chemical reactor. *Industrial and Engineering Chemistry Research*, 39(6):1554–1564, June 2000.

[32] A. Karimi and I. Landau. Robust adaptive control of a flexible transmission system using multiple models. *IEEE Transactions on Control Systems Technology*, 8(2):321–331, March 2000.

Chapter 6

CONSTRAINED CONTROL: POLYTOPIC TECHNIQUES*

D. Q. Mayne

Imperial College of Science, Technology and Medicine, London.

Dedicated to Larry Ho with affection, respect, and gratitude for the inspiration he has abundantly provided to all in our field

Abstract

There is a resurgence of interest in the control of dynamic systems with hard constraints on states and controls and many significant advances have been made. A major reason for the success of model predictive control, which, with over 2000 applications, is the most widely used modern control technique, is precisely its ability to handle effectively hard constraints. But there are also other important developments. The solution of the *constrained* linear quadratic regulator problem has been characterized permitting, at least in principle, explicit determination of the value function and the optimal state feedback controller. Maximal output admissible sets have been effectively harnessed to provide easily implementable regulation and control of constrained linear systems. The solution of the robust, constrained time-optimal control problem has also been characterized. A common feature of all these advances is their reliance on polytopic techniques. Knowledge of robust controllability sets is required in model predictive control of constrained dynamic systems; these sets are polytopes when the system being controlled is linear and the constraints polytopic. In other problems, such as robust time optimal control and (unconstrained) ℓ_1 optimal control, the value function itself is polytopic. Partitioning of the state space into polytopes is required for the characterization of the solution of the constrained linear quadratic regulator problem for which the value function is piecewise quadratic and the optimal control piecewise affine. It is possible that polytopic computation may become as useful a tool for the

*Research supported by Engineering and Physical Sciences Research Council, UK

control engineer as numerical linear algebra so successfully employed in H_2 and H_∞ control.

1. Introduction

Dynamic programming provides, in principle, a constructive method for obtaining optimal controllers but its use for this purpose has been largely restricted to H_2 and H_∞ optimal control where the system being controlled is linear and the cost quadratic. In these problems the value function is quadratic and, therefore, simply parameterized and computed. Despite wide use, the H_2 and H_∞ methodologies do not address important requirements (such as hard constraints on control and states) that are present in nearly every application. The absence, for a considerable period, of effective techniques for handling hard constraints motivated the development of many ad-hoc techniques. A typical procedure for enforcing control constraints is to include a saturator prior to the actuator together with an anti-windup mechanism to mitigate the adverse effect of the added non-linearity; stability is not guaranteed.

The necessity for anti-windup procedures when constraints are naively enforced highlights the fact that *nonlinear* control is required for an effective solution to the problem of constrained control, even when the system being controlled is nonlinear. This requirement (for non-linear control) meant that an entirely new approach was needed. Interestingly, not one, but several new methodologies emerged.

One of the most successful of these new methodologies is *model predictive control.* Forsaking off-line determination of the optimal control law (via dynamic programming) because of its difficulty, model predictive control determines the current control action by solving *on-line*, at each sampling instant, a finite-horizon open-loop optimal control problem in which the initial state is the current state of the plant; optimization yields an optimal control sequence and the first control action in this sequence is applied to the plant. Model predictive control therefore differs considerably, in *implementation*, from conventional control that uses a pre-computed control law. Implicitly, however, it employs a control law that can, in principle, be determined off-line via dynamic programming; the difference from conventional control lies merely in the method of implementation. However, the difference is important in practice. Since solving a sequence of *open-loop* optimal control problems (one for each state encountered) is often possible, despite the presence of constraints, non-linearity or time-variation, model predictive control provides an extremely effective and simple way of realizing optimal control. Given its almost unique ability to handle hard constraints and nonlinearity, model predictive control has been widely employed; over 2000 applications to linear plants and a rapidly increasing number of applications to nonlinear plants have been reported [30]. Practicality requires that the optimal control problem solved on-line has a finite terminal time; the resultant controller is therefore not necessarily stabilizing. Many applications are to stable plants for which a long horizon automatically yields stability; this property delayed the development of an adequate analysis which is now fortunately available and reviewed in, inter alia, [1, 24].

But there have been many other contributions to the control of constrained dynamic systems. We mention a few that are germane to this paper. Optimal feedback controllers for constrained and robust constrained linear time-optimal problems are described in [15, 4, 25]. In [13] and [19] a sequence of linear controllers, ranging from 'cautious' to 'aggressive', are chosen. A higher level controller selects, at each sampling instant, the controller to be employed, yielding an interesting and useful form of hybrid control. Stability is ensured and the most aggressive controller is ultimately selected and maintained. This approach was successfully extended to cope with bounded disturbances in [18, 20]; related work appears in [7]. Reference governors [12] permit the extension of this methodology to the tracking problem; a related approach, employing Lyapunov functions parameterized by the set-point for the nonlinear tracking problem, is presented in [26]. Indeed, it has been shown, at least for linear systems [8, 9], that the domain of attraction of a stabilizing controller is also the tracking domain and that tracking controllers may be constructed from the stabilizing controller. The solution to the constrained linear quadratic optimal control problem has recently been characterized [3, 33]; this characterization, which is surprisingly late in the history of optimal control, permits off-line computation of the optimal control law thereby providing a serious competitor to model predictive control when the plant is linear.

What binds these seemingly disparate approaches to the control of *linear* constrained systems is their use of polytopic techniques. In model predictive control, the controllability sets (knowledge of which are required to ensure safe use of this methodology) are polytopes. These sets are also the level sets of the value function in (robust) time-optimal control of discrete-time systems. Maximal output admissible sets, employed to construct stabilizing and tracking controllers in [13, 18, 19, 20] are polytopes. The value function for the *constrained* linear quadratic optimal control problem is piecewise quadratic and the optimal controller piecewise affine; the regions that demarcate the 'pieces' are polytopes. In all these examples, constructive techniques exist for controller design. Polytopic techniques provide an alternative procedure for feedback controller design, formerly restricted almost exclusively to linear quadratic optimal control; these techniques are also able to address the practically significant problem of hard constraints on states and controls. The main objective of this paper is to explore the potential provided by polytopic techniques to obtain feedback solutions for problems other than the conventional linear quadratic optimal control problem, an aim shared in part in [34] where it is shown that the level sets for the ℓ_1 unconstrained optimal control problem are polytopic (indeed, for this problem, the value function is completely specified by one level set).

Conventional linear quadratic control provides both a simple (dynamic programming) recursion for computing the (time-varying) value function and control law for finite horizon optimal control problems but, also, relatively simple computation of the (stationary) value function and control law for the infinite horizon problem. While the examples below show that it is sometimes possible to compute (again using dynamic programming) the value function and control law for *non-quadratic*, finite horizon optimal control problems, there is, as yet, no indication that solutions of the infinite horizon problem are possible. Hence, the solutions to non-quadratic optimal control problems will only be useful if these problems can be formulated in such a way as to ensure stability of the optimal controller. The problem of obtaining stabilizing controllers from the solution to finite horizon optimal control problems has been much studied in the model predictive control literature (for the obvious reason that it is impractical to solve infinite horizon optimal control problems on line). We review the main results obtained in this literature in §2.

To achieve our objective of showing that interesting non quadratic optimal control problems may be solved using polytopic techniques, we examine, in §3, three examples, namely: ℓ_1 optimal control, (robust) time-optimal control, and constrained linear quadratic control. Then, in §4, we present an interesting example of stabilizing, rather than optimizing, control, for constrained linear system for two cases, disturbance-free and bounded disturbance.

2. Stability from finite horizon optimal control

2.1 Preliminaries

Consider control of a nonlinear discrete-time system described by

$$x^+ = f(x, u) \tag{1}$$

where x is the current state (assumed known), u the current control, and x^+ the successor state. The system is subject to hard constraints on control and state:

$$u(k) \in \mathbb{U}, \quad x(k) \in \mathbb{X} \tag{2}$$

for all $k \in Z := \{0, 1, 2, \ldots\}$. A finite horizon optimal control problem is one defined over a finite interval 0 to i, say, where i is any non-negative integer; in this formulation 0 is the initial time and i the terminal time. By time invariance, this is equivalent to the problem defined over the interval $N-i$ to N; here N is the fixed terminal time, and i is time-to-go. For each i, Let $\mathbf{u}$ denote the control sequence $\{u(0), u(1), \ldots, u(i-1)\}$, and let $x^{\mathbf{u}}(k; x, 0)$ denote the solution of (1) at time k if the initial state is x at time 0 and the control sequence $\mathbf{u}$ is applied to the system. The finite horizon optimal control problem (on interval 0 to i) is the minimization (with respect to $\mathbf{u}$) of

$$V_i(x, \mathbf{u}) = \sum_{k=0}^{i-1} \ell(x(k), u(k)) + F(x(i)) \tag{3}$$

(with $x(k) := x^{\mathbf{u}}(k; x, 0)$) subject to the constraints

$$u(k) \in \mathbb{U}, \quad x(k) \in \mathbb{X}, \quad x(i) \in X_f \tag{4}$$

These constraints are, via the difference equation (1), an implicit constraint on the control sequence $\mathbf{u}$; the constraints may be expressed as $\mathbf{u} \in \mathcal{U}_i(x)$ where

$$\mathcal{U}_i(x) := \{\mathbf{u} \mid u(k) \in \mathbb{U}, x(k) \in \mathbb{X}, \ k = 0, \ldots, i-1, x(i) \in X_f\} \tag{5}$$

(with $x(k) := x^{\mathbf{u}}(k; x, 0)$, $k = 0, 1, \ldots, i$). The optimal control problem $\mathbf{P}_i(x)$ is, therefore, minimization of $V_i(x, \mathbf{u})$ subject to the state-dependent constraint $\mathbf{u} \in \mathcal{U}_i(x)$:

$$\mathbf{P}_i(x): \qquad \min_{\mathbf{u}} \{V_N(x, \mathbf{u}) \mid \mathbf{u} \in \mathcal{U}(x)\} \tag{6}$$

A control $\mathbf{u}$ is said to be admissible (for state x) if it satisfies $\mathbf{u} \in \mathcal{U}_i(x)$. The solution to $\mathbf{P}_i(x)$, if it exists, is

$$\mathbf{u}^0(x) = \{u^0(0; x), u^0(1; x), \ldots, u^0(i-1; x)\} \tag{7}$$

The associated optimal state trajectory is

$$\mathbf{x}^0(x) = \{x^0(0; x), x^0(1; x), \ldots, x^0(N; x)\} \tag{8}$$

where $x^0(0; x) = x$. The value function for the optimal control problem $\mathbf{P}_i$ is

$$V_i^0(x) = V_i(x, \mathbf{u}^0(x)) \tag{9}$$

and the associated optimal control law is $\kappa_i(\cdot)$ defined by

$$\kappa_i(x) := u^0(0; x) \tag{10}$$

Dynamic programming yields the solution not only to P_N say, but to the whole sequence of problems $\{P_i \mid i = 1, \ldots, N\}$ through the recursion:

$$\begin{aligned} V_i^0(x) &= \min_{u \in \mathbb{U}} \{\ell(x,u) + V_{i-1}^0(f(x,u)) \mid f(x,u) \in X_{i-1}\} && (11)\\ \kappa_i(x) &= \arg\min_{u \in \mathbb{U}} \{\ell(x,u) + V_{i-1}^0(f(x,u)) \mid f(x,u) \in X_{i-1}\} && (12)\\ X_i &= \{x \mid \exists\, u \in \mathbb{U} \text{ such that } f(x,u) \in X_{i-1}\} \cap \mathbb{X} && (13) \end{aligned}$$

with boundary conditions

$$V_0^0(\cdot) = F(\cdot), \quad X_0 = X_f \tag{14}$$

Conventionally, dynamic programming does not include the recursion (13) for the sets X_i; X_i is usually implicitly defined as the domain of the value function V_i^0. However, it is preferable to compute these sets explicitly through the recursion (13); for each i, X_i is a controllability set (the set of states that can be steered to X_f in i steps or less by a control satisfying the state and control constraints) and the set in which P_i is feasible:

$$X_i = \{x \mid \exists\, \mathbf{u} \in \mathcal{U}_i(x)\} \tag{15}$$

(Conventional) optimal control (on the interval 0 to N) employs the time-varying control law:

$$u(k) = \kappa_{N-k} x(k) \tag{16}$$

for $k = 0, 1, \ldots, N-1$; the control is undefined for $k \geq N$. Model predictive control (with horizon N), on the other hand, *implicitly* employs the *time-invariant* control law

$$u(k) = \kappa_N x(k) \tag{17}$$

for all k (in fact, at state x, model predictive control applies the control $u_N^0(0; x)$, obtained by solving on-line an open-loop optimal control problem, but, by (11), this is precisely $\kappa_N(x)$). This establishes the relevance of model predictive control to our objective; we wish to employ polytopic techniques to solve non-quadratic optimal control problems, but are likely to succeed only if the problem is defined over a finite interval 0 to N. The resultant controller κ_N is neither necessarily optimal nor stabilizing. We have, therefore, to formulate P_N so that κ_N does possess these properties. This is precisely the problem studied in the model predictive control literature.

2.2 Stabilizing ingredients

We extract from the model predictive control literature conditions on the formulation of P_N that ensure that κ_N is stabilizing and optimal for a *modified* infinite horizon optimal control problem. The following hypotheses are assumed to be satisfied in the sequel:

A0: $f(\cdot)$, $F(\cdot)$ are continuous, $f(0,0) = 0$, $\ell(\cdot)$ is positive definite, $\ell(0,0) = 0$, $\mathbb{X}$ and X_f are closed, $\mathbb{U}$ is compact, each of the sets $\mathbb{X}, X_f, \mathbb{U}$ contains the origin in its interior, and the point-to-set map $x \mapsto \mathcal{U}_N(x)$ is continuous in X_N.

A set $X \subset \mathbb{X}$ is said to be control invariant if, for any $x \in X$, there exists a $u \in \mathbb{U}$ such that $f(x,u) \in X$. Under these assumptions, the sets $X_0, \ldots X_N$ are compact and satisfy $X_0 \subset X_1 \ldots \subset X_N$. By construction, for each $i \in \{1, \ldots, N\}$, the set X_i is control invariant (any $x \in X_i$ can be steered into $X_{i-1} \subset X_i$ by a control $u \in \mathbb{U}$) and, by 10) and (15), $f(x, \kappa_i(x)) \in X_{i-1}$ for all $x \in X_i$ so that X_i is positively invariant (for the closed loop system $x^+ = f(x, \kappa_i(x))$ (a set $X \subset \mathbb{X}$ is said to be control invariant if, for any $x \in X$, there exists a $u \in \mathbb{U}$ such that $f(x,u) \in X$). The value functions $V_0^0(\cdot), \ldots, V_N^0(\cdot)$ are well-defined, continuous, and have domains $X_0, \ldots, X_N$ respectively. Finally, the control laws $\kappa_i(\cdot)$, $i = 0, \ldots, N-1$, are continuous.

As is done in the model predictive control literature, we employ $V_N^0(\cdot)$ as a candidate Lyapunov function (recalling that we are dealing with a constrained optimal control problem). It follows from our assumptions that $V_N^0(\cdot)$ is positive definite with $V_N^0(x) = 0$ if and only if $x = 0$. For any function $\gamma : \mathbb{R}^n \to \mathbb{R}$ we define $\overset{*}{\gamma} : \mathbb{R}^n \times \mathbb{R}^m \to \mathbb{R}$ (the analog of $\dot{\gamma}$ in continuous time) by

$$\overset{*}{\gamma}(x,u) := \gamma(f(x,u)) - \gamma(x) \tag{18}$$

A simple calculation shows

$$[\overset{*}{V_N^0} + \ell](x, \kappa_N(x)) = [V_N^0 - V_{N-1}^0](f(x, \kappa_N(x)) \tag{19}$$

so that $V_N^0(\cdot)$ has the desirable property $[\overset{*}{V_N^0} + \ell](x, \kappa_N(x)) \leq 0$ for all $x \in X_N$ if $V_N^0(x) \leq V_{N-1}^0(f(x, \kappa_N(x))$. To find conditions under which the latter is true we make use of an monotonicity argument first employed (in the context of the linear quadratic optimal control problem) in [6] and extended to the nonlinear case in [21] and [24] (where the explicit dependence on the constraint sets is specified).

Lemma 2 *(Monotonicity of the value function)*
If $V_1^0(x) \leq V_0^0(x) = F(x)$ for all $x \in X_0 = X_f$, then

$$V_i^0(x) \leq V_{i-1}^0(x) \quad \forall\, x \in X_{i-1},\; i \in \{1, \ldots, N\}. \tag{20}$$

To prove this [24], assume $V_i^0(x) \leq V_{i-1}^0(x)$ for all $x \in X_{i-1}$; then

$$\begin{aligned} [V_{i+1}^0 - V_i^0](x) &\leq \ell(x, \kappa_i(x)) + V_i^0(f(x, \kappa_i(x))) \\ &\quad -[\ell(x, \kappa_i(x)) + V_{i-1}^0(f(x, \kappa_i(x)))] \\ &\leq [V_i^0 - V_{i-1}^0](f(x, \kappa_i(x))) \leq 0 \end{aligned}$$

for all $x \in X_i$ since $x \in X_i$ implies $f(x, \kappa_i(x)) \in X_{i-1}$. We are now in a position to formulate P_N to ensure that κ_N is stabilizing. Since $V_0^0 = F$,

$$[V_1^0 - V_0^0](x) = \min_u \{[\overset{*}{F} + \ell](x, u) \mid u \in \mathbb{U},\; f(x, u) \in X_f\} \tag{21}$$

for all $x \in X_f$. By (19), (21) and the monotonicity lemma $[\overset{*}{V_N^0} + \ell](x, \kappa_N(x)) \leq 0$ for all $x \in X_N$ if the ingredients F and X_N of P_N satisfy the two conditions [11, 22, 24] :

A1: X_f is control invariant, $X_f \subset \mathbb{X}$.

A2: $\min_{u \in \mathbb{U}} \{[\overset{*}{F} + \ell](x, u) \mid f(x, u) \in X_f\} \leq 0, \quad \forall\, x \in X_f$.

The essential requirement is that F is a *local* Control Lyapunov function (CLF) [14] in which case X_f may be chosen to be a level set of F. A global (rather than a local) CLF (e.g. V_∞^0) would be a preferable, but impractical, choice. There exist many examples of (F, X_f) [16, 31, 32, 10, 22, 24] that satisfy **A1** and **A2**. We require one further result [14]:

Lemma 3 *Suppose **A1, A2** hold. Then*

$$V_N^0(x) \leq F(x) \quad \forall x \in X_f \tag{22}$$

that is easily proven. We make one more assumption before stating the main the main result of this section:

A3: (i) $\ell(x, u) = (1/2)[|x|_Q^2 + |u|_R^2$ where Q and R are positive definite; (ii) there exists a positive constant d such that $F(x) \leq d|x|^2$ for all $x \in X_f$.

Theorem 1 *Suppose **A1, A2** hold. Then the origin is exponentially stable with a domain of attraction X_N.*

This result follows from the positive invariance of X_N (for $x^+ = f(x, \kappa_N(x)))$ and the fact $\ell(x,u) \geq c|x|^2$ for some $c > 0$, all x, so that

$$c|x|^2 \leq V_N^0(x) \leq d|x|^2, \quad \forall x \in X_f$$

$$\overset{*}{V_N^0}(x, \kappa_N(x)) \leq -c|x|^2, \quad \forall x \in X_N.$$

There exists a finite time in which any $x \in X_N$ is steered to a level set of $V_N^0(\cdot)$ contained in X_f after which exponential convergence to the origin obtains.

Finally, we note [29, 21] that the Hamilton-Jacobi difference equation may be written as a *fake* Hamilton-Jacobi algebraic equation:

$$V_N^0(x) = L(x, \kappa_N(x)) + V_N^0(f(x, \kappa_N(x)) \tag{23}$$

for all $x \in X_N$ where

$$L(x,u) := \ell(x,u) + [V_{N-1}^0 - V_N^0](f(x, \kappa_N(x)) \geq \ell(x,u) \tag{24}$$

if **A1, A2** hold. Thus, the *finite-horizon* controller $\kappa_N(\cdot)$ *is* optimal for the *infinite-horizon* optimal control problem obtained by replacing N by ∞ and $\ell(\cdot)$ by $L(\cdot)$.

3. Non-quadratic optimal control

3.1 Unconstrained non-quadratic optimal control

An interesting example of non-quadratic optimal control is described in [34]. The system is unconstrained and linear, described by:

$$x^+ = Ax + Bu \tag{25}$$

To specify the cost, we introduce the function $(\cdot)_{\max} : \mathbb{R}^p \to \mathbb{R}$ defined by:

$$(y)_{\max} := \max\{y_i \mid i = 1, 2, \ldots, p\} \tag{26}$$

Then, for each i in $\{1, \ldots, N\}$,

$$V_i(x, \mathbf{u}) := \sum_{k=1}^{i-1} \ell(x(k), u(k)) + F(x(i)) \tag{27}$$

(with $x(k) = x^{\mathbf{u}}(k; x, 0)$ and $\mathbf{u} := \{u(0), u(1), \ldots, u(i-1)\})$) where

$$\ell(x,u) := (Cx + Du)_{\max} \tag{28}$$

$$F(x) := (F'x)_{\max} \tag{29}$$

There exists a choice of C and D such that $\ell(x,u) = |x|_\infty + |u|_\infty$ (or $|x|_1 + |u|_1$). The level sets of $\ell(x,u)$ and $F(x)$ are polyhedrons (in (x,u) space and x space, respectively); moreover $F(x)$ (and any function of this form) varies linearly along rays emanating from the origin:

$$F(\alpha x) = \alpha F(x) \tag{30}$$

for any $\alpha \geq 0$. Consequently, the function F is completely characterized by its level set $\mathcal{F}$ defined by

$$\mathcal{F} := \{x \mid F(x) \leq 1\} \tag{31}$$

To proceed we need to specify a few operations. For any two matrices $M \in \mathbb{R}^{r,p}$ and $N \in \mathbb{R}^{s,p}$

$$(Mz)_{\max} + (Nz)_{\max} = (Pz)_{\max} \tag{32}$$

where $P \in \mathbb{R}^{(r+s),p}$ is a matrix each row of which is the sum of a row of M and a row of N; this is called the pairing operation in [34]. The second operation is projection. Let $\mathcal{P}$ be a polyhedron in (x,u) space ($\mathbb{R}^{n \times m}$) and let its matrix representation be $P = [P_1, P_2]$ where $P_1 \in \mathbb{R}^{n \times n}$:

$$\mathcal{P} := \{(x,u) \mid (P_1 x + P_2 u)_{\max} \leq 1\} \tag{33}$$

Then $\mathcal{S} := \text{Proj}_X \mathcal{P}$ (the orthogonal projection of $\mathcal{P}$ onto x space) satisfies

$$\mathcal{S} = \{x \mid \exists\ u \text{ such that } (x,u) \in \mathcal{P}\} \tag{34}$$

$\mathcal{S}$ is also a polyhedron, with matrix representation S, i.e.

$$\mathcal{S} = \{x \mid (Sx)_{\max} \leq 1\} \tag{35}$$

It is an interesting fact [34], not difficult to show, that

$$\min_u (P_1 x + P_2 u)_{\max} = (Sx)_{\max} \tag{36}$$

We now have the tools needed to implement the dynamic programming recursion for this optimal control problem:

$$V_i^0(x) = \min_u \{\ell(x,u) + V_{i-1}^0(Ax + Bu)\} \tag{37}$$

$$V_0^0(x) = F(x) \tag{38}$$

Now, $V_0^0(x) = (M_0 x)_{\max}$ where $M_0 = F'$. Suppose, then, that $V_{i-1}^0 = (M_{i-1}x)_{\max}$. Then

$$V_i^0(x) = \min_u \{(Cx + Du)_{\max} + (M_{i-1}(Ax + Bu)_{\max} \tag{39}$$

$$= \min_u \{(L_1 x + L_2 u)_{\max}\} \text{ (by pairing)} \tag{40}$$

$$= (M_i x)_{\max} \text{ (by projection)} \tag{41}$$

By induction, each V_i^0 is a max function and the sequences of optimal value functions $\{V_i^0\}$ and optimal control laws $\{\kappa_i\}$, defined by

$$\kappa_i(x) := \arg\min_u (L_1 x + L_2 u)_{\max}, \tag{42}$$

are yielded by (39-42). For any N, the time-invariant controller $u = \kappa_N(x)$ is not necessarily stabilizing. As shown in [34], κ_N is stabilizing for sufficiently large N. Alternatively, for given N, the approach described in §2 may be employed. X_f is chosen to be $\mathbb{R}^n$ so that a global Control Lyapunov function is required for F. Polytopic methods may be employed. Choose W to be the set $\{x \mid (x)_\infty \leq 1\}$ and $u = Kx$ to be a minimum time controller ($A_K^n = 0$,where $A_K := A + BK$)) and let $\mathcal{F}$ be defined by

$$\mathcal{F} := W \oplus A_K W + \ldots \oplus A_K^{n-1} W \tag{43}$$

where $\oplus$ denotes set addition. Then $A_K \mathcal{F} \oplus W = \mathcal{F}$ so that $x \in \mathcal{F}$ implies $A_K x \in \mathcal{F} \ominus W$ where $\ominus$ denotes Minkowski (Pontryagin) set subtraction [25, 20] defined by:

$$X \ominus Y := \{x \mid x + Y \subset X\} \tag{44}$$

Let F denote the matrix representation of $\mathcal{F}$, i.e F satisfies:

$$\mathcal{F} := \{x \mid (Fx)_{\max} \leq 1\} \tag{45}$$

Since W is compact and contains the origin in its interior, so does $\mathcal{F}$. Clearly $\max\{F(x) \mid x \in \mathcal{F}\} = 1$ and $\max\{F(x) \mid x \in \mathcal{F} \ominus W\} = \gamma < 1$. We deduce from the contractive property ($A_K x \in \mathcal{F} \ominus W$ for all $x \in \mathcal{F}$) and the linearity of F along rays from the origin that

$$F(A_K x) - F(x) \leq -(1-\gamma) F(x) \tag{46}$$

for all x. By scaling (replacing F by αF, $\alpha > 0$, we obtain an F satisfying $F(A_K x) - F(x) \leq -\ell(x, Kx)$ so that F satisfies **A2** (with $X_f = \mathbb{R}^n$):

$$\min_u [\overset{*}{F} + \ell(x,u)] \leq 0 \tag{47}$$

The results of §2 may be modified to suit non-quadratic ℓ and the other hypotheses are easily satisfied yielding (global) asymptotic stability.

3.2 (Robust) time-optimal control

In the example just considered, the value function is a max function characterized by a single level set that is a polytope. In our next example, (robust) time-optimal control of constrained, linear, discrete-time systems, the value function is characterized by a nested sequence of polytopes whose boundaries are isochrones. For time optimal control to the origin, we consider a linear system, described by (25), with state and control constraints (2) that are polytopic:

$$\mathbb{U} := \{u \mid A_u u \le a_u\}, \quad \mathbb{X} := \{x \mid A_x x \le a_x\} \tag{48}$$

With $X_f = \{0\}$, the controllability set X_i, defined in (15), is the set of states that may be steered to the origin in i steps or less while satisfying all constraints; for each i, the isochrone for time i is the boundary of the set X_i. The sets $\{X_i\}$ may be calculated by the recursion (13) and (14) which, in the case considered here, reduces to

$$X_i = \{x \mid \exists u \in \mathbb{U} \text{ such that } Ax + Bu \in X_{i-1}\} \cap \mathbb{X} \tag{49}$$

with boundary condition

$$X_0 = \{0\} \tag{50}$$

When A is invertible, the projection operation in (49) may be replaced [25] by:

$$X_i = \{A^{-1}X_{i-1} \oplus (-A^{-1}B\mathbb{U}\} \cap \mathbb{X} \tag{51}$$

where $\oplus$ denotes set addition. A time optimal control law is easily obtained: for each x, determine the least integer i such that $x \in X_i$ and compute (using linear programming) any control $u \in \mathbb{U}$ such that $Ax + Bu \in X_{i-1}$.

A simple modification yields robust controllable sets [4, 25], required when a bounded disturbance is present, i.e. when the system satisfies

$$x^+ = Ax + Bu + w \tag{52}$$

with $w \in W$ where W is compact and contains the origin in its interior. It is no longer possible to steer the state to the origin; instead we define a target set $X_f \subset \mathbb{X}$ that is robust control invariant: for each $x \in X_f$ there exists a $u \in \mathbb{U}$ such that $Ax + Bu + w \in X_f$ for all $w \in W$. Hence, once the state is steered to X_f it may be kept there by an admissible control. There are several ways of obtaining a robust control invariant set. A simple method, employed in [25], is to choose $u = Kx$ to be a minimum-time (unconstrained) controller for $x^+ = Ax + Bu$, and to define X_f by

$$X_f = W \oplus A_K W \oplus \ldots \oplus A^{n-1} W \tag{53}$$

where $A_K := A + BK$. Then $A_K X_f + W = X_f$ (since $A^n = 0$), yielding robust control invariance. The robust controllable sets X_i are then yielded by the recursion (when A is invertible)

$$\begin{aligned} X_{i+1} &= \{A^{-1}(X_i \ominus W) \oplus (-A^{-1}B\mathbb{U})\} \cap \mathbb{X} && (54)\\ X_0 &= X_f && (55) \end{aligned}$$

Projection may be used, as above, when A is not invertible. A robust minimum time controller (that minimizes (with respect to control laws) the maximum (with respect to disturbance sequences) time to steer any state x to X_f is obtained as above: for each x, determine the least integer i such that $x \in X_i$ and compute (using linear programming) any control $u \in \mathbb{U}$ such that $Ax + Bu \in X_{i-1} \ominus W$.

3.3 Constrained linear quadratic control

The structure of the solution to the constrained linear quadratic (linear) regulator problem $\mathbf{P}_\mathbf{N}(x)$ was discovered surprisingly late [33, 2] in the history of optimal control. The essential results are that the optimal control law $\kappa_N(\cdot)$ is piecewise affine and that the value function is piecewise quadratic (piecewise affine if $\ell(\cdot)$ is affine rather than quadratic). The regions in which the control is affine and the cost quadratic are polyhedrons. Perhaps even more surprisingly, the value function $V_N^0(\cdot)$ is continuously differentiable [28]; indeed, the gradient of the value function is piecewise affine and continuous. We assume that the system is described by (25) and the cost by (27) where ℓ and F are quadratic with ℓ strictly positive definite. The system is subject, as before, to constraints on controls and states and the sets $\mathbb{U}$ and X_f are polytopes, and the set $\mathbb{X}$ polyhedral. Under these and our previous assumptions, the value function $V_N^0(\cdot)$ is continuous. Since $V_N(\cdot)$ is strictly convex in $\mathbf{u}$ and $\mathcal{U}(\cdot)$ is continuous, $\mathbf{u}^0(\cdot)$ and $\kappa_n(\cdot)$ are continuous. For each $x \in X_N$, the set $\mathcal{U}(x)$ is a polytope, expressible as $\{\mathbf{u} \mid C\mathbf{u} \leq c + Dx\}$. At each $x \in X_N$, let $I(x)$ denote the active constraints, i.e. $i \in I(x)$ if and only if $C_i\mathbf{u}^0(x) = c_i + D_i x$ where the subscript a_i denotes the ith row of of a. For each $x \in X_N$, there exists a region $\{y \mid I(y) = I(x)\}$ in which the active set I is constant ($I = I(x)$). This region may be found as follows: for given x, solve the equality constrained optimization problem

$$P_N^I(x): \quad \min_{\mathbf{u}}\{V(x, \mathbf{u}) \mid C_i\mathbf{u} = c_i + D_i x, i \in I\} \tag{56}$$

to obtain the value function $V_N^0(\cdot)$ and optimal control sequence $\mathbf{u}^0(\cdot)$ satisfying

$$V_N^0(x) = (1/2)x'P^I x + p^I x + q^I \tag{57}$$

$$\mathbf{u}^0(x) = K^I x + k^I \tag{58}$$

These expressions (for the value function and optimal control) are valid in the set R^I defined by

$$R^I := \{x \in X_N \mid C_i(K^I x + k^I)x = c_i + D_i x, i \in I, \\ C_i(K^I x + k^I) \leq c_i + D_i x, i \in I^c, Px \leq q\} \quad (59)$$

The last constraint ensures positivity of the multipliers at the solution of $P_N^I(x)$ for any $\in R^I$. The regions R^I are polyhedral and within these regions, which partition X_N, the value function is quadratic and the optimal control affine.

To show that $V_x^0(\cdot) := (\partial/\partial x)V_N^0(\cdot)$ is continuous we use the argument employed in [23]. Consider the simple case when there is merely one constraint $C\mathbf{u} \leq c + Dx$ (C, c and D each have one row). Then X_N is partitioned into two regions R^{I_1}, R^{I_2} (with $I_1 = \emptyset, I_2 = \{1\}$) separated by the hyperplane $H := \{x \mid C(K^{I_1}x + k^{I_1}) = c + Dx\}$. Consider an arbitrary sequence $\{x_i\}$ in the interior of R^{I_1} ($C(K^{I_1}x_i + k^{I_1}) < c + Dx_i$) converging to $x^* \in H \cap X_N$. Then along this sequence $V_\mathbf{u}(x_i, \mathbf{u}^0(x_i)) \equiv 0$ and $V_x^0(x_i) = V_x(x_i, \mathbf{u}^0(x_i))$. So, by the continuity of $V_x(\cdot)$, $V_\mathbf{u}(\cdot)$ and $\mathbf{u}^0(\cdot)$, $V_x^0(x_i) \to V_x(x^*, \mathbf{u}^0(x^*))$ as $i \to \infty$ and $V_\mathbf{u}(x^*, \mathbf{u}^0(x^*)) = 0$. Now consider an arbitrary sequence $\{x_i\}$ in R^{I_2} satisfying $C(K^{I_2}x_i + k^{I_2}) = c + Dx_i$; along this sequence $V_\mathbf{u}(x_i, \mathbf{u}^0(x_i))$ is not necessarily zero (because of the equality constraint), so that $V_x^0(x_i) = V_x(x_i, \mathbf{u}^0(x_i)) + V_\mathbf{u}(x_i, \mathbf{u}^0(x_i))K^{I_2}$ which converges to $V_x(x^*, \mathbf{u}^0(x^*)) + V_\mathbf{u}(x^*, \mathbf{u}^0(x^*))K^{I_2}$. But $V_\mathbf{u}(x^*, \mathbf{u}^0(x^*)) = 0$ so that $V_x^0(x_i) \to V_x(x^*, \mathbf{u}^0(x^*))$, which is the same limit obtained for the sequence in R^{I_1}. Hence $V_x^0(\cdot)$ is continuous. This argument may be generalized.

Constrained optimal control may be implemented by storing the inequalities that define each region, testing the inequalities to find in which region the current state x lies, and setting the current control $u = K_1^I x + k_1^I$ if $x \in R^I$ where K_1^I, k_1^I are, respectively, the first m rows of K^I, k^I. If the regions can be reliably computed, this procedure may prove to be a serious competitor for model predictive control, at least in those applications where it is difficult to perform the online optimization in a time short compared with the response time of the plant.

3.4 Nonlinear systems

The discussion above has been limited to determination of optimal controllers for linear systems with polyhedral constraints. However, nonlinear systems (described by $x^+ = f(x, u)$) may be arbitrarily closely approximated by piecewise affine systems. For such systems, (x, u) space is partitioned into polyhedrons P_i, $i = 1, \ldots, I$; in each polyhedron P_i, f is approximated by $A_i x + B_i u + c_i$. It is then possible [28, 17] to express many sets of interest, such as the controllability sets X_i, as the *union* of a set of convex sets where the convex sets may be computed using using formulae similar to those employed above. The union of a set of convex sets is not, of course, necessarily convex, so that non-convex controllability and reachability sets, and sets characterizing constrained nonlinear quadratic control, may be suitably approximated.

4. Stabilizing control

There is an even larger literature dealing with polytopic techniques for computing stabilizing control; we merely scratch the surface of this literature to glimpse its contribution. Many references to this literature may be found in [17, 20]. Consider the problem of controlling the linear system (25) subject to the hard constraint

$$y = Cx + Du \in \mathbb{Y} \tag{60}$$

where Y is polyhedral. A stabilizing controller that satisfies this constraint is sought. One method, that seems particularly useful, has been proposed in [13] where the concept of the *maximal output admissible set* is introduced. Suppose $u = Kx$ is a (stabilizing) controller for system (25; the maximal output admissible set for the closed-loop system $x^+ A_K x$, $y = C_K x$ (with $A_K = A + BK$, $C_K = C + DK$) is

$$\mathcal{O}_\infty := \{x \mid C_K A_K^i x \in \mathbb{Y},\ i = 0, 1, 2, \ldots\} \tag{61}$$

This set is is a polytope (if A_K is stable and (C, A) observable) and is positively invariant for $x^+ = F_K x$; any trajectory starting in $\mathcal{O}_\infty$ remains in $\mathcal{O}_\infty$ and satisfies the control and state constraints. The set $\mathcal{O}_\infty$ is the largest positively invariant set lying in the set $\{x \mid C_K x \in Y\}$. For all i let $\mathcal{O}_i := \{x \mid C_K F_K^j x \in \mathbb{Y},\ j = 0, 1, \ldots, i\}$. Then $\{\mathcal{O}_i\}$ satisfies the recursion

$$\begin{aligned}
\mathcal{O}_0 &= \{x \mid C_K x \in \mathbb{Y}\} \\
\mathcal{O}_{i+1} &= \{x \mid C_K A_K^j x \in \mathbb{Y}, j = 0, 1, \ldots, i+1\} \\
&= \{x \mid A_K x \in \mathcal{O}_i\} \cap \mathcal{O}_0 \\
&= A_K^{-1} \mathcal{O}_i \cap \mathcal{O}_0
\end{aligned}$$

if A_K is invertible. The sequence is terminated at t if $\mathcal{O}_{t+1} = \mathcal{O}_t$, in which case $\mathcal{O}_\infty = \mathcal{O}_t$; finite termination occurs if A_K is stable. If this is the case, $\mathcal{O}_\infty$ is a polyhedron (a polytope if (C, A) is observable). An effective method for controlling constrained systems, described in [13], utilizes a set of controllers $K_1, \ldots, K_I$, ranging from 'cautious' (with 'large' $\mathcal{O}_\infty$) to 'aggressive' (with 'small' $\mathcal{O}_\infty$) together with logic for switching controllers; this logic employs the maximal output admissible sets $\mathcal{O}^i_\infty, i = 1 \ldots I$ associated with controller K_i, $i = 1 \ldots I$; K_i is employed while the state is in $\mathcal{O}^i_\infty$ until the state enters the output admissible set associated with a more aggressive controller when the more aggressive controller is selected. As time increases, the controller selected becomes increasingly more 'aggressive' until, finally, the most aggressive controller is selected and maintained. The sets $\mathcal{O}^i_\infty, i = 1 \ldots I$ are not necessarily nested.

The concept of maximal output admissible set may be extended to deal with bounded disturbances [20]. Suppose the system is, now,

$$x^+ = Ax + Bu + w \tag{62}$$

where $w \in W$ and W is compact and contains the origin in its interior. For all t, let

$$\begin{aligned} \mathcal{A}_t &:= \{W \oplus A_K W \oplus \ldots \oplus A_K^{t-1} W\} && (63)\\ \mathcal{O}_t &:= \{x \mid C_K[A_K^i x \oplus \mathcal{A}_i] \subset Y,\ i = 1, \ldots, t\} && (64)\\ &= \{x \mid C_K A_K^i x \in Y \ominus \mathcal{A}_i,\ i = 1, \ldots, t\} && (65) \end{aligned}$$

Both $\mathcal{A}_t$ and $\mathcal{O}_t$ have limits, $\mathcal{A}_\infty$ and $\mathcal{O}_\infty$ respectively, as $i \to \infty$ if A_K is asymptotically stable; $\mathcal{O}_\infty$, the maximal disturbance invariant set, is robust positively invariant for $x^+ = A_K x + w$; any trajectory starting in $\mathcal{O}_\infty$ remains in $\mathcal{O}_\infty$ and satisfies the control and state constraints for *all* possible realizations of the disturbance sequence. If the set $\hat{Y} := Y \ominus C_K \mathcal{A}_\infty$ is not empty, then $\hat{\mathcal{O}}_t := \{x \mid C_K A_K^i x \in \hat{Y},\ i = 0, 1, \ldots, t\}$ is an inner approximation to $\mathcal{O}_t$. The inner approximation is achieved by replacing the original constraint set Y by $\hat{Y} = Y \ominus C_K \mathcal{A}_\infty$, and employing the conventional (disturbance-free) theory of maximal output admissible sets.

5. Conclusion

The previous subsections show that a variety of interesting control problems may be solved using classical polytopic operations such as addition, Minkowski (Pontryagin) subtraction, intersection, linear transformation and projection together with a few other simple algebraic operations such as pairing. Related work on bounded estimation that yields polytopes, in which the unknown variable (for example the state or unknown vector of parameters of a system) is known to lie, should also be mentioned [5, 27, 36]. Usefulness of these results depends crucially on our ability to carry out these computations. Fortunately some recent developments hold out the promise that these computations can be reliably executed, if not exactly then approximately with approximations that do not destroy the major objectives of the computations.

The simplest representation of a polytope P is often the inequality representation ($P = \{x \mid Mx \leq m\}$). This representation is, for example, suited to the computation of maximal output admissible sets [20] since these sets are defined by inequalities and the additional operations required, such as Pontryagin subtraction, are easily implemented using the inequality representation and linear programming. However, many other operations that are more generally required, such as set addition, projection and finding the convex hull of a set of points, either require, or are much more easily computed, using the vertex representation ($P = \text{convex hull}\{x_1, \ldots, x_I\}$). It appears necessary to have procedures that permit computation of a vertex representation from an inequality representation and vice versa; such procedures are provided, for example, in the software *cdlib* developed by Fukuda at ETH, Zurich. A new version of a MATLAB tool-box (Geometric Bounding Tool-Box) that provides all these operations is currently being developed [35]. In this tool-box, polytopes are specified both by equality and inequality constraints and by vertices. The incorporation of equality constraints permits consideration of polytopes that lie in affine manifolds. Consistency checks ensure validity of the specification. Addition is performed iteratively using a routine that computes the convex hull of a set of points. The tool-box employs procedures that prevent the formation of numerically ill-conditioned polytopes and automatically restricts complexity to that required for a specified accuracy. Because the complexity of the polytopes required may be very high (the complexity of the controllability set X_i is exponential in i) it is important to have techniques that yield adequate approximations; for example, the approximate controllability sets must be inner approximations to be useful. The Geometric Bounding Tool-Box provides this facility.

The number of examples of optimal control problems that may be solved using polytopic techniques, while still small, is perhaps sufficient to encourage further research. These techniques permit the solution of problems, such as constrained linear quadratic optimal control, that are important for applications, and hold out the promise of providing a useful method for approximating the solution of constrained, nonlinear, optimal control problems.

References

[1] Frank Allgöwer, Thomas A. Badgwell, Joe S. Qin, James B. Rawlings, and Stephen J. Wright. Nonlinear predictive control and mov-

ing horizon estimation - an introductory overview. In Paul M. Frank, editor, *Advances in Control: highlights of ECC'99*, pages 391–449, London, 1999. Springer.

[2] A. Bemporad, F. Borelli, and M. Morari. The explicit solution of constrained LP-based receding horizon control. In *Proceedings of the 39th IEEE Conference on Decision and Control*, page 632, Sydney, 2000.

[3] A. Bemporad, M. Morari, V. Dua, and E. Pistikopoulos. The explicit linear quadratic regulator for constrained systems. Technical Report AUT99-16, Automatic Control Laboratory, ETH-Swiss Federal Institute of Technology, 1999.

[4] D. P. Bertsekas and I. B. Rhodes. On the minimax reachability of target sets and target tubes. *Automatica*, 7:233–247, 1971.

[5] D. P. Bertsekas and I. B. Rhodes. Recursive state estimation for a set-membership description of uncertainty. *IEEE Transactions on Automatic Control*, 16:117–128, 1971.

[6] R. R. Bitmead, M. Gevers, and V. Wertz. *Adaptive Optimal Control—The Thinking Man's GPC*. Prentice Hall Int., 1990.

[7] F. Blanchini. Control synthesis for discrete-time linear systems with control and state bounds in the presence of disturbances. In *Proceedings 30th IEEE Conference on Decision and Control*, pages 3464–3467, 1990.

[8] Franco Blanchini and Stefano Miani. Any domain of attraction for a linear constrained system is a tracking domain of attraction. *SIAM Journal of Control and Optimization*, 38(3):971–994, 2000.

[9] Franco Blanchini and Stefano Miani. Any domain of attraction for a linear constrained system is a tracking domain of attraction. In *Proceedings 39th IEEE Conference on Decision and Control*, Sydney, 2000.

[10] D. Chmielewski and V. Manousiouthakis. On constrained infinite-time linear quadratic optimal control. *Systems & Control Letters*, 29:121–129, 1996.

[11] G. De Nicolao, L. Magni, and R. Scattolini. Stability and robustness of nonlinear model predictive control. In Frank Allgöwer and Alex Zheng, editors, *Nonlinear Model Predictive Control*, pages 3–22. Birkhäuser Verlag, Basle, 2000.

[12] E. C. Gilbert, I. Kolmanovsky, and K. T. Tan. Discrete-time reference governors and the nonlinear control of systems with state and control constraints. *Journal of Robust and Nonlinear Control*, 5:487–504, 1995.

[13] E. G. Gilbert and K. T. Tan. Linear systems with state and control constraints: the theory and application of maximal output admissible sets. *IEEE Transactions on Automatic Control*, AC-36:1008–1020, 1991.

[14] Ali Jadbabaie, Jie Yu, and John Hauser. Stabilizing receding horizon control of nonlinear systems: a control Lyapunov function approach. In *Proceedings of American Control Conference*, 1999.

[15] S. S. Keerthi and E. G. Gilbert. Computation of minimum-time feedback control laws for systems with state-control constraints. *IEEE Transactions on Automatic Control*, AC-32:432–435, 1987.

[16] S. S. Keerthi and E. G. Gilbert. Optimal, infinite horizon feedback laws for a general class of constrained discrete time systems: Stability and moving-horizon approximations. *Journal of Optimization Theory and Applications*, 57:265–293, 1988.

[17] Eric. C. Kerrigan. *Robust constraint satisfaction: invariant sets and predictive control.* PhD thesis, University of Cambridge, 2000.

[18] I. Kolmanovsky and E. C. Gilbert. Maximal output admissible sets for discrete-time systems with disturbance inputs. In *Proceedings of the American Control Conference*, Seattle, June 1995.

[19] I. Kolmanovsky and E. C. Gilbert. Multi-modal regulators for systems with state and control constraints and disturbance inputs. In A. S. Morse, editor, *Control using logic-based switching: Lecture Notes in Control and Information Sciences*, pages 118– 127. Springer-Verlag, 1996.

[20] I. Kolmanovsky and E. C. Gilbert. Theory and computation of disturbance invariant sets for discrete-time linear systems. *Mathematical Problems in Engineering*, 4:317–367, 1998.

[21] L. Magni and R. Sepulchre. Stability margins of nonlinear receding-horizon control via inverse optimality. *Systems & Control Letters*, 32:241–245, 1997.

[22] D. Q. Mayne. Nonlinear model predictive control: challenges and opportunities. In Frank Allgöwer and Alex Zheng, editors, *Nonlinear Model Predictive Control*, pages 23–44. Birkhäuser Verlag, Basle, 2000.

[23] D. Q. Mayne. Control of constrained dynamic systems. *European Journal of Control*, 23, 2001.

[24] D. Q. Mayne, J. B. Rawlings, C. V. Rao, and P. O. M. Scokaert. Constrained model predictive control: stability and optimality. *Automatica*, 36:789–814, June 2000. Survey paper.

[25] D. Q. Mayne and W. R. Schroeder. Robust time-optimal control of constrained linear systems. *Automatica*, 33(12):2103–2118, 1997.

[26] Robert H. Miller, Ilya Kolmanovsky, Elmer C. Gilbert, and Peter D. Wahabaugh. Constrained linear systems: a case study. *IEEE Control Systems Magazine*, pages 23–32, February 2000.

[27] S. H. Mo and J. P. Norton. Parameter-bounding identification algorithms for bounded-noise records. *Proceedings of the IEE*, 135 Part D:127–132, 1987.

[28] M. Morari. Mathematical programming approach to hybrid systems, analysis and control. In *25 Years of Nonlinear Control at Ècole des Mines de Paris*, http://cas.ensmp.fr/25ans, 2001.

[29] M. A. Poubelle, R. R. Bitmead, and M. Gevers. Fake algebraic Riccati techniques and stability. *IEEE Transactions on Automatic Control*, AC–31:379–381, 1988.

[30] S. J. Qin and T. A. Badgwell. An overview of nonlinear model predictive control applications. In Frank Allgöwer and Alex Zheng, editors, *Nonlinear Model Predictive Control*, pages 369–392. Birkhäuser Verlag, Basle, 2000.

[31] J. B. Rawlings and K. R. Muske. Stability of constrained receding horizon control. *IEEE Transactions on Automatic Control*, AC-38(10):1512–1516, 1993.

[32] P. O. M. Scokaert and J. B. Rawlings. Infinite horizon linear quadratic control with constraints. In *Proceedings of the 13th IFAC triennial world congress*, volume M, pages 109–114, San Francisico, June 1996.

[33] Marià M. Seron, José A. De Doná, and Graham C. Goodwin. Global analytical model predictive control with input constraints. In *Proceedings of the 39th IEEE Conference on Decision and Control*, Sydney, Australia, December 2000.

[34] Jeff S. Shamma and Dapeng Xiong. Linear non-quadratic optimal control. *IEEE Transactions on Automatic Control*, 42:875–879, 1997.

[35] S. M. Veres. Numerical error control in polytope computations. *Journal of Optimization Theory and Applications*, 2001. Accepted.

[36] S. M. Veres and J. P. Norton. Predictive self-tuning control by parameter bounding and worst-case design. *Automatica*, 29(4):911–928, 1993.

Chapter 7

AN INTRODUCTION TO CONSTRAINED CONTROL

Graham C. Goodwin
eegcg@ee.newcastle.edu.au

Tristan Pérez
tristan@ee.newcastle.edu.au

José A. De Doná
eejose@ee.newcastle.edu.au

School of Electrical Engineering and Computer Science,
The University of Newcastle, NSW, Australia

Abstract **An ubiquitous problem in control system design is that the system must operate subject to various constraints. Although the topic of constrained control has a long history in practice, there have been recent significant advances in the supporting theory. In this chapter, we give an introduction to constrained control. In particular, we describe contemporary work which shows that the constrained optimal control problem for discrete-time systems has an interesting geometric structure and a simple local solution. We also discuss issues associated with the output feedback solution to this class of problems, and the implication of these results in the closely related problem of anti-windup. As an application, we address the problem of rudder roll stabilization for ships.**

Introduction

Dealing with constraints is an essential part of practical control system design. This is a consequence of two facts:

(i) Practical control systems have, in almost all cases, an associated set of constraints. For instance, inputs take values between a maximum and a minimum, and states and outputs must lie within certain ranges. These constraints are mainly present due to constitutive and structural relationships existing between the components of the control system; and in some cases also due to the need to maintain safe operating conditions.

(ii) It is generally true that high performance is associated with the application of larger input signals. Thus, operating conditions are likely to be on, or near, the constraints.

As an example, consider the problem of rudder roll stabilization of ships. The prime function of the rudder is to maintain the ship's heading. However, the rudder also imparts a rolling moment to the ship. Thus, in some cases, the rudder can be used to achieve roll stabilization to reduce the motion induced by the waves. Since the rolling moment induced by the rudder is relatively small, it can be appreciated that large rudder displacements will be called upon especially under heavy sea conditions and/or low forward speeds of the vessel. Of course, real rudders must operate subject to constraints on both their total displacement (typically ± 20 degrees) and slew rate (typically ± 10 degrees per second). Thus, input constraints are present. On the other hand, it is also desirable to keep the yaw or heading of the ship within certain ranges of deviation; therefore, there may also be state constraints associated with the yaw motion.

Most of control theory deals with unconstrained problems; nevertheless, as argued above it is usually desirable to operate systems on or near the constraints. Thus, there is an on-going interest in developing tools to address constrained control problems.

The strategies to deal with constrained control can be broadly classified into four approaches, which we denote:

- Cautious
- Serendipitous
- Remedial
- Tactical

In the cautious case, the control gains are reduced so the constraints are avoided. In the serendipitous case, large control gains are used so the constrains are met sometimes, but there is no particular action to protect the system in the event of constraint violation. In the remedial case, special features (such as anti-windup protection) are added to prevent dangerous situations when the constraints are met. Finally, in the tactical case the constraints are included into the control design *ab initio*.

The remedial strategies have long been used in practice due to their simplicity. The associated control schemes are usually called "anti-windup" schemes. These strategies perform relatively well; however, advances in theoretical aspects as well as electronics have contributed to the applicability of the alternative and more sophisticated tactical strategies. Within the latter class, perhaps the most widely known are the so-called Model Predictive Control (MPC) strategies.

In this chapter, we give an introduction to the topic of constrained control for discrete-time systems. In particular, we describe contemporary work on constrained optimal control for this class of systems and the implication of this in the closely related problem of anti-windup. As an application, we address the problem of rudder roll stabilization for ships.

The rest of the chapter is organized as follows. In section 1, we summarize the fundamental ideas behind the remedial and tactical approaches to deal with constrained control systems. In section 2, we present the elements of MPC and show how the problem can be posed as a quadratic programming optimization problem for the case of linear systems. In section 3, we develop the geometric structure of the solution using a particular form of MPC and a short prediction horizon. In section 4, we analyze a particularly simple form of the solution that holds locally and leads to a connection to anti-windup schemes; we also refer to methods that allow a generalization of the results we present. In section 5, we address the issue of output feedback by analyzing of optimality of strategy that uses the so-called Certainty Equivalence principle. This yields a better understanding of the connection between MPC and anti-windup and also implications on integral action in MPC. In section 6, we review the ingredients that are generally used to establish stability of MPC. In order to do this, we make assumptions that lead to a simple proof and we refer to more general results in the literature. Although simple, the results presented in this section capture the essential idea behind more general stability proofs and help to explain recent contributions in this area. In section 7, we present a case study in which we apply MPC to a particular problem of practical interest: rudder roll stabilization for ships. Finally, in section 8, we summarize and discuss the results.

1. Remedial and Tactical Strategies for Constrained Control

1.1 Remedial Strategies

If constraints are ignored in a design, then it is not surprising that difficulties arise when encountered. A well known example is that of "integral-windup". This term describes the fact that the integral term in a simple PID controller will continue to grow when the controller output hits a constraint. This produces a degradation in the performance of the control loop. Schemes that address the above problem by mitigating the effect of windup are generally called anti-windup schemes. Many such schemes are described in the literature, see for example the survey paper [1].

One way of thinking about the problem of windup is that the states of the control law have not been properly "informed" when the input signal meets the constraint limit. This suggests that one approach to mitigate the problem could be to rearrange the controller so that its internal states are "fed" by the true constrained input signals. This can be achieved, for instance, by implementing the controller transfer function $C(s)$ as shown in figure 7.1, where c_∞ is the high-frequency gain of $C(s)$ (which we assume to be nonzero, *i.e.*, $C(s)$ is assumed bi-proper), and where '*limit*' is a static nonlinear function that represents the relevant constraint on the plant input, *e.g.*, magnitude, rate, a combination of both, *etc.*

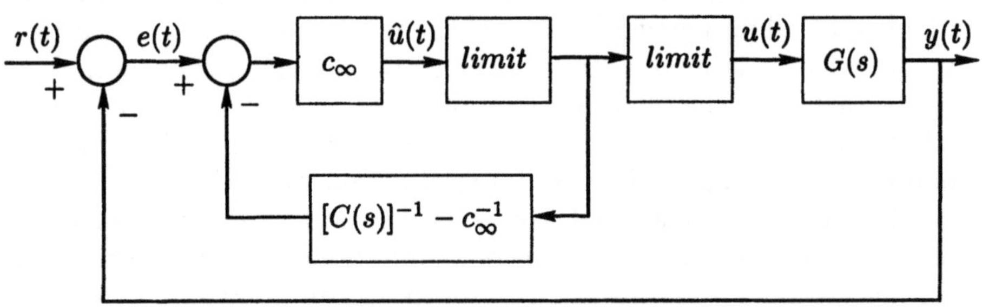

Figure 7.1. Simplified anti-windup control loop.

Two observations immediately follow from this figure:

(i) If one replaces the function '*limit*' by a unity gain, then the transfer function from the error $e(t)$, to the controller output, $u(t)$ is

$$F(s) = \frac{c_\infty}{1 + c_\infty[C(s)]^{-1} - c_\infty c_\infty^{-1}} = C(s).$$

Thus, in the absence of constraints, the above implementation preserves the original linear feedback control law, $C(s)$.

(ii) The internal states of the control law are driven by the constrained input signal.

Remedial strategies of the kind illustrated above appear in most practical implementations of control algorithms. Because of the importance of these ideas in practice, there has been substantial research interest in these schemes.

Summarizing, we see that the essential idea of this class of control scheme is to ensure that the internal states of the control law are "informed" regarding the constraints on the input signal. This can also be given a state variable interpretation [2]. In the state space framework, the dynamics of the control law are typically only associated with the state observer which forms part of the control law. Hence, the appropriate extension of the ant-wind-up scheme described above to this form of controller is to simply ensure that the corresponding observer is "informed" about the constrained input. This leads to the anti-windup configuration shown in figure 7.2. Here, the state estimates will typically also include disturbance states (for example, integral action is achieved by including an appropriate estimator for a constant disturbance [3].)

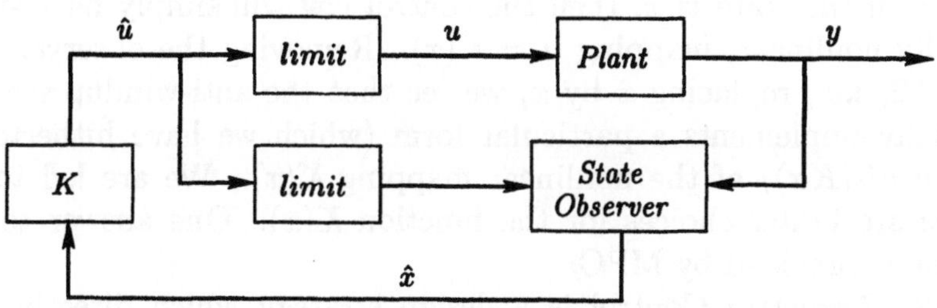

Figure 7.2. Anti-wind up scheme for state variable control.

It is interesting to consider the linear time-invariant case. Here the observer output is simply a linear function of the plant input, u, and the plant output, y. Thus, in transfer form, the state estimate $\hat{x}$ can be expressed as

$$\hat{X}(s) = T_1(s)Y(s) + T_2(s)U(s)$$

with appropriate transfer functions $T_1(s)$ and $T_2(s)$.

We see that the scheme of figure 7.1 is a special case of the state variable scheme of figure 7.2, where $u = limit(\hat{u}) = limit(K\hat{x})$, and

$$\begin{aligned} K\hat{X}(s) &= [K * T_1(s)]Y(s) + [K * T_2(s)]U(s)\,, \\ K * T_1(s) &= -c_\infty\,, \\ K * T_2(s) &= 1 - c_\infty[C(s)]^{-1}\,. \end{aligned}$$

1.2 Tactical Strategies

The schemes discussed above have focused on ensuring that the controller/observer operates correctly in the presence of constraints. This can be seen as a remedial viewpoint since the underlying design strategy begins from an unconstrained perspective. On the other hand, the alternative, tactical viewpoint begins by making a complete reappraisal of the control system design procedure. Specifically, the aim is to optimize the overall performance subject to satisfaction of the appropriate constraints.

To "set the scene" for this alternative viewpoint, we shall temporarily remove the focus on the observer *per se.* Instead, we will, for the moment, assume that the state vector, x, is directly measured.

We also assume that, all components of the problem (model, performance index and constraints), are time invariant. Under these conditions, it follows that the optimal control policy will also be time invariant, *i.e.*, if the state is x, then the control law will simply be a static, generally nonlinear, mapping $u = \mathcal{K}(x)$. Removing the observer from figure 7.2, *i.e.*, replacing $\hat{x}$ by x, we see that the anti-windup schemes essentially implements a particular form (which we have hitherto denoted $limit(Kx)$) of the nonlinear mapping $\mathcal{K}(x)$. We are led to ask if there are better choices for the function $\mathcal{K}(x)$. One answer to this question is provided by MPC.

Model Predictive Control is a design strategy which directly aims at optimizing performance (measured by some cost function) subject to the satisfaction of appropriate constraints. Thus MPC fits into the framework of traditional constrained optimal control — see, for example, [4]. MPC adds a novel aspect based on the use of a receding horizon optimization technique in order to make computation of the control law viable. The core idea of this scheme is to compute, on-line, the solution to an open-loop optimal control problem over some future horizon. The first part of this solution is then applied to the plant and a new state is measured (or estimated). The optimization problem is then repeated over the same-length horizon but with the starting time advanced by one time-unit. Thus, MPC uses on-line constrained optimization to evaluate the nonlinear function $u = \mathcal{K}(x)$ based on the current state, x (or its estimate). This is the main difference from conventional control which uses a pre-computed law, and it can be inferred that the *raison d'être* for model predictive control is its ability to handle control problems in which off-line computations of a control law are deemed difficult or impossible [5].

The key elements of an MPC algorithm are as follows:

- a cost function;
- a set of input and state constraints;
- a method of predicting the future response of the system;
- a prediction horizon over which predictions of the future response are made;
- a control horizon over which the plant input is assumed to change;
- a constraint horizon over which constraints are considered;
- an algorithm for optimizing the cost function subject to the constraints.

We have argued that MPC is, in essence, an on-line procedure utilizing optimization for computing an appropriate nonlinear mapping $u = \mathcal{K}(x)$ from the current state, x, to the input, u. Significant work has been devoted to the associated constrained optimization problem — see for example [6], [7] and [8]. This raises the question as to whether it would be feasible to compute the function $\mathcal{K}(\cdot)$, off-line, for all possible states and then to use these precomputed values, on-line, via some form of table look-up. This would be a particularly attractive option if $\mathcal{K}(\cdot)$ took a simple form, *i.e.,* if it were finitely parameterized.

It is well known that, in the case of linear unconstrained optimal control problems with quadratic cost function, the resultant function, $\mathcal{K}(\cdot)$ always has a finite parameterization. When constraints are added, the situation becomes more complex. Indeed, it is not clear, a-priori, that $\mathcal{K}(\cdot)$ has a finite parameterization; and for this reason, the default option has been to compute $\mathcal{K}(\cdot)$ on-line at the current state.

There have been several recent developments which give additional insight into the MPC solution for constrained linear systems. It has been shown ([9], [10], [11]) that the function $\mathcal{K}(\cdot)$ has a finite parameterization in the presence of constraints. This parameterization turns out to be much simpler than might, *a-priori*, be thought possible. Besides, under non-trivial conditions the finitely parameterized solution reveals a strong connection to the anti-windup strategies previously mentioned.

2. Elements of Model Predictive Control

Many formulations of MPC have been proposed in the literature. They differ essentially on the choice of model and performance index, selection of tuning parameters and derivation of stability conditions. In this section, we describe the essential elements of MPC within a general formulation that encompasses many of the existing MPC formulations. We then specialize to linear models.

2.1 MPC Strategy

MPC uses *receding horizon* optimization, which can be summarized as follows:

(i) At the time instant k, measure the state $x(k)$.

(ii) Given the event $(k, x(k))$, solve the open-loop optimal control problem to find the sequence of control actions that minimize the associated cost function; consider the optimization over some future (finite) interval or horizon taking into account the *current* and *future* constraints.

(iii) Update the control by applying the first element of the solution sequence to the optimization problem as the actual control action $u(k)$.

(iv) Repeat the whole procedure at the time instant $k+1$.

When $x(k)$ is not directly measured, then one can obtain an output feed back policy by replacing $x(k)$ by an estimate provided by some form of observer. The latter topic is taken up in section 5. For the moment, we assume that $x(k)$ is measured.

Specifically, in a general nonlinear setting, the method is as follows. Given a model

$$x(\ell+1) = f(x(\ell), u(\ell)), \tag{1}$$

the MPC control at event (x, k) is computed by solving a constrained optimal control problem:

$$\mathcal{P}_N(x): \qquad V_N^{\text{OPT}}(x) = \min_{\mathbf{u} \in \mathcal{U}_N} V_N(x, \mathbf{u}),$$

where

$$\mathbf{u} = \{u(k), u(k+1), \ldots, u(k+N-1)\},$$

$$V_N(x, \mathbf{u}) = \sum_{\ell=k}^{k+N-1} L(x(\ell), u(\ell)) + F(x(k+N)),$$

and $\mathcal{U}_N$ is the constraint set for $\mathbf{u}$ which encompasses the constraints over the entire interval $[k, k+N-1]$:

$$\begin{aligned} u(\ell) &\in \mathbb{U} \qquad \ell = k, k+1, \dots, k+N-1\,, \\ x(\ell) &\in \mathbb{X} \qquad \ell = k, k+1, \dots, k+N\,, \end{aligned}$$

together with the terminal constraint $x(k+N) \in \mathbb{X}_f$.

Usually, the input constraint set $\mathbb{U} \subset \mathbb{R}^m$ is convex and compact, the state constraint set $\mathbb{X} \subset \mathbb{R}^n$ is convex and closed, and the terminal constraint set $\mathbb{X}_f$ is a set that can be appropriately selected to achieve stability. In the above formulation, the model and cost are time invariant. Hence, one obtains a time-invariant feedback control law, and in particular, we can set $k = 0$ without loss of generality. Then at event (x, k) we solve:

$$\mathcal{P}_N(x): \qquad V_N^{\text{OPT}}(x) = \min_{\mathbf{u} \in \mathcal{U}_N} V_N(x, \mathbf{u})\,, \tag{2}$$

where

$$\mathbf{u} = \{u(0), u(1), \dots, u(N-1)\}\,, \tag{3}$$

$$V_N(x, \mathbf{u}) = \sum_{\ell=0}^{N-1} L(x(\ell), u(\ell)) + F(x(N))\,, \tag{4}$$

subject to

$$u(\ell) \in \mathbb{U} \qquad \ell = 0, 1, \dots, N-1\,, \tag{5}$$

$$x(\ell) \in \mathbb{X} \qquad \ell = 0, 1, \dots, N\,, \tag{6}$$

$$x(N) \in \mathbb{X}_f\,. \tag{7}$$

Let the minimizing control sequence be

$$\mathbf{u}^{\text{OPT}}(x) = \{u^{\text{OPT}}(0; x), u^{\text{OPT}}(1; x), \dots, u^{\text{OPT}}(N-1; x)\}\,; \tag{8}$$

then the actual control applied at time k is the first element of this sequence, *i.e.*,

$$u = u^{\text{OPT}}(0; x)\,. \tag{9}$$

The above procedure is then repeated endlessly.

The above MPC strategy *implicitly* defines a time-invariant control policy $\mathcal{K}_N : \mathbb{X} \to \mathbb{U}$ of the form

$$\mathcal{K}_N(x) = u^{\text{OPT}}(0; x)\,. \tag{10}$$

We will expand on the above skeleton description of MPC as the chapter evolves. For example, we will treat linear constrained problems in some detail in the sequel, and show that when the system model is linear, the performance index quadratic and the constraint sets are polyhedral, the optimal control problem $\mathcal{P}_N(\cdot)$ is a quadratic programming problem.

2.2 Specialization for Linear Models with Quadratic Cost

In this subsection, we take a special case, namely linear systems with quadratic cost function. This will allow us to refine the core idea of MPC. Let us assume that the system is described by the following linear, time-invariant model:

$$x(\ell+1) = Ax(\ell) + Bu(\ell), \tag{11}$$
$$y(\ell) = Cx(\ell), \tag{12}$$

where $x(\ell) \in \mathbb{R}^n$, $u(\ell) \in \mathbb{R}^m$, and $y(\ell) \in \mathbb{R}^m$. We assume that (A, B, C) is stabilizable and detectable.

So as to illustrate the principles involved, we go beyond the set-up described in subsection 2.1 to include reference tracking and disturbance rejection. Thus, we consider the problem of tracking a constant setpoint y_s and rejecting a time-varying output disturbance $\{d(\ell)\}$, *i.e.*, we wish to regulate, to zero, the error

$$e(\ell) = y(\ell) + (d(\ell) - y_s). \tag{13}$$

It will be convenient in the sequel to make no special distinction between the output disturbance and the setpoint, so we define an "equivalent" output disturbance d_e as the external signal

$$d_e(\ell) = d(\ell) - y_s. \tag{14}$$

We also assume knowledge of the external signal d_e and the current state measurement $x(0) = x$. Our aim is to find the M-move control sequence $\mathbf{u} = \{u(0), u(1), \ldots, u(M-1)\}$ that minimizes the finite-horizon cost:

$$V_{N,M}(x, \mathbf{u}) = [x(N) - x_s]^T Q_f [x(N) - x_s] + \sum_{\ell=0}^{N-1} e^T(\ell) Q e(\ell) + \sum_{\ell=0}^{M-1} [u(\ell) - u_s]^T R [u(\ell) - u_s], \tag{15}$$

where $Q \geq 0$, $R > 0$, $Q_f \geq 0$. In (15), N is the prediction horizon and $M \leq N$ is the control horizon. We assume that $d_e(\ell)$ in (14) contains time-varying components as well as a constant (steady-state) component, denoted by $\bar{d}_e$. Then, from (14),

$$\bar{d}_e = \bar{d} - y_s, \tag{16}$$

where $\bar{d}$ is the constant component of the output disturbance $d(\ell)$.

In equation (15), we then let u_s and x_s be the stead- state values of u and x:

$$u_s = -[C(I-A)^{-1}B]^{-1}\bar{d}_e\,, \tag{17}$$

$$x_s = (I-A)^{-1}Bu_s\,. \tag{18}$$

and (for simplicity) we have assumed that 1 is not an eigenvalue of A; however, this assumption can be relaxed–see [12]).

The minimization of (15) is performed on the assumption that the control reaches its steady state value after M steps, that is $u(\ell) = u_s$, $\forall \ell \geq M$. We also assume that the setpoint y_s and the corresponding input and state setpoints u_s and x_s are feasible, *i.e.*, they satisfy the required constraints.

2.3 The Finite Horizon Optimal Control Problem as a Quadratic Programming Problem

In the presence of constraints on the input and output, the dynamic optimization problem described in the subsection 2.2 can be transformed into a *non-dynamic constrained quadratic programming problem.*

To see this, we start by writing[1], from (11) and using $Bu_s = (I-A)x_s$:

$$\mathbf{x} - \mathbf{x}_s = \Gamma\,\mathbf{u} + \Omega\,x - \overline{\mathbf{x}}_s\,, \tag{19}$$

where

$$\mathbf{x} = \begin{bmatrix} x(1) \\ x(2) \\ \vdots \\ x(N) \end{bmatrix}; \quad \mathbf{x}_s = \begin{bmatrix} x_s \\ x_s \\ \vdots \\ x_s \end{bmatrix}; \quad \mathbf{u} = \begin{bmatrix} u(0) \\ u(1) \\ \vdots \\ u(M-1) \end{bmatrix}; \quad \Omega = \begin{bmatrix} A \\ A^2 \\ \vdots \\ A^N \end{bmatrix}; \tag{20}$$

$$\Gamma = \begin{bmatrix} B & 0 & \dots & 0 & 0 \\ AB & B & \dots & 0 & 0 \\ \vdots & \vdots & \ddots & \vdots & \vdots \\ A^{M-1}B & A^{M-2}B & \dots & AB & B \\ A^{M}B & A^{M-1}B & \dots & A^2B & AB \\ \vdots & \vdots & \ddots & \vdots & \vdots \\ A^{N-1}B & A^{N-2}B & \dots & \dots & A^{N-M}B \end{bmatrix}; \quad \overline{\mathbf{x}}_s = \begin{bmatrix} x_s \\ x_s \\ \vdots \\ x_s \\ Ax_s \\ \vdots \\ A^{N-M}x_s \end{bmatrix}.$$

[1]With a slight abuse of notation, we use bold letters to denote both sequences and "piles" of vectors.

Then, using (19), (20) and

$$
\begin{aligned}
\mathbf{Q} &= \mathrm{diag}[C^T QC, \dots, C^T QC, Q_f]\,, \\
\mathbf{R} &= \mathrm{diag}[R, \dots, R]\,, \\
\mathbf{u}_s &= \begin{bmatrix} u_s^T & u_s^T & \dots & u_s^T \end{bmatrix}^T, \\
\mathbf{D} &= \begin{bmatrix} d_e(0)^T - \bar{d}_e^T & d_e(1)^T - \bar{d}_e^T & \dots & d_e(N-1)^T - \bar{d}_e^T & 0 \end{bmatrix}^T, \\
Z &= \mathrm{diag}[C^T Q, C^T Q, \dots C^T Q]\,,
\end{aligned}
\tag{21}
$$

we can express (15) as

$$
\begin{aligned}
V_{N,M}(x, \mathbf{u}) =& e^T(0)Qe(0) + (\mathbf{x} - \mathbf{x}_s)^T \mathbf{Q}(\mathbf{x} - \mathbf{x}_s) + (\mathbf{u} - \mathbf{u}_s)^T \mathbf{R}(\mathbf{u} - \mathbf{u}_s) \\
&+ 2(\mathbf{x} - \mathbf{x}_s)^T Z\mathbf{D} + \mathbf{D}^T \mathrm{diag}[Q, Q, \dots, Q]\mathbf{D} \\
=& \overline{V} + \mathbf{u}^T W \mathbf{u} + 2\mathbf{u}^T (Fx + H)\,.
\end{aligned}
\tag{22}
$$

In (22), $\overline{V}$ is independent of $\mathbf{u}$ and

$$
W = \Gamma^T \mathbf{Q} \Gamma + \mathbf{R}\,, \quad F = \Gamma^T \mathbf{Q}\, \Omega\,, \quad H = -\Gamma^T \mathbf{Q}\, \overline{\mathbf{x}}_s - \mathbf{R}\mathbf{u}_s + \Gamma^T Z\mathbf{D}\,.
$$

From (22) it is clear that, if the design is *unconstrained*, $V_{N,M}(x, \mathbf{u})$ is minimized by taking

$$
u_{\mathrm{UC}}^{\mathrm{OPT}} = -W^{-1}(Fx + H)\,. \tag{23}
$$

where the subscript "UC" stands for unconstrained. In fact, even in the constrained case, the optimal solution has the form (23) in a region that contains the operating point. This will be discussed in more detail later on. Note that, if the setpoint for the output is zero and there are no external disturbances, then the matrix H in (23) is zero. The first element of (23), $u_{\mathrm{UC}}^{\mathrm{OPT}}(0; x)$, has a linear time-invariant feedback structure of the form

$$
u_{\mathrm{UC}}^{\mathrm{OPT}}(0; x) = -K(x - x_s) + u_s\,, \tag{24}
$$

where K is such that $(A - BK)$ is Hurwitz if the terminal state weighting Q_f in (15) is selected adequately–see [13].

Next we introduce constraints into the problem formulation. Magnitude and rate constraints on the plant *input* and *output* can be expressed as follows

$$
\begin{aligned}
u_{\min} \le u(\ell) \le u_{\max}\,; &\qquad \ell = 0, 1, \dots, M-1\,, \\
y_{\min} \le y(\ell) \le y_{\max}\,; &\qquad \ell = 1, 2, \dots, N-1\,, \\
\delta u_{\min} \le u(\ell) - u(\ell - 1) \le \delta u_{\max}\,; &\qquad \ell = 0, 1, \dots, M-1\,.
\end{aligned}
\tag{25}
$$

These constraints can be written as *linear* constraints on $\mathbf{u}$ of the form $L\mathbf{u} \leq K$.

Example 2.1 (Constraint handling) *As an illustration for the single-input single-output case, L and K take the following form:*

$$L = \begin{bmatrix} I \\ -I \\ \Phi \\ -\Phi \\ E \\ -E \end{bmatrix}, \qquad K = \begin{bmatrix} \mathbf{u}_{max} \\ \mathbf{u}_{min} \\ \mathbf{y}_{max} \\ \mathbf{y}_{min} \\ \delta\mathbf{u}_{max} \\ \delta\mathbf{u}_{min} \end{bmatrix},$$

where I is the $M \times M$ identity matrix (M is the control horizon). Φ is the following $(N-1) \times M$ matrix:

$$\Phi = \begin{bmatrix} CB & 0 & \dots & 0 & 0 \\ CAB & CB & \dots & 0 & 0 \\ \vdots & \vdots & \ddots & \vdots & \vdots \\ CA^{M-1}B & A^{M-2}B & \dots & CAB & CB \\ CA^{M}B & CA^{M-1}B & \dots & CA^2B & CAB \\ \vdots & \vdots & \ddots & \vdots & \vdots \\ CA^{N-2}B & CA^{N-3}B & \dots & \dots & CA^{N-M-1}B \end{bmatrix}.$$

E is the following $M \times M$ matrix:

$$E = \begin{bmatrix} 1 & 0 & \dots & 0 \\ -1 & 1 & \dots & 0 \\ & \ddots & \ddots & \\ 0 & \dots & -1 & 1 \end{bmatrix}$$

and

$$\mathbf{u}_{max} = \begin{bmatrix} u_{max} \\ \vdots \\ u_{max} \end{bmatrix}, \qquad \mathbf{u}_{min} = \begin{bmatrix} -u_{min} \\ \vdots \\ -u_{min} \end{bmatrix},$$

$$\delta\mathbf{u}_{max} = \begin{bmatrix} u(0) + \delta u_{max} \\ \delta u_{max} \\ \vdots \\ \delta u_{max} \end{bmatrix}, \qquad \delta\mathbf{u}_{min} = \begin{bmatrix} -u(0) - \delta u_{min} \\ -\delta u_{min} \\ \vdots \\ -\delta u_{min} \end{bmatrix},$$

$$\mathbf{y}_{max} = \begin{bmatrix} y_{max} - CAx - d(1) \\ \vdots \\ y_{max} - CA^{N-1}x - d(N-1) \\ \sum_{i=0}^{N-M-2} CA^i B u_s \end{bmatrix},$$

$$\mathbf{y}_{min} = \begin{bmatrix} -y_{min} + CAx + d(1) \\ \vdots \\ -y_{min} + CA^{N-1}x + d(N-1) \\ \sum_{i=0}^{N-M-2} CA^i B u_s \end{bmatrix},$$

where x is the initial state, and u_{max}, u_{min}, δu_{max}, δu_{min}, y_{max}, y_{min} are the constraint limits defined in (25). ◦

The optimal solution

$$\mathbf{u}^{\text{OPT}}(x) = [(u^{\text{OPT}}(0;x))^{\text{T}}, (u^{\text{OPT}}(1;x))^{\text{T}}, \ldots, (u^{\text{OPT}}(M-1;x))^{\text{T}}]^{\text{T}}$$

can be numerically computed via the following static optimization problem:

$$\text{QP:} \qquad \mathbf{u}^{\text{OPT}}(x) = \arg \min_{L\mathbf{u} \leq K} \mathbf{u}^{\text{T}} W \mathbf{u} + 2\mathbf{u}^{\text{T}}(Fx + H)\,. \tag{26}$$

This problem is a convex problem due to the quadratic cost and linear constraints. Also, standard numerical procedures (called Quadratic Programming algorithms or QP for short) are available to solve this optimization—see for example [14].

3. Geometric Structure of MPC

In the previous section we have seen that for linear systems, the solution of MPC can be posed as a Quadratic Programming optimization problem. We can also see from the previous development that embedded in the solution of QP, there exists an implicit mapping $u = \mathcal{K}(x)$. This raises the question as to whether it would be feasible to compute the function $\mathcal{K}(\cdot)$, off-line, for all possible states and then to use these precomputed values, on-line via some form of table look-up. This would be a particularly attractive option if $\mathcal{K}(\cdot)$ took a simple form, *e.g.*, if it were finitely parameterized.

In this section we outline a procedure by which finitely-parameterized solutions with interesting geometric structure of general MPC cases can be obtained. In order to introduce the reader to the finite parameterized solution of MPC, we consider the solution for the case of short prediction horizon, $N = 2$, and particular choices of the MPC elements as follows.

The system considered is (11) with scalar input $u \in \mathbb{R}$ and state measurement. *i.e.*, $y = x$. The optimal control problem to be solved at each sampling instant is

$$\mathcal{P}_2(x): \qquad V_2^{\text{OPT}}(x) = \min_{\mathbf{u}\in\mathcal{U}_2} V_2(x,\mathbf{u}), \tag{27}$$

where $\mathbf{u} = \{u(0), u(1)\}$, and

$$V_2(x,\mathbf{u}) = \sum_{\ell=0}^{1}[x^{\text{T}}(\ell)Qx(\ell) + u^{\text{T}}(\ell)Ru(\ell)] + x^{\text{T}}(2)Sx(2). \tag{28}$$

The terminal cost $F(x) = x^{\text{T}}Sx$ in (28) is the value function for the associated *unconstrained* infinite horizon control problem, that is, the problem defined by (2) with $N = \infty$, $\mathbf{u} = \{u(0), u(1), \dots\}$, $L(x,u) = x^{\text{T}}Qx + u^{\text{T}}Ru$, $F(x) \equiv 0$ and no constraints on the inputs or states (we will comment more on this choice in section 6). The latter problem is a standard linear quadratic (LQ) optimal control problem whose solution is well known to be achieved by the feedback control law $u = -Kx$, where

$$K \triangleq \bar{R}^{-1}B^{\text{T}}SA, \qquad \bar{R} \triangleq R + B^{\text{T}}SB, \tag{29}$$

and S satisfies the discrete-time algebraic Riccati equation (DARE)

$$S = A^{\text{T}}SA + Q - K^{\text{T}}\bar{R}K. \tag{30}$$

The controller $u = -Kx$ is also optimal for the unconstrained *finite-horizon* control problem defined by (2) with arbitrary finite N, with terminal cost $F(x) = x^{\text{T}}Sx$, and with no constraints on the inputs and states.

The choice of a quadratic function as terminal cost yields a quadratic performance index (28), and the latter makes it possible to derive the finitely-parameterized MPC solution explicitly. At this stage, the choice of the matrix S as the solution to (30) is not essential, although it will prove useful for stability purposes as we discuss in section 6.

The control constraint set in (5) is then

$$\mathbb{U} \triangleq [-\Delta, \Delta], \quad \Delta > 0, \tag{31}$$

that is, the magnitude of the control input is restricted to be less than or equal to the positive number Δ. Also for simplicity, we impose no constraints on the system states, *i.e.*, $\mathbb{X} = \mathbb{X}_f = \mathbb{R}^n$ in (6) and (7).

Let the control sequence that minimizes (28) be

$$\mathbf{u}^{\text{OPT}}(x) = \{u^{\text{OPT}}(0;x), u^{\text{OPT}}(1;x)\}. \tag{32}$$

Then the MPC law is given by the first element of (32), that is,

$$\mathcal{K}_2(x) = u^{\text{OPT}}(0;x). \tag{33}$$

In this section, we will derive the solution of $\mathcal{P}_2$ defined in (27) using standard *dynamic programming* arguments [15]. To this end, we need to introduce some definitions.

The *partial value functions* (or, *optimal costs to go* at each step of the dynamic programming algorithm), are defined by

$$\begin{aligned} V_0^{\mathrm{OPT}}(x(2)) &= x^{\mathrm{T}}(2)Sx(2)\,, \\ V_1^{\mathrm{OPT}}(x(1)) &= \min_{u(1)\in\mathbb{U}} [x^{\mathrm{T}}(1)Qx(1) + u^{\mathrm{T}}(1)Ru(1) + x^{\mathrm{T}}(2)Sx(2)]\,, \\ V_2^{\mathrm{OPT}}(x) &= \min_{\mathbf{u}\in\mathbb{U}^2} V_2(x,\mathbf{u})\,, \end{aligned} \tag{34}$$

where $x(0) = x$ and $V_2(x, \mathbf{u})$ is given in (28).

Dynamic programming leads to:

$$V_{2-\ell}^{\mathrm{OPT}}(x) = \min_{u\in\mathbb{U}} x^{\mathrm{T}}Qx + u^{\mathrm{T}}Ru + V_{2-(\ell+1)}^{\mathrm{OPT}}(Ax+Bu)\,, \qquad \mathit{for} \quad \ell = 0,1, \tag{35}$$

where u and x denote, $u = u(\ell)$ and $x = x(\ell)$, respectively.

The saturation function $\mathrm{sat}_\Delta(\cdot)$ is defined for the saturation level Δ as

$$\mathrm{sat}_\Delta(u) \triangleq \begin{cases} \Delta & \text{if } u > \Delta\,, \\ u & \text{if } |u| \le \Delta\,, \\ -\Delta & \text{if } u < -\Delta\,. \end{cases} \tag{36}$$

The following result reveals an interesting geometric strucuture of MPC law (33).

Theorem 1 *The MPC law* (33) *has the form*

$$\mathcal{K}_2(x) = \begin{cases} -\mathrm{sat}_\Delta(Gx + h) & \mathit{if}\, x \in X_1^- \\ -\mathrm{sat}_\Delta(Kx) & \mathit{if}\, x \in X_1 \\ -\mathrm{sat}_\Delta(Gx - h) & \mathit{if}\, x \in X_1^+ \end{cases} \tag{37}$$

where K is as in (29)*, the gain $G \in \mathbb{R}^{1\times n}$ and the constant $h \in \mathbb{R}$ are given by*

$$G \triangleq \frac{K + KBKA}{1 + (KB)^2}\,, \qquad h \triangleq \frac{KB}{1 + (KB)^2}\Delta\,, \tag{38}$$

and the sets (X_1^-, X_1, X_1^+) are defined by

$$X_1^- \triangleq \{x \mid KA_Kx < -\Delta\}\,, \tag{39}$$

$$X_1 \triangleq \{x \mid |KA_Kx| \le \Delta\}\,, \tag{40}$$

$$X_1^+ \triangleq \{x \mid KA_Kx > \Delta\}\,, \tag{41}$$

with $A_K \triangleq A - BK$.

Proof. We start from the last partial value function (the first equation in (34)), at time $\ell = N = 2$, and solve the problem backwards in time by using the principle of optimality (35).

The partial value function $\mathbf{V_0^{opt}}$ ($\boldsymbol{\ell} = \mathbf{N} = \mathbf{2}$). By definition, the partial value function at time $\ell = N = 2$ is

$$V_0^{\text{OPT}}(x) \triangleq F(x) = x^T S x, \quad \forall x \in \mathbb{R}^n .$$

The partial value function $\mathbf{V_1^{opt}}$ ($\boldsymbol{\ell} = \mathbf{N} - \mathbf{1} = \mathbf{1}$). By the principle of optimality, for all $x \in \mathbb{R}^n$

$$\begin{aligned} V_1^{\text{OPT}}(x) &= \min_{u \in \mathbb{U}} \left\{ x^T Q x + u^T R u + V_0^{\text{OPT}}(Ax + Bu) \right\} \\ &= \min_{u \in \mathbb{U}} \left\{ x^T Q x + u^T R u + (Ax + Bu)^T S (Ax + Bu) \right\} \\ &= \min_{u \in \mathbb{U}} \left\{ x^T S x + \bar{R}(u + Kx)^2 \right\} , \end{aligned} \tag{42}$$

where $\bar{R}$ is defined in (29). In deriving (42) we have made use of (30). It is clear that the *unconstrained* ($u \in \mathbb{R}$) optimal control is given by $u = -Kx$. From the convexity of the value function it then follows that the *constrained* ($u \in \mathbb{U}$) optimal control law is given by

$$u^{\text{OPT}}(1; x) = \text{sat}_\Delta(-Kx) = -\text{sat}_\Delta(Kx), \quad \forall x \in \mathbb{R}^n ,$$

and the partial value function at time $\ell = N - 1 = 1$ is

$$V_1^{\text{OPT}}(x) = x^T S x + \bar{R}\left[Kx - \text{sat}_\Delta(Kx)\right]^2 , \quad \forall x \in \mathbb{R}^n .$$

The (partial) value function $\mathbf{V_2^{opt}}$ ($\boldsymbol{\ell} = \mathbf{N} - \mathbf{2} = \mathbf{0}$). By the principle of optimality, for all $x \in \mathbb{R}^n$

$$\begin{aligned} V_2^{\text{OPT}}(x) &= \min_{u \in \mathbb{U}} \left\{ x^T Q x + u^T R u + V_1^{\text{OPT}}(Ax + Bu) \right\} \\ &= \min_{u \in \mathbb{U}} \Big\{ x^T Q x + u^T R u + (Ax + Bu)^T S (Ax + Bu) \\ &\qquad + \bar{R}\left[K(Ax + Bu) - \text{sat}_\Delta(K(Ax + Bu))\right]^2 \Big\} \\ &= \min_{u \in \mathbb{U}} \Big\{ x^T S x + \bar{R}(u + Kx)^2 \\ &\qquad + \bar{R}\left[KAx + KBu - \text{sat}_\Delta(KAx + KBu)\right]^2 \Big\} . \end{aligned} \tag{43}$$

Let us denote the terms in (43), $f_1(u) \triangleq (u + Kx)^2$, and $f_2(u) \triangleq [KAx + KBu - \text{sat}_\Delta(KAx + KBu)]^2$. Notice that the function $f_2(u)$ has a 'cup' shape formed by three zones: (i) Half-parabola corresponding to the case $KAx + KBu < -\Delta$, (ii) A flat zone corresponding to the case $|KAx + KBu| \leq \Delta$, and, (iii) Half-parabola corresponding to the case $KAx+KBu > \Delta$. We can then analyse the three different cases in (37).

Case (a) $[x \in X_1^-]$: In this case, the minimiser of $f_1(u)$—*i.e.*, $u = -Kx$—is such that $KAx + KB(-Kx) = KA_Kx < -\Delta$ (see (39)), *i.e.*, $u = -Kx$ is in zone (i) of function $f_2(u)$. Then, the minimum of $f_1(u) + f_2(u)$ (situated between the minimum of $f_1(u)$, at $u = -Kx$, and the minimum of $f_2(u)$) will also fall in zone (i), and we conclude that the value function is

$$V_2^{\text{OPT}}(x) = \min_{u \in \mathbb{U}} \left\{ x^T S x + \bar{R}(u + Kx)^2 + \bar{R}\,[KAx + KBu + \Delta]^2 \right\},$$

whose *unconstrained* minimum is easily found to be at $u = -(Gx+h)$, with G and h as given in (38). The result in (37), for this case, follows from the convexity of $f_1(u) + f_2(u)$.

Case (b) $[x \in X_1]$: This case corresponds to the situation where $u = -Kx$ is in zone (ii) of $f_2(u)$ and hence, the *unconstrained* minimum of $f_1(u) + f_2(u)$ occurs at $u = -Kx$. The result in (37) follows from the convexity of the function.

Case (c) $[x \in X_1^+]$: The result follows from a similar analysis to the case $x \in X_1^-$.

■

To summarize, the MPC implicitly defined feedback control laws for horizons $N = 1$ and $N = 2$ are then given by

- For $N = 1$

$$u_k = \mathcal{K}_1(x_k) = \mu_1^{\text{OPT}}(x_k) = -\text{sat}_\Delta(Kx_k) \qquad \forall x_k \in \mathbb{R}^n. \tag{44}$$

- For $N = 2$

$$u_k = \mathcal{K}_2(x_k) = \mu_0^{\text{OPT}}(x_k) = \begin{cases} -\text{sat}_\Delta(Gx_k + h) & \text{if } x_k \in X_1^- \\ -\text{sat}_\Delta(Kx_k) & \text{if } x_k \in X_1 \\ -\text{sat}_\Delta(Gx_k - h) & \text{if } x_k \in X_1^+ \end{cases} \tag{45}$$

Remark 3.1 *Although simple, the case we have considered here illustrates the geometric structure of the solution. This result can be generalized for the case of an arbitrary horizon N and more general constraints using, for example, multi-parametric QP as in [9] or the geometric properties of QP as in [11]. Both approaches use the idea of partitioning the* *state space into regions in which a local controller (that is a continuous and piecewise affine function of the state, see [9]) is used.* ∘

Remark 3.2 *An interesting feature of theorem 1 is that for $x \in X_1$, the MPC solution assumes a* very simple *form, i.e.,* $u = -\text{sat}_\Delta(Kx)$, *which is a clipped version of the unconstrained infinite horizon solution. This local result represents a link to the anti-wind-up strategies presented in section 7; hence, for $x \in X_1$, MPC and anti-wind-up are equivalent control policies; or similarly, anti-wind-up is optimal in a restricted region of the state space.* ∘

Remark 3.2 naturally leads to the question as to whether this simple local form of the solution of MPC, $u = -\text{sat}_\Delta(Kx)$, can be generalized for an arbitrary horizon N, and whether its region of validity is of large size and positive invariant under this control law. These issues are addressed in the following section.

4. Simplified Solution of MPC in a Local Region

In this section we obtain a simplified MPC result that holds in a *restricted region* of the state space. This solution turns out to be equivalent to a class of anti-windup algorithms in the restriction.

4.1 Preliminaries

For clarity of exposition, we present here some notation and preliminary results. In particular various sets in $\mathbb{R}^n$ are defined. These sets make possible the characterization of the state-space regions in which the simple solution holds.

Consider the discrete-time system (11) with state measurement, *i.e.*, $y = x$ where $x \in \mathbb{R}^n$ and $u \in \mathbb{R}$ are the current state and control. The pair (A, B) is assumed to be stabilizable, and the control is assumed to satisfy the constraint $u \in \mathbb{U}$ where $\mathbb{U} \triangleq [-\Delta, \Delta]$, $\Delta > 0$.

For some initial time $i \in \{0, 1, \ldots, N-1\}$, let $\mathbf{u}_i$ denote the *partial control sequence* $\mathbf{u}_i \triangleq \{u(i), u(i+1), \ldots, u(N-1)\}$., where $\mathbf{u} \in \mathbb{U}^N$ ($\mathbf{u}_i \in \mathbb{U}^{N-i}$) denotes the case when each element in the sequence satisfies $u(k) \in \mathbb{U}$, $k = 0, \ldots, N-1$ ($k = i, \ldots, N-1$).

The solution of (11) at time $k \geq i$ when the initial state at time i is $x(i) = x$, and the control sequence is $\mathbf{u}_i$, is denoted $x^{\mathbf{u}_i}(k; x, i)$. To simplify notation, the initial time is dropped when it is zero, *i.e.*, $x^{\mathbf{u}}(k; x) = x^{\mathbf{u}_0}(k; x, 0)$. The *finite-horizon optimal control problem* considered is

$$\mathcal{P}_N(x) : \begin{cases} V_N^{\text{OPT}}(x) & = \min_{\mathbf{u}\in\mathbb{U}^N} V_N(x, \mathbf{u}) , \\ \mathbf{u}^{\text{OPT}} & = \arg \min_{\mathbf{u}\in\mathbb{U}^N} V_N(x, \mathbf{u}) , \end{cases} \tag{46}$$

where

$$V_N(x, \mathbf{u}) = \sum_{k=0}^{N-1} [x^{\text{T}}(k)Qx(k) + u^{\text{T}}(k)Ru(k)] + x^{\text{T}}(N)Sx(N) , \tag{47}$$

and $x(k) \triangleq x^{\mathbf{u}}(k; x)$, $k = 0, 1, \ldots, N$. We assume that $Q \geq 0$ and $R > 0$, and S satisfies the algebraic Riccati equation (30). For each $i = 0, 1, 2, \ldots, N-1$, the *partial value function* (or, *optimal cost to go*), is defined by

$$V_{N-i}^{\text{OPT}}(x) = \min_{\mathbf{u}\in\mathbb{U}^{N-i}} V_{N-i}(x, \mathbf{u}_i) , \tag{48}$$

where the *partial cost* $V_{N-i}(\cdot, \cdot)$ is defined by

$$V_{N-i}(x, \mathbf{u}_i) = \sum_{k=i}^{N-1} [x^{\text{T}}(k)Qx(k) + u^{\text{T}}(k)Ru(k)] + x^{\text{T}}(N)Sx(N) ,$$

with $x(k) \triangleq x^{\mathbf{u}_i}(k; x, i)$, $k = i, i+1, \ldots, N$. We refer to V_{N-i}^{OPT} as the partial value function (or, just the value function) 'at time i', meaning that the (partial) value function 'starts at time i'. We also define

$$V_0^{\text{OPT}}(x) \triangleq x^{\text{T}}Px .$$

We shall consider in the sequel both the linear controller $u = -Kx$ and the nonlinear controller $u = -\text{sat}_\Delta(Kx)$. To this end we define the mappings $\phi_l, \phi_{nl} : \mathbb{R}^n \to \mathbb{R}^n$ as

$$\phi_l(x) \triangleq A_K x, \qquad A_K \triangleq A - BK, \tag{49}$$

$$\phi_{nl}(x) \triangleq Ax - B\text{sat}_\Delta(Kx), \tag{50}$$

so that when the linear (nonlinear) controller is employed, the closed-loop system satisfies $x^+ = \phi_l(x)$ ($x^+ = \phi_{nl}(x)$), where x^+ is the sucessor state of x when the control law $u = -Kx$ ($u = -\text{sat}_\Delta(Kx)$) is applied in (11). In the sequel we will use $\circ$ to denote operator concatenation, *i.e.*, $(a \circ b)(x) \triangleq a(b(x))$ and denote $a^i(x) \triangleq (a \circ a^{i-1})(x)$, $i = 1, 2, \ldots$, where $a^0(x) \triangleq x$. We also define the functions $\delta_i : \mathbb{R}^n \to \mathbb{R}$ as

$$\delta_i(x) \triangleq Kx - \text{sat}_{\Delta_i}(Kx), \quad i = 1, 2, \ldots, N , \tag{51}$$

where the saturation bounds Δ_i are defined as

$$\Delta_i \triangleq \left(1+\sum_{k=0}^{i-2}|KA^kB|\right)\Delta, \quad i=1,2,\ldots,N. \tag{52}$$

In summations, it is to be understood that $\sum_{k=k_1}^{k_2}(\cdot) = 0$ whenever $k_2 < k_1$, so that, in (52) we have

$$\Delta_1 = \Delta, \quad \Delta_2 = \Delta_1 + |KB|\Delta, \ldots, \Delta_{i+1} = \Delta_i + |KA^{i-1}B|\Delta, \ldots.$$

We define, for future use, the sets $X_i \subseteq \mathbb{R}^n$:

$$X_i \triangleq \left\{x \mid \delta_i\left(A^{i-1}\phi_l(x)\right) = 0\right\}, \qquad i=1,2,\ldots,N-1. \tag{53}$$

Denote

$$\bar{K}_i \triangleq KA^{i-1}A_K, \qquad i=1,2,\ldots,N-1.$$

Then, the sets X_i are given by the set of linear inequalities:

$$X_i = \left\{x \mid \bar{K}_i x \le \Delta_i, \; -\bar{K}_i x \le \Delta_i\right\}, \quad i=1,2,\ldots,N-1. \tag{54}$$

We also define, for future use, the sets Y_i, $Z_i \subseteq \mathbb{R}^n$:

$$\begin{aligned} &Y_0 \triangleq Y_1 \triangleq \mathbb{R}^n, \\ &Y_i = \textstyle\bigcap_{j=1}^{i-1} X_j, \quad i=2,3,\ldots,N, \qquad (55)\\ &Z_0 \triangleq Z_1 \triangleq \mathbb{R}^n, \\ &Z_i \triangleq \left\{x \mid \phi_{nl}^k(x) \in Y_{i-k}, \; k=0,1,\ldots,i-2\right\}, \; i=2,3,\ldots,N, \qquad (56) \end{aligned}$$

so that

$$\begin{aligned} Z_2 &= Y_2, \\ Z_3 &= \{x \mid x \in Y_3, \; \phi_{nl}(x) \in Y_2\}, \\ Z_4 &= \{x \mid x \in Y_4, \; \phi_{nl}(x) \in Y_3, \; \phi_{nl}^2(x) \in Y_2\}, \end{aligned}$$

and so on.

We have the following properties of these sets:

Proposition 1

(i) $Y_{i+1} = Y_i \cap X_i$, $i=1,2,\ldots,N-1$.

(ii) The set sequence $\{Z_i \mid i = 0,1,\ldots,N\}$ *is monotonically non-increasing (with respect to inclusion), i.e.* $Z_{i+1} \subseteq Z_i$, $i=0,1,\ldots,N-1$.

(iii) $Z_{i+1} = Y_{i+1} \cap \{x \mid \phi_{nl}(x) \in Z_i\}$, $\quad i=0,1,\ldots,N-1$.

Proof. See [10] ∎

4.2 Regional Solution to the Fixed-horizon Optimal Control Problem

Here we present the a regional result to the fixed horizon optimal control problem.

For a horizon $N \geq 1$, consider the set

$$\begin{aligned} Z_N &= \mathbb{R}^n, \quad \text{if } N = 1, \\ Z_N &= \left\{ x \,\middle|\, \phi_{nl}^k(x) \in Y_{N-k},\ k = 0, 1, \ldots, N-2 \right\}, \ \text{if } N \geq 2, \end{aligned} \tag{57}$$

then we have:

Theorem 2 *Given the fixed-horizon optimal control problem $\mathcal{P}_N$ defined in (46), where x denotes the initial state $x = x(0)$ of system (11) with state measurement, then for all $x \in Z_N$ the minimum cost is*

$$V_N^{\mathrm{OPT}}(x) = x^T S x + \bar{R} \sum_{k=1}^{N} \delta_k (A^{k-1} x)^2, \tag{58}$$

and, for all $x \in Z_N$ the optimal control sequence $\mathbf{u}^{\mathrm{OPT}}(x)$ that attains this minimum is

$$u^{\mathrm{OPT}}(k; x) = \mathrm{sat}_\Delta(-Kx(k)) = -\mathrm{sat}_\Delta(Kx(k)), \tag{59}$$

for $k = 0, 1, \ldots, N-1$, where $x(k) = \phi_{nl}^k(x)$.

Proof. See [10]. ∎

The above result is interesting since it shows that the simple policy (59) provides a solution to the linear quadratic MPC problem in a region of the state space *for an arbitrary horizon.* In the sequel, we will explore various aspects of this solution, including refinement of the set in which the result holds. We first present a simple example for which the solution (59) holds globally.

Example 4.1 (Scalar System with Cheap Control) *Consider a scalar system $x(k+1) = ax(k) + bu(k)$, $x(0) = x$, with $b \neq 0$ and weights $Q = 1$, $R = 0$, in the cost function (47). Such a design, with no weight on the control input, is a limiting case of the controller considered, known as* cheap control.

For this case, the unconstrained *optimal control is* $u = -Kx$, *where* K *computed from* (29) *is* $K = a/b$. *Now, notice that, with* $K = a/b$, *the gain* A_K *computed from* (49) *is* $A_K = 0$ *and, hence, the sets in* (53)–(56) *are:* $X_i \equiv \mathbb{R}^n$, $Y_i \equiv \mathbb{R}^n$, $Z_i \equiv \mathbb{R}^n$, *for all* i. *It then follows from theorem 2 that the optimal control sequence for all* $x \in \mathbb{R}$ *in this case is*

$$u^{\text{OPT}}(k;x) = \text{sat}_\Delta(-ax(k)/b) = -\text{sat}_\Delta(ax(k)/b), \tag{60}$$

for $k = 0, 1, \ldots, N-1$, *where* $x(k) = \phi_{nl}^k(x)$. *Note that here the result* (60) *holds* globally *in the state space. This is particular to this simple problem.* ∘

4.3 Invariant Sets Associated with the Regional Solution

We have seen above that the simple control law (59) solves the finite-horizon constrained linear quadratic problem in a special region of the state space. However, before we can use this control law as the solution to the MPC problem, we need to extend the results to the receding-horizon formulation of the problem. In particular, in order to guarantee that the MPC mapping

$$\mathcal{K}_N(x) = u^{\text{OPT}}(0;x), \tag{61}$$

is regionally given by (59), it is essential to know if future states remain in the region in which the result holds or whether they are driven outside this region. Clearly, in the former case, we can implement the MPC algorithm as in (59) without further consideration. We thus proceed to examine the conditions under which the state remains in the region Z_N where (59) applies. Towards this end we define:

Definition 4.1 *The set* $\mathcal{S} \subset \mathbb{R}^n$ *is said to be* positively invariant *for the system* $x(k+1) = \phi(x(k))$ *if for all* $x(0) = x \in \mathcal{S}$ *the solution* $x(k) = \phi^k(x) \in \mathcal{S}$ *for* $k > 0$, *i.e., if* $\phi(\mathcal{S}) \subseteq \mathcal{S}$. *In particular, in our case, the mapping* $\phi_{nl} : \mathbb{R}^n \to \mathbb{R}^n$ *represents the time evolution of the dynamic system* (11) *under the control law* $u = -\text{sat}_\Delta(Kx)$, *and such a set* $\mathcal{S}$ *constitutes a positively invariant set for the system under this particular nonlinear control.* ∘

Definition 4.2 *Define, for* $N \geq 2$, *the set* $\bar{Z}_N$ *as*

$$\bar{Z}_N \triangleq \left\{x \mid \phi_{nl}^k(x) \in Y_N,\ k = 0, 1, 2, \ldots\right\}. \tag{62}$$

∘

From its definition, it is clear that $\bar{Z}_N$ is the *maximal positively invariant set* in Z_N and in Y_N for the closed-loop system $x^+ = \phi_{nl}(x)$. Notice that $\bar{Z}_N \subseteq Z_N \subseteq Y_N$ (see Proposition 1 (i) and (iii)).

We have from theorem 2 that, if the initial state $x(0) = x \in Z_N$ then $\mathcal{K}_N(x(0)) = u^{\text{OPT}}(0; x(0)) = -\text{sat}_\Delta(Kx(0))$. But, in general, since Z_N is not positively invariant under $\phi_{nl}(\cdot)$, after this control is applied, there is no guarantee that the successor state $x(1) = \phi_{nl}(x(0))$ will stay in Z_N. (Notice from the definition of Z_N that, for example, $x(0) \in Z_N \Rightarrow x(1) = \phi_{nl}(x(0)) \in Y_{N-1} = \bigcap_{i=1}^{N-2} X_i$, but that there is no guarantee that $x(1) = \phi_{nl}(x(0)) \in X_{N-1}$.) Hence, in general, $\mathcal{K}_N(x(1)) = u^{\text{OPT}}(0; x(1)) \neq -\text{sat}_\Delta(Kx(1))$. So, in order that the solution (59) can be applied to the receding-horizon control problem we must ensure that all successor states belong to Z_N. In this connection, we present the following theorem:

Theorem 3 *For all $x \in \bar{Z}_N$ the MPC law $\mathcal{K}_N$ (61) is given by*

$$\mathcal{K}_N(x) = \text{sat}_\Delta(-Kx) = -\text{sat}_\Delta(Kx)\,. \tag{63}$$

Proof. The proof of the theorem follows from the fact that $\bar{Z}_N \subseteq Z_N$ and that $\bar{Z}_N$ is positively invariant under the control $\mathcal{K}_N(x) = -\text{sat}_\Delta(Kx)$. Then, for all states in $\bar{Z}_N$ the future trajectories of the system will be such that $x \in \bar{Z}_N \subseteq Z_N$, and from theorem 2 we conclude that $\mathcal{K}_N(x) = u^{\text{OPT}}(0; x) = \text{sat}_\Delta(-Kx) = -\text{sat}_\Delta(Kx)$. ■

Notice that, if the set $\bar{Z}_N$, in which the theorem is valid, were small enough, such that the control sequence $\{u = -\text{sat}_\Delta(Kx)\}$ stayed unsaturated along the system trajectories, then the result of Theorem 3 would be *trivial*, since it would readily follow from the result of the *unconstrained case* (see [16]). Also, if this set were such that *only* the 'first' control in the sequence $\{u = -\text{sat}_\Delta(Kx)\}$ stayed saturated, then the result would also be trivial (although this is not as evident—see [10]). We shall see in the sequel that $\bar{Z}_N$ is considerably bigger than both of these trivial cases.

We first introduce:

Definition 4.3 (Gilbert and Tan, [17]) *The* maximal output admissible set $\mathcal{O}_\infty$,

$$\mathcal{O}_\infty \triangleq \{x \mid |KA_K^i x| \leq \Delta, \quad i = 0, 1, \ldots\} \tag{64}$$

is the maximal positively invariant set for the system $x^+ = \phi_l(x) = A_K x$ in which input constraints are avoided.

Theorem 4 *The set $\bar{Z}_N$ constitutes a (nonlinear) positively invariant extension of the maximal output admissible set, i.e., $\mathcal{O}_\infty \subseteq \bar{Z}_N$.*

Proof. See [30] ∎

We have seen that the set $\bar{Z}_N$ contains the maximal positively invariant set in which the control constraints are avoided. Although a complete characterization of the set $\bar{Z}_N$ is not currently known, it has been shown in [30] that this set is finitely parameterized, and examples show that, in general, the set $\bar{Z}_N$ is considerably larger than the set $\mathcal{O}_\infty$. The following example illustrate this point.

Example 4.2 *In this example, the relative size of the sets $\mathcal{O}_\infty$, Y_N and $\bar{Z}_N$ is illustrated. Consider the system $x^+ = Ax + Bu$, $y = Cx$ with*

$$A = \begin{bmatrix} 1 & 0 \\ 0.4 & 1 \end{bmatrix}, \quad B = \begin{bmatrix} 0.4 \\ 0.08 \end{bmatrix}, \quad C = \begin{bmatrix} 0 & 1 \end{bmatrix},$$

which is the zero-order hold discretization, with a sampling period of 0.4 sec., of the double integrator $\dot{x}_1 = u$, $\dot{x}_2 = x_1$, $y = x_2$. We take: $Q = I_{2\times 2}$ and $R = 0.25$.

In Figure 7.3 we show the set Y_N, the maximal output admissible set

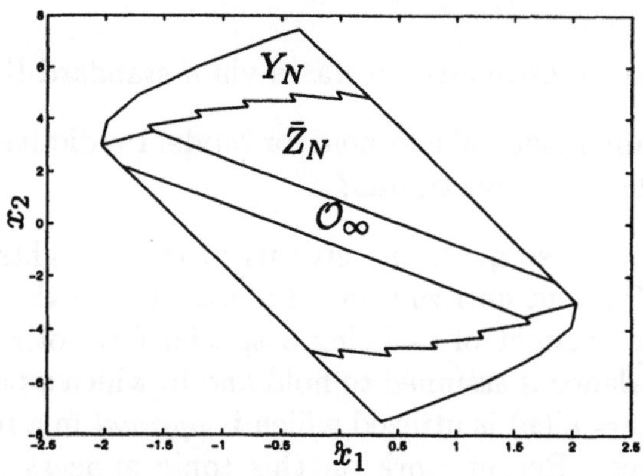

Figure 7.3. Set boundaries for the example. (In the case of $\bar{Z}_N$ an *estimate* obtained *numerically* is shown. Note that $\mathcal{O}_\infty \subset \bar{Z}_N \subset Y_N$.)

$\mathcal{O}_\infty$ and an 'estimate' of the set $\bar{Z}_N$. The estimate of $\bar{Z}_N$ was obtained numerically by gridding the set Y_N and checking the condition in (62) for each point of the grid. As proven in [30], the set $\bar{Z}_N$ is finitely determined and, in this example, the test $\phi^k_{n\ell}(x) \in \bar{Y}_N$ (cf. (62)) is stopped for k such that $\phi^k_{n\ell}(x) \in \mathcal{O}_\infty$. As the figure shows, the new terminal constraint set $\bar{Z}_N$ is considerably larger than $\mathcal{O}_\infty$. ○

As mentioned in section 7, an alternative line of attack on the problem of input constraints (which we have termed "remedial") grouped under the name of anti-windup techniques was developed beginning from a different perspective. Consider now that the original compensator is the static state feedback $u = -Kx$, *i.e.*, the controller has no dynamics. Particularizing the unified framework presented in [18] (which encompasses many anti-windup algorithms) to this case, one easily obtains that the corresponding anti-windup strategy is equivalent to saturating the control signal, that is, $u = -\text{sat}_\Delta(Kx)$. Comparing with the MPC control law (63), we conclude that for all $x \in \bar{Z}_N$, the MPC and anti-windup strategies have the identical parameterization $u = -\text{sat}_\Delta(Kx)$. That is, the MPC and anti-windup control laws coincide in the region where MPC has the simple finite parameterization (63).

Up to this point in our discussion about MPC, we have assumed that the full state vector is measured. When this is not the case, then it seems natural to replace the true state by a state estimate. For example, in [19] it is described an MPC type algorithm incorporating a Kalman filter to estimate both system and disturbance states. This strategy is commonly called Certainty Equivalence since the estimated states are used in the control law as if they were the true states. This raises several questions, *e.g.*,

- Is it optimal to estimate the states via a standard Kalman filter?
- Does Certainty Equivalence hold for Model Predictive Control with linear models and constraints?

The answers to these questions give us further insights into the the link between MPC and anti-windup. Indeed, we see that anti-windup strategies can be thought of as being a special form of MPC in which Certainty Equivalence is assumed to hold and in which a particular form of the mapping $u = \mathcal{K}(x)$ is utilized which is *optimal* in a restricted area of the state space. Recent work on this topic appears in [20] and is reviewed in the following section.

5. Output Feedback Model Predictive Control

In the early sections of the chapter we have assumed that the full state vector is measured; however, due to the presence of uncertainties such as model error or disturbances, and costly or impossible measurements, the true value of the state is not generally available to implement control laws. When this is the case, then a simple solution is to to invoke the so-called Certainty Equivalence Principle (CE), *i.e.*, estimate the state via an observer and use these estimates in a control law that results from the assumption that there is no uncertainty in the knowledge of the state. This control strategy is well known to be optimal in the absence of constraints [22], [23]. However, it is not clear whether this is true when the constraints are added.

In this section, we investigate the issues related to the optimality of observer based MPC with CE by using Stochastic Dynamic Programming to find the optimal control policy for the finite horizon linear input constrained control problem with output measurements and stochastic disturbances. Due to the constraints, the solution potentially has elements of "probing" and "caution" [21]. In our analysis, we find that:

- The use of CE does not, in general, yield an optimal control policy for the associated open-loop constrained optimal control problem. There is an element of caution in the final policy, *i.e.*, the optimal control law has to be modified to take into account the uncertainty associated with the estimated state.
- CE gives an optimal regional solution, but this has no application to the receding horizon formulation of MPC since the region is not guaranteed to be positively invariant under the CE control law.
- There exist cases and conditions in which the CE solution can approximate the optimal solution. Indeed, CE is near optimal for low power disturbance and measurement noise and also if a particular condition is satisfied. In the latter case, the result is independent of the disturbance and noise power.

5.1 The Stochastic MPC Problem

We consider the following time-invariant linear system [2]

$$\begin{aligned} x_{k+1} &= A\, x_k + B\, u_k + w_k\,, \\ y_k &= C\, x_k + v_k\,, \end{aligned} \tag{65}$$

where $x \in \mathbb{R}^n$, $u \in \mathbb{R}$, $y \in \mathbb{R}$. The disturbances, w_k and v_k, are modelled as independent equally-distributed Gaussian random vectors of appropriate dimension, with zero mean and covariance matrices R_w and R_v respectively.

We assume that at the event k, the past data is available. Let us denote the information available to the controller at the event k as $I^k = \{y_k, y_{k-1}, \ldots, y_0, u_{k-1}, u_{k-2}, \ldots u_0\}$ *for* $k = 0, \ldots, N-1$, with $I^0 = \{y_0\}$.

We then consider the class of control laws (or polices) that consist of

[2]To simplify notation in the section, we denote $x(k)$ as x_k.

a finite sequence of functions $\mathbf{u} = \{\mu_0, \mu_1, \ldots, \mu_{N-1}\}$, where each function μ_k maps the information I^k into $\mathbb{R}$, and such that $\mu_k(I^k) \in \mathbb{U} = [-\Delta, \Delta]$ for $k = 0, \ldots, N-1$. Such control laws are termed admissible. Hence, consider the following problem.

Definition 1 (SFHOCP) *We define the Stochastic Finite-Horizon Optimal Control Problem (SFHOCP) with incomplete state information as that of finding the admissible control law* $\mathbf{u}^{\mathrm{OPT}} = \{\mu_0^{\mathrm{OPT}}, \mu_1^{\mathrm{OPT}}, \ldots, \mu_{N-1}^{\mathrm{OPT}}\}$ *that minimizes the cost functional:*

$$V_N(\mathbf{u}) = \mathop{\mathrm{E}}_{x_0, w_k, v_k} \left\{ x_N^{\mathrm{T}} S x_N + \sum_{k=0}^{N-1} x_k^{\mathrm{T}} Q x_k + \mu_k(I^{k-1})^{\mathrm{T}} R \mu_k(I^{k-1}) \right\} \quad (66)$$

subject to the system and measurement equation (65).

5.2 Solution via Stochastic Dynamic Programming

Standard Stochastic Dynamic Programming (SDP) for the problem with imperfect information gives—see [24]:

- For $i = N-1$,

$$\begin{aligned} & V_1^{\mathrm{OPT}}(I^{N-1}) \\ & = \min_{u_{N-1} \in \mathbb{U}} \mathop{\mathrm{E}}_{x_{N-1}, w_{N-1}} \{ \\ & (Ax_{N-1} + Bu_{N-1} + w_{N-1})^{\mathrm{T}} S (Ax_{N-1} + Bu_{N-1} + w_{N-1}) \\ & + x_{N-1}^T Q x_{N-1} + u_{N-1}^T R u_{N-1} | I^{N-1}, u_{N-1} \}; \end{aligned} \quad (67)$$

- For $i = N-2, \ldots, 0$,

$$\begin{aligned} & V_{N-i}^{\mathrm{OPT}}(I^i) \\ & = \min_{u_i \in \mathbb{U}} \mathop{\mathrm{E}}_{x_i, w_i, y_{i+1}} \{ x_i^T Q x_i + u_i^T R u_i + V_{N-(i+1)}^{\mathrm{OPT}}(I^i, y_{i+1}, u_i) | I^i, u_i \}. \end{aligned} \quad (68)$$

It is evident from the SDP algorithm, (67) and (68), that the estimates of the state or conditional means $E\{x_k | I^k\}$ are necessary to solve the problem. The quality of the estimates can be described by the statistics of the estimation error $x_k - E\{x_k | I^k\}$. Before proceeding with the solution of the problem, let us note that due to the linearity of the system and measurement equation, the quality of the estimate 'is not' influenced by the choice of the control [24].

Remark 5.1 *The above fact establishes that there is a partial separation and the problem of estimation can be studied independently from the control policy being adopted.* ○

The implementation of the optimal estimator $E\{x_k|I^k\}$ under the assumption that the initial state x_0, w_k, and v_k are Gaussian, is given by the Kalman filter [26]. Thus,

$$E\{x_k|I^k\} = \hat{x}_{k|k-1}$$

$$cov\{x_k|I^k\} = P_k .$$

The conditional expectation then satisfies

$$\hat{x}_{k+1|k} = A\hat{x}_{k|k-1} + Bu_k + L_k[y_k - C\hat{x}_{k|k-1}] ,$$
$$\hat{x}_{k_0|k_0-1} = \hat{x}_0 ,$$

where

$$L_k = AP_kC^T[CP_kC^T + R_v]^{-1} ,$$

and P_k is the covariance of the estimation error given by the following recursive equation:

$$P_{k+1} = [A - L_kC]P_kC^T + R_w , \qquad P_{k_0} = P_0 .$$

We next state the main result of this section.

5.3 Optimal Control Policy

Theorem 5 *For $N = 2$, the solution of the SFHOCP with incomplete information at the event k is of the form:*

$$\{\mu_k^{\text{OPT}}(I^k), \mu_{k+1}^{\text{OPT}}(I^{k+1})\} = \{\mu_k^{\text{OPT}}(\hat{x}_{k|k-1}, P_k), \mu_{k+1}^{\text{OPT}}(\hat{x}_{k+1|k})\} , \tag{69}$$

where,

$$\mu_{k+1}^{\text{OPT}}(\hat{x}_{k+1|k}) = -\text{sat}_\Delta(K\hat{x}_{k+1|k}) \quad \forall \hat{x}_{k+1|k} \in \mathbb{R}^n , \tag{70}$$

$$\begin{aligned}\mu_k^{\text{OPT}}(\hat{x}_{k|k-1}, P_k) &= \arg\min_{u\in\mathbb{U}}\{\\ g_1(\hat{x}_{k|k-1}, u) &+ g_2(\Delta, \eta(\hat{x}_{k|k-1}, u), \sigma^2(P_k))\} \qquad \forall \hat{x}_{k|k-1} \in \mathbb{R}^n .\end{aligned} \tag{71}$$

The function $\text{sat}_\Delta(\cdot)$ is defined in (36) and

$$g_1(\hat{x}_{k|k-1}, u) \triangleq [u + K\hat{x}_{k|k-1}]^2 \tag{72}$$

$$\begin{aligned}g_2(\Delta, \eta(\hat{x}_{k|k-1}, u), \sigma^2(P_k)) &\triangleq [\sigma^2 + (\eta + \Delta)^2]\ \Phi(\eta, \sigma^2, -\Delta)\\ &+[\sigma^2 + (\eta - \Delta)^2](1 - \Phi(\eta, \sigma^2, \Delta))\\ -\frac{(\eta + \Delta)\sigma}{\sqrt{2\pi}} \exp(\frac{-(\Delta + \eta)^2}{2\sigma^2}) &- \frac{(\Delta - \eta)\sigma}{\sqrt{2\pi}} \exp(\frac{-(\Delta - \eta)^2}{2\sigma^2}) ,\end{aligned} \tag{73}$$

where

$$\begin{aligned}\eta &= \eta(\hat{x}_{k|k-1}, u) = K(A\hat{x}_{k|k-1} + Bu)\,,\\ \sigma^2 &= \sigma^2(P_k) = K(AP_kA^{\mathrm{T}} + R_w)K^{\mathrm{T}}\,.\end{aligned} \tag{74}$$

and the function $\Phi(\eta, \sigma^2, x)$ is the Standard Normal Distribution Function given by

$$\Phi(\eta, \sigma^2, x) = \int_{-\infty}^{x} \frac{1}{\sigma\sqrt{2\pi}} \quad e^{\frac{-(z-\eta)^2}{2\sigma^2}} dz\,.$$

Proof. We apply standard Stochastic Dynamic Programming (*cf.*, (67) and (68))

- For $N-1$, we have

$$\begin{aligned}&V_1^{\text{OPT}}(I^{N-1}) =\\ &\min_{u\in\mathbb{U}}\{(AE\{x_{N-1}|I^{N-1}\} + Bu)^T S(AE\{x_{N-1}|I^{N-1}\} + Bu)\\ &+ trace(Scov\{Ax_{N-1} + R_w|I^{N-1}\})\\ &+ E\{x_{N-1}|I^{N-1}\}^T QE\{x_{N-1}|I^{N-1}\}\\ &+ u^{\mathrm{T}}Ru + trace(Qcov\{x_{N-1}|I^{N-1}\})\}\end{aligned}$$

where we have used the fact that if z is Gaussian $E\{z^{\mathrm{T}}Hz\} = E\{z\}^{\mathrm{T}}HE\{z\} + trace(Hcov\{z\})$– see for example [22]– and that $E\{w_{N-1}|I^{N-1}\} = E\{w_{N-1}\} = 0$. Furthermore, using the algebraic Riccati equation (30) we have

$$\begin{aligned}V_1^{\text{OPT}}(I^{N-1}) = &\min_{u\in\mathbb{U}}\{E\{x_{N-1}|I^{N-1}\}^{\mathrm{T}} SE\{x_{N-1}|I^{N-1}\}\\ &+ \tilde{R}[u + KE\{x_{N-1}|I^{N-1}\}]^2\}\\ &+ trace(Scov\{Ax_{N-1} + R_w|I^{N-1}\})\\ &+ trace(Qcov\{x_{N-1}|I^{N-1}\})\}\end{aligned} \tag{75}$$

It is clear that the *unconstrained* minimum of (75) is attained when $u = -KE\{x_{N-1}|I^{N-1}\}$; and from the convexity of (75), it then follows that for the *constrained* case the optimal control is given by

$$u = -\text{sat}_{\Delta}(KE\{x_{N-1}|I^{N-1}\})\,. \tag{76}$$

Using the estimate given by the Kalman filter, we obtain

$$\mu_{k+1}^{\text{OPT}}(I^{k+1}) = \mu_{k+1}^{\text{OPT}}(\hat{x}_{k+1|k}) = -\text{sat}_{\Delta}(K\hat{x}_{k+1|k}) \qquad \forall \hat{x}_{k+1|k} \in \mathbb{R}^n\,.$$

Also

$$cov\{x_{N-1}|I^{N-1}\} = P_{N-1}$$

and finally, since x_i and w_i are independent then

$$\begin{aligned} cov\{Ax_{N-1} + Bu + w_{N-1}|I^{N-1}, u\} &= cov\{Ax_{N-1} + w_{N-1}|I^{N-1}, u\} \\ &= AP_{N-1}A^{\mathrm{T}} + R_w . \end{aligned}$$

By replacing (76) into (75), the optimal value function at $N-1$ is then

$$\begin{aligned} V_1^{\mathrm{OPT}}(I^{N-1}) =& E\{x_{N-1}|I^{N-1}\}^{\mathrm{T}} S E\{x_{N-1}|I^{N-1}\} \\ &+ \tilde{R}[KE\{x_{N-1}|I^{N-1}\} - \mathrm{sat}_\Delta(KE\{x_{N-1}|I^{N-1}\})]^2\} \\ &+ trace(S cov\{Ax_{N-1} + R_w|I^{N-1}\}) \\ &+ trace(Q cov\{x_{N-1}|I^{N-1}\}) \end{aligned} \tag{77}$$

- For $i = N-2$, we have

$$\begin{aligned} V_2^{\mathrm{OPT}}(I^{N-2}) =& \min_{u\in\mathbb{U}} E\{x_{N-2}^T Q x_{N-2} + u^T R u \\ &+ V_1^{\mathrm{OPT}}(I^{N-2}, y_{N-1}, u)|I^{N-2}, u\} \\ =& \min_{u\in\mathbb{U}} E\{x_{N-2}^T Q x_{N-2} + u^T R u \\ &+ (Ax_{N-2} + Bu + w_{N-2})^T S (Ax_{N-2} + Bu + w_{N-2}) \\ &+ \tilde{R}[K(Ax_{N-2} + Bu + w_{N-2}) \\ &- \mathrm{sat}_\Delta(K(Ax_{N-2} + Bu + w_{N-2}))]^2 \\ &+ trace(S(AP_{N-1}A^{\mathrm{T}} + R_w)) \\ &+ trace(QP_{N-1})|I^{N-2}, u\} . \end{aligned} \tag{78}$$

Proceeding in a similar fashion as in the stage $i = N-1$, we obtain

$$\begin{aligned} V_2^{\mathrm{OPT}}(I^{N-2}) =& E\{x_{N-2}|I^{N-2}\}^T S E\{x_{N-2}|I^{N-2}\} \\ &+ \tilde{R} \min_{u\in\mathbb{U}} [g_1(E\{x_{N-2}|I^{N-2}\}, u) \\ &+ g_2(\Delta, \eta(E\{x_{N-2}|I^{N-2}\}, u), \sigma^2(cov\{x_{N-2}|I^{N-2}\}))] \\ &+ trace(S(AP_{N-1}A^{\mathrm{T}} + R_w)) + trace(QP_{N-1}) \\ &+ trace(S(AP_{N-2}A^{\mathrm{T}} + R_w)) + trace(QP_{N-2}) \end{aligned} \tag{79}$$

Where we have denoted:

$$g_1(E\{x_{N-2}|I^{N-2}\},u) = [u + KE\{x_{N-2}|I^{N-2}\}]^2\} \tag{80}$$

and

$$\begin{aligned} &g_2(\Delta, \eta(E\{x_{N-2}|I^{N-2}\},u), \sigma^2(cov\{x_{N-2}|I^{N-2}\})) = \\ &E\{[K(Ax_{N-2} + Bu + w_{N-2}) \\ &- \text{sat}_\Delta(K(Ax_{N-2} + Bu + w_{N-2}))]^2|I^{N-2},u\} \end{aligned} \tag{81}$$

After integration, we can write this last expression as a function of the Standard Normal Distribution Function $\Phi(\eta, \sigma^2, \Delta)$, *i.e.*,

$$\begin{aligned} &g_2(\Delta, \eta(E\{x_{N-2}|I^{N-2}\},u), \sigma^2(cov\{x_{N-2}|I^{N-2}\})) = \\ &[\sigma^2 + (\eta+\Delta)^2]\ \Phi(\eta, \sigma^2, -\Delta) + [\sigma^2 + (\eta-\Delta)^2](1 - \Phi(\eta,\sigma^2,\Delta)) \\ &- \frac{(\eta+\Delta)\sigma}{\sqrt{2\pi}} \exp(\frac{-(\Delta+\eta)^2}{2\sigma^2}) - \frac{(\Delta-\eta)\sigma}{\sqrt{2\pi}} \exp(\frac{-(\Delta-\eta)^2}{2\sigma^2}), \end{aligned}$$

where

$$\begin{aligned} \eta &= \eta(E\{x_{N-2}|I^{N-2}\},u) = E\{K(Ax_{N-2} + Bu + w_{N-2})|I^{N-2}\} \\ &= K(AE\{x_{N-2}|I^{N-2}\} + Bu), \end{aligned}$$

$$\begin{aligned} \sigma^2 &= \sigma^2(cov\{x_{N-2}|I^{N-2}\}) = cov\{K(Ax_{N-2} + Bu + w_{N-2})|I^{N-2}\} \\ &= K(Acov\{x_{N-2}|I^{N-2}\}A^{\mathrm{T}} + R_w)K^{\mathrm{T}}. \end{aligned}$$

Using the estimates given by the Kalman filter, we obtain that the optimal control that minimizes (79) is given by:

$$\begin{aligned} \mu_k^{\text{OPT}}(I^k) = \mu_k^{\text{OPT}}(\hat{x}_{k|k-1}, P_k) = \arg\min_{u \in \mathbb{U}} \{ & g_1(\hat{x}_{k|k-1}, u) \\ & + g_2(\Delta, \eta(\hat{x}_{k|k-1}, u), \sigma^2(P_k))\}. \end{aligned} \tag{82}$$

This completes the proof. ■

The stochastic output feedback MPC solutions for the horizons $N = 1$ and $N = 2$ are then given by

- For $N = 1$

$$u_k = \mathcal{K}_1(\hat{x}_{k|k-1}) = \mu_1^{\text{OPT}}(\hat{x}_{k|k-1}) = -\text{sat}_\Delta(K\hat{x}_{k|k-1}) \quad \forall \hat{x}_{k|k-1} \in \mathbb{R}^n. \tag{83}$$

- For $N = 2$

$$\begin{aligned} u_k = \mathcal{K}_2(\hat{x}_{k|k-1}, P_k) = \arg\min_{u \in \mathbb{U}} \{ & g_1(\hat{x}_{k|k-1}, u) \\ & + g_2(\Delta, \eta(\hat{x}_{k|k-1}, u), \sigma^2(P_k))\} \quad \forall \hat{x}_{k|k-1} \in \mathbb{R}^n. \end{aligned} \tag{84}$$

Remark 5.2 *We see from (44) and (83) that the use of Certainty Equivalence in the MPC formulation is optimal for the case of horizon $N = 1$, but from (45) and (84) we see that this is not so for the case of horizon $N = 2$; and therefore, neither will be for greater horizons, since the control law starts to depend upon information regarding the state uncertainty from the second step of recursion of Dynamic Programming. Consequently, the use of Certainty Equivalence in the MPC methodology does not, in general, yield an optimal control policy.*

In addition, the control (84) cannot be written in an explicit form since it depends upon the Standard Normal Distribution function. Hence some form of numerical search is necessary to evaluate it given $\hat{x}_{k|k-1}$ and P_k. This can be seen from (81) which shows that the function g_2 obtained by considering the expected value of $y = f(z)$, where

$$f(z) = [z - \text{sat}_\Delta(z)]^2 = \begin{cases} (z-\Delta)^2 & \textit{if } z > \Delta\,, \\ 0 & \textit{if } |z| \le \Delta\,, \\ (z+\Delta)^2 & \textit{if } z < -\Delta\,, \end{cases} \tag{85}$$

and $z = K(Ax_k + Bu + w_k)$ is normal (given u and I^k) with conditional mean $\eta = E\{z|I^k, u\}$ and covariance $\sigma^2 = cov(z|I^k, u)$ given in (74). ∘

The results of Theorem 5 lead us to ask whether CE could be optimal, or near optimal, for $N > 1$ in particular cases.

5.4 Conditions under which CE yields optimal and near optimal control policies

We first analyze the optimal solution in a restricted region of the state space.

Corollary 5.1 *For $(N = 2)$ in the SFHOCP with incomplete state information, CE gives a regional optimal solution,* i.e.,

$$\mu_{k+1}^{\text{OPT}}(I^{k+1}) = \mu_{k+1}^{\text{OPT}}(\hat{x}_{k+1|k}) = -\text{sat}_\Delta(K\hat{x}_{k+1|k}) \quad \forall \quad \hat{x}_{k+1|k} \in \mathbb{R}^n\,, \tag{86}$$

$$\mu_k^{\text{OPT}}(I^k) = \mu_k^{\text{OPT}}(\hat{x}_{k|k-1}) = -\text{sat}_\Delta(K\hat{x}_{k|k-1}) \quad \text{if} \quad \hat{x}_{k|k-1} \in X_2\,, \tag{87}$$

where, the set X_2 is defined by

$$X_2 = \{x | K(A - BK)x = 0\}\,. \tag{88}$$

Proof. To see this, we can follow the proof of Theorem 5, in which it was shown that

$$\begin{aligned}V_2^{\text{OPT}}(I^{N-2}) =&\hat{x}_{N-2}^{\mathrm{T}} S \hat{x}_{N-2} \\ &+ \tilde{R} \min_{u \in \mathbb{U}} [g_1(\hat{x}_{N-2}, u) + g_2(\Delta, \eta(\hat{x}_{N-2}, u), \sigma^2(P_{N-2}))] \\ &+ trace(S(AP_{N-1}A^{\mathrm{T}} + R_w)) + trace(QP_{N-1}) \\ &+ trace(S(AP_{N-2}A^{\mathrm{T}} + R_w)) + trace(QP_{N-2}).\end{aligned} \tag{89}$$

From the definition of X_2 in (88), η (*cf.*, (74)) and (89) it follows that

$$\begin{aligned}\mu_{N-2}^{\text{OPT}}(I^{N-2}) = \mu_{N-2}^{\text{OPT}}(\hat{x}_{N-2}) = \arg\min_{u \in \mathbb{U}} \{g_1(\hat{x}_{N-2}, u) + K_2\} \\ \text{if} \quad \hat{x}_{N-2} \in X_2 \,.\end{aligned} \tag{90}$$

Where $K_2 = g_2(\Delta, 0, \sigma^2)$. Therefore, the unconstrained minimum of (90) is $u = -K\hat{x}_{N-2}$; and it is clear from the convexity of the function (90) that the constrained minimum is attained by $u = -\text{sat}_\Delta(K\hat{x}_{N-2})$. This completes the proof. ■

Remark 5.3 *The above result establishes that there is a condition associated with a region in the state space, namely $\hat{x}_{N-2} \in X_2$, that if satisfied, then CE yields an optimal control policy for the SFHOCP with $N = 2$. The result, however, cannot be directly used in receding horizon MPC since it requires the region in which the CE solution is optimal be positively invariant. This is very unlikely to happen since X_2 is a line through the origin in the state space.* ○

We next study the conditions under which CE gives a near optimal control policy.

For the case of reduced noise we have the following corollary.

Corollary 5.2 *As the variance of the noise reduces, the sub-optimal Certainty Equivalence solution of the constrained finite-horizon ($N = 2$) optimal control problem approximates the optimal solution,* i.e.,

$$u_k = \mathcal{K}_2(\hat{x}_{k|k-1}) = \mu_0^{\text{OPT}}(\hat{x}_{k|k-1}) \approx \begin{cases} -\text{sat}_\Delta(G\hat{x}_{k|k-1} + h) & \textit{if}\, \hat{x}_{k|k-1} \in X_1^- \\ -\text{sat}_\Delta(K\hat{x}_{k|k-1}) & \textit{if}\, \hat{x}_{k|k-1} \in X_1 \\ -\text{sat}_\Delta(G\hat{x}_{k|k-1} - h) & \textit{if}\, \hat{x}_{k|k-1} \in X_1^+ \end{cases} \tag{91}$$

Proof. If R_w and R_v are small, then the covariance of the estimation error P is small. Using the fact that if z has mean $E\{z\} = \eta$ and its probability density function $f_z(z)$ is negligible outside the interval$(\eta - \epsilon, \eta + \epsilon)$, and in this interval $g(z) \approx g(\eta)$, then the expected value of $y = f(z)$ can be approximated by

$$E\{y\} = \int_{-\infty}^{\infty} f(s)f_z(z)dz \approx \int_{\eta-\epsilon}^{\eta+\epsilon} f(\eta)f_z(z)dz \approx f(\eta)\,.$$

The result then follows by defining $z \triangleq K(Ax_{N-2}+Bu+w_{N-2})$, defining $f(z)$ as in (85) and using the estimates given by the Kalman filter. ■

Figure 7.4 Shows $g_2(\eta) = E\{f(z)\}$ and its derivative as a function of

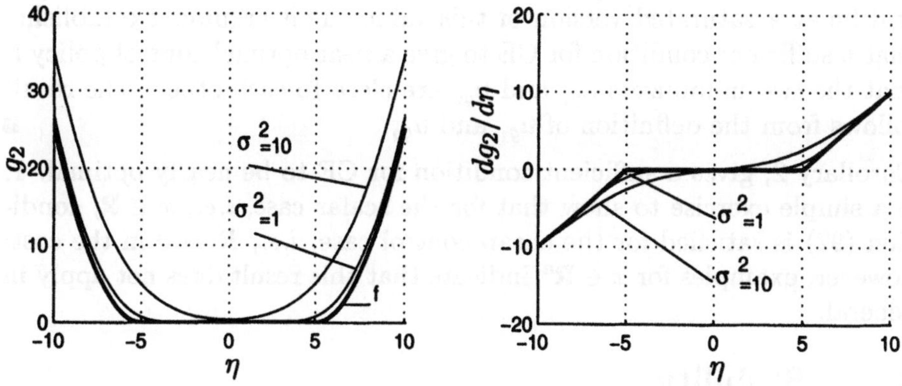

Figure 7.4. $g_2(\Delta, \eta, \sigma^2)$ and $dg_2(\Delta, \eta, \sigma^2)/d\eta$ for different values of σ^2 and for $\Delta = 5$.

the mean $E\{z\} = \eta$ for different values of σ^2 and also (in thick line) $f(\eta)$ (*cf.*, (85)) and its derivative. In this figure, we see a graphical representation of the result of corollary 5.2.

The function g_2 presents some properties that can be exploited to find other conditions under which the use CE is near optimal. In this regard, we present the following proposition.

Proposition 2 *The function g_2 is convex with respect of η (and thus with respect of u) and also has its unique minimum at $\eta = 0$ independently of the value of σ^2.*

Proof. See [25] ■

Finally, we have the following result.

Corollary 5.3 *CE gives a near optimal control policy if*

$$K - (KB)^{-1}KA \to 0. \tag{92}$$

Proof. Using the result of proposition 2 and the definition of η (*cf.*, (74)), we find that the control that minimizes g_2 is

$$u_{g_2} = -(KB)^{-1}(KA)\hat{x}\,.$$

On the other hand, the quadratic function g_1 has its minimum for

$$u_{g_1} = -K\hat{x}\,.$$

Since both g_1 and g_2 are scalar and convex, then it follows that unconstrained minimum of (71) will be between u_{g_1} and u_{g_2}. Also, from the convexity of $g_1 + g_2$ it then follows that the constrained minimum of (71) will be at a saturated version of this value. This argument establishes that a sufficient condition for CE to give a near optimal control policy is that the two minimizers u_{g_1} and u_{g_2} are close to each other. The result follows from the definition of u_{g_1} and u_{g_2}. ■

Corollary 2, gives a sufficient condition for CE to be nearly optimal. It is a simple exercise to show that for the scalar case, *i.e.*, $x \in \mathbb{R}$, condition (92) is satisfied for the cheap control case, *i.e.*, $R \to 0$ in the cost; however, examples for $x \in \mathbb{R}^n$ indicate that this result does not apply in general.

6. Stability

An important property of any control law is whether or not it stabilizes the system. Note that "optimality" does not, by itself, guarantee stability—see for example comments made in [26] and later in [27]. Thus stability needs to be independently established. This is often a non-trivial task. It is interesting to note in this regard, that the absence of a formal proof of stability has historically not impeded practical use of various control schemes. Indeed, constrained MPC has been used in industry for three decades, yet formal proofs of stability have only appeared in recent times.

A great deal of literature about MPC has focused on establishing sufficient conditions for stability. This has developed, essentially, in a number of different proposals for model predictive control with their specific properties and ingredients that guarantee stability of the algorithms. Recently, in [5] it has been compiled a survey that comprehensively presents the state-of-the-art in this area of on-going research activity. This survey distills the common ingredients that characterize the different forms of model predictive control reported in the literature.

Since the stability analysis of MPC is not a trivial issue, in this section of the chapter we give the reader a feel for the essence of the stability properties of MPC. Thus, we establish stability under simplifying assumptions leading to a particular forms of MPC that essentially captures the core of more general form of stability proof—like the one presented in [5].

A standard technique to analyze MPC stability, proposed originally in [29] and later generalized in [28], is to use the value function of the finite-horizon optimal control problem as a Lyapunov function for the closed-loop system.

Several 'ingredients' of the finite-horizon optimal control problem presented in section 2.1 directly affect closed-loop stability; these are: the terminal cost $F(x)$, the terminal constraint set $\mathbb{X}_f$, and a local controller $\mathcal{K}_f(x)$ employed in $\mathbb{X}_f$ that ensures existence of feasible solutions to the optimal control problem. Since different specifications of $(F, \mathbb{X}_f, \mathcal{K}_f)$ lead to different formulations of MPC (see [5]) , to introduce the reader to the core idea of the general stability proofs, we establish stability under the following simplifying assumptions:

A1 The terminal constraint set is the origin, i.e., $\mathbb{X}_f = 0$.

A2 $L(x,u)$ and $F(x)$ in (4) are positive definite functions of their arguments.

The following theorem establishes exponential stability of the closed-loop system.

Theorem 6 *Consider the system* (1) *controlled by the receding-horizon MPC algorithm* (2)–(9) *and subject to Assumptions A1 and A2 above. Then the closed-loop system is asymptotically stable.*

Proof. Under regularity assumptions, the value function $V_N^{\text{OPT}}(x)$ in (2) is positive definite and proper. It is therefore a Lyapunov function candidate for the problem.

We recall that at event (x,k) MPC solves (2) subject to (5)–(7) to obtain the optimal open-loop control sequence (8). Then the actual control applied at time k is the first value of this sequence, i.e.,

$$u = \mathcal{K}_N(x) = u^{\text{OPT}}(0; x)\,.$$

Let $x(1) = f(x, \mathcal{K}_N(x))$, and let $x(N)$ be the terminal state resulting from the application of (8) to (1). Note that $x(N) = 0$ by Assumption A1.

A *feasible* (but not necessarily optimal) control sequence for the second step in the receding horizon computation $\mathcal{P}_N(x(1))$ is then

$$\tilde{\mathbf{u}}(x) = \{u^{\text{OPT}}(1;x), u^{\text{OPT}}(2;x), \ldots, u^{\text{OPT}}(N-1;x), 0\}. \tag{93}$$

The increment of the Lyapunov function, upon using the true MPC optimal input and then moving from $x(0) = x$ to $x(1) = f(x, \mathcal{K}_N(x))$ satisfies

$$V_N^{\text{OPT}}(x(1)) - V_N^{\text{OPT}}(x) = V_N(x(1), \mathbf{u}^{\text{OPT}}(x(1))) - V_N(x, \mathbf{u}^{\text{OPT}}(x)). \tag{94}$$

However, $\mathbf{u}^{\text{OPT}}(x(1))$ is optimal, so we know that

$$V_N(x(1), \mathbf{u}^{\text{OPT}}(x(1))) \leq V_N(x(1), \tilde{\mathbf{u}}(x)), \tag{95}$$

where $\tilde{\mathbf{u}}(x)$ is the feasible sequence defined in (93). Using (95) in (94), and the facts that (i) $\tilde{\mathbf{u}}(x)$ shares $N-1$ terms in common with $\mathbf{u}^{\text{OPT}}(x)$, and that (ii) $\mathbf{u}^{\text{OPT}}(x)$ leads to $x(N) = 0$ and so $\tilde{\mathbf{u}}(x)$ leads to $x(N+1) = 0$, we obtain

$$\begin{aligned} V_N^{\text{OPT}}(x(1)) - V_N^{\text{OPT}}(x) &\leq V_N(x(1), \tilde{\mathbf{u}}(x)) - V_N(x, \mathbf{u}^{\text{OPT}}(x)) \\ &= -L(x, \mathcal{K}_N(x)). \end{aligned}$$

Since $L(x,u)$ is positive definite in both arguments (by Assumption A2), global asymptotic stability follows from Lyapunov arguments. ■

Remark 6.1 *Conventionally, the terminal cost function is chosen, when the system being controlled is linear, to be the value function of the infinite horizon unconstrained optimal control problem,* i.e., $F(x) = x^{\text{T}} S x$ *where S is the solution of the associated Algebraic Riccati Equation; and the terminal constraint set is chosen to be the output admissible set [17] for the closed-loop system using the control law $\mathcal{K}_f(x) = -Kx$. Recent results reported in [30], use the region in which the finite parameterization given in section 3 is valid to establish enlarged terminal sets that guarantee the stabilizing properties of MPC and facilitate the on-line solution of the optimization problem.* ○

7. Applications to Rudder Roll Stabilization for Ships

In this section we present an application of MPC to a case of practical interest: Rudder Roll Stabilization for Ships (RRS). This section is based on recent work reported in [31]

Besides controlling the heading of a ship, it is also desirable to reduce the rolling motion produced by the waves so as to prevent cargo damage and improve crew efficiency and passenger comfort. Conventional methods for ship roll stabilization include water tanks, stabilizing fins, and bilge keels; however, RRS is attractive because no extra equipment needs to be added to the ship and it performs similarly to other methods [32].

From the control design point of view, the main limiting factor of the RRS problem is the highly non linear behavior of the mechanical devices that command the rudder. These devices essentially impose constraints on the maximum slew rate and excursion attainable for the rudder movement. This, makes, in some cases, traditional linear control design infeasible since the generated control signals are typically too large to be realistic [33]. In order to deal with the saturation and slew rate problems, methods such as automatic gain control, see [33], or gain scheduling algorithms, see [34], have been proposed. All this methods fall, according to our classification, into "remedial" strategies. Therefore, we propose the use of a tactical strategy, *i.e.*, MPC, to address the problem of RRS.

In order to describe the motion of a ship, six independent coordinates are necessary. The first three coordinates and their time derivatives correspond to the position and translational motion while the other three coordinates and their time derivatives correspond to orientation and rotational motion description. For marine vehicles, the six different motion components are called: *surge, sway, heave, roll, pitch,* and *yaw*. Accordingly, the most generally used notation for theses quantities are: x, y, z, ϕ, θ, and ψ respectively, while their time derivatives are denoted u, v, w, p, q, and r respectively. Figure 7.5 shows the 6 coordinate definitions and the most generally adopted reference frame to which they are referred. Apart from the mentioned magnitudes, figure 7.5 also shows the definition and positive convention for the *Rudder Angle*, which we denote δ.

The common method of modelling the ship motion is based on the use of Newton's equations in surge, sway and roll. Using this approach, a non linear state space model is obtained in the variables $x \triangleq [v\ r\ p\ \phi\ \psi]^{\mathrm{T}}$ and from this non-linear model, a linearized state space model can be derived, see for example [35].

Undesired variations of the roll and yaw angles are mainly generated by the action of waves and usually modeled as output disturbances [35]. The complete model of the ship dynamics including the output equation

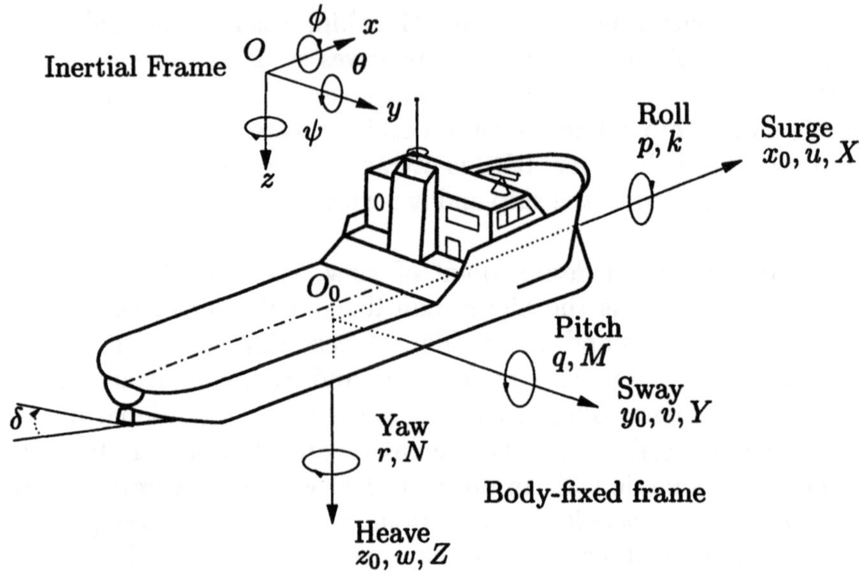

Figure 7.5. **Standard notation and sign conventions for ship motion description.**

is of the form:

$$\begin{aligned}\dot{x} &= A\,x + B\,\delta\,,\\ y &= C\,x + d_{wave}\,,\end{aligned} \tag{96}$$

Typically, only the roll and yaw angles are measured; thus, $y \triangleq [\phi\ \psi]^T$.

The parameters of this model depend upon the forward speed of the ship; however, this is of not of much concern since the model remains quasi-constant over a reasonable speed range. This allows the use of gain scheduling to update the model according to different operating speeds.

The wave-produced disturbances can be characterized in terms of the frequency spectrum of the sea elevation, for instance, using the Pierson-Moskowits spectrum, see [36] . This frequency spectrum can be simulated by a series of added sinusoidal signals with random phases, and the ship motion is calculated using the response amplitude operators. This way of modelling the wave disturbances is accurate; however, for ship motion control system design, the wave disturbances are generally modelled using filtered white noise [32]. In the latter case, the filter used to approximate the spectrum is usually a second order filter of the form

$$H(s) = \frac{K_w\ s}{s^2 + 2\ \xi\ \omega_e\ s + \omega_e^2} \tag{97}$$

in which K_w is a constant describing the wave intensity, ξ is a damping coefficient, and ω_e is the encounter frequency, which depends on the peak frequency of the wave elevation spectrum, forward speed of the ship, and the angle between the heading and the predominant direction of the wave—for more details see [32], [36].

The design objectives for RRS are

- Increase the damping and reduce roll amplitude
- Control the heading of the ship
- Satisfy rudder magnitude and rate constraints.

For merchant ships the rudder rate is within 3 to 4 deg/sec while for military vessels, it is in the range of 3 to 8 deg/sec. The maximum rudder excursion is in the range of 20 to 35 deg. In order to achieve better roll reduction, new steering machinery systems have been designed, and the rudder rate has been increased to 15-20 deg/sec, [34].

In the sequel, we present some simulation results using the algorithm given in sections 2.2–2.3 .

Example 7.1 (MPC via QP) *The parameters of the continuous time model for a constant forward speed of a naval vessel of 9 m/sec are taken from [35]:*

$$A = \begin{bmatrix} -0.1795 & -0.8404 & 0.2115 & 0.9665 & 0 \\ -0.0159 & -0.4492 & 0.0053 & 0.0151 & 0 \\ 0.0354 & -1.5594 & -0.1714 & -0.7883 & 0 \\ 0 & 0 & 1 & 0 & 0 \\ 0 & 1 & 0 & 0 & 0 \end{bmatrix}, \quad B = \begin{bmatrix} 0.2784 \\ -0.0334 \\ -0.0894 \\ 0 \\ 0 \end{bmatrix}$$

and the model used in the MPC algorithm is a zero-order hold equivalent for the sampling interval 0.5 sec. Details of the MPC controller parameters are:

$$Q = \begin{bmatrix} 90 & 0 \\ 0 & 3 \end{bmatrix} \quad \text{and} \quad R = [0.1]. \tag{98}$$

The Matrix S is calculated from (30) *using* (98). *The prediction horizon is taken* $N = 20$ *and the control horizon* $M = 18$.

Since only roll, ϕ, *and yaw,* ψ, *are assumed to be measured, a Kalman filter is employed to estimate the 5 system states and the 2 disturbance states in a composite model,* i.e., $[\hat{x}^{\mathrm{T}} \quad \hat{x}_d^{\mathrm{T}}]^{\mathrm{T}}$, *where* $\hat{x}_d$ *is the estimate of the state of the disturbance model* (97) *converted into state space model* (A_d, B_d, C_d). *The prediction of the disturbance over the horizon N, is done by the optimal n-step-ahead predictor [37],* i.e., *running the disturbance model without input and using the value given by the Kalman*

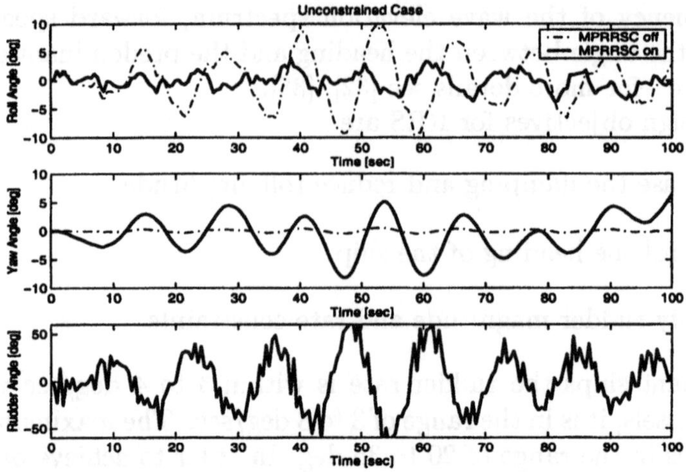

Figure 7.6. **Ship Motion with no constraints on the rudder.**

filter as initial condition:

$$\begin{bmatrix} \hat{d}(1) \\ \hat{d}(2) \\ \vdots \\ \hat{d}(N-1) \end{bmatrix} = \begin{bmatrix} C_d A_d \\ C_d A_d^2 \\ \vdots \\ C_d A_d^{N-1} \end{bmatrix} \hat{x}_d(0)$$

The simulation results are shown in figures 7.6 and 7.7. ∘

Figure 7.6, shows the results for the case in which no constrains are applied either to the magnitude or the slew rate of the rudder. In this case, a roll reduction of 66% is achieved. This reduction is the standard deviation of the stabilized roll angle relative to no stabilization [32]. The price paid for this reduction is an increase in the standard deviation of the yaw angle. This behavior is due to a fundamental trade-off arising from the single input two outputs characteristic of the plant; analysis of this fundamental trade-off can be found in [38]. Notice, also, that the rudder magnitude reaches angles of 60 degrees and has a very fast slew rate.

Figure 7.7, shows the results for the case that the rudder is constrained to maximum angle of 30 deg. and maximum rate of 15 deg/sec. In this case a roll reduction of 59% is achieved. Notice that the rudder movement satisfies the constraints via the application of the MPC algorithm. Note that Certainty Equivalence was employed here to obtain the results even though we know from section 5 that it is not strictly optimal.

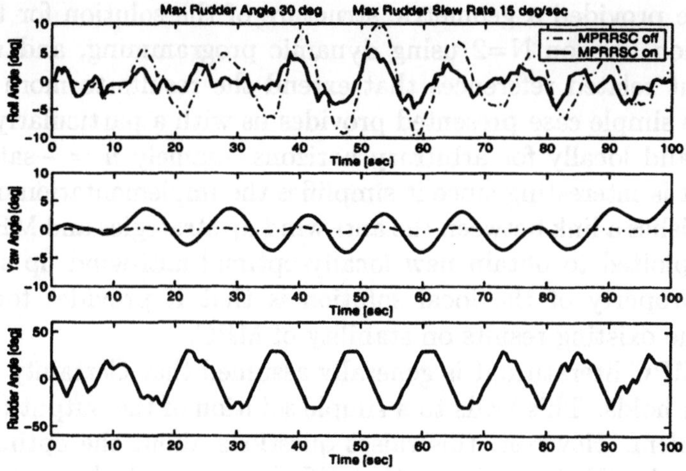

Figure 7.7. **Ship motion when the rudder is constrained to maximum angle of 30 deg. and maximum rate of 15 deg/sec.**

Remark 7.1 ***Although not shown in this example, a constraint on the states so as to prevent the Yaw angle from exceeding a certain range of values can be easily incorporated into the QP optimization problem.*** ∘

8. Summary and Discussion

In this chapter, we have reviewed some recent results on constrained control. We have mentioned that there are basically four approaches to deal with constraints. One of these approaches (which we have called remedial) includes the so-called anti-windup strategies, which aim at mitigating the deteriorating effects produced by the constraints and have long been used in practice. In these strategies, the constrains are handled in an ad-hoc manner and they are not incorporated as an integral part of the design. Another approach (which we have termed tactical) incorporates model based optimization strategies which aim at optimizing performance in the presence of constraints. These strategies have gained considerable popularity during the last few decades. Among the later, the most widely used method is Model Predictive Control (MPC).

We have described the elements of MPC within a general formulation that encompasses many of the existing MPC versions and we have made specialization to linear models with quadratic cost function. In this framework, we have shown that the open-loop optimization problem associated with solution of the MPC can be posed as non-dynamic constrained quadratic programme.

We have provided a geometric structure of the solution for the case of prediction horizon N=2 using dynamic programming, and we have pointed out related references that extend the results to more general cases. The simple case presented provides us with a particularly simple solution valid locally for arbitrary horizons—namely $u = -\text{sat}_{\Delta}(Kx)$. This result is interesting since it simplifies the implementation of MPC, and establishes a link between the anti-wind up strategies and MPC, that can be exploited to obtain new locally-optimal anti-wind up schemes. Another property of the local solution is that it provides tools that improve the existing results on stability of MPC.

In the MPC literature it is generally assumed that Certainty Equivalence (CE) holds. This leads to a simple solution of the output feedback MPC problem. However, this raises questions about the optimality of this approach. We have shown that CE does not hold in general since the optimal solution should incorporate information about the uncertainty of the state. Although not optimal, we have also shown that under certain conditions CE is nearly optimal.

Finally, we have presented a practical application case, in which some of the results presented here could be applied.

References

[1] D. Bernstein and A. Michel. A chronological bibliography on saturating actuators. *International Journal of Robust and Nonlinear Control*, 5:375–380, 1995.

[2] K. Walgama and J. Sternby. The inherent observer property in a class of anti-windup compensators. *International Journal of Control*, 52:705–724, 1990.

[3] Kwakernaak, H. and R. Sivan. *Linear Optimal Control systems.* Wiley, 1972.

[4] E. Lee and L. Markus. *Foundations of Optimal Control Theory.* Wiley New York, 1967.

[5] D. Q. Mayne, J. B. Rawlings, C. V. Rao, and P. O. M. Scokaert. Constrained model predictive control: Stability and optimality. *Automatica*, 36:789–814, 2000.

[6] D. Q. Mayne. Optimization in model based control. In *Proceedings of the IFAC symposium on dynamics and control of chemical reactors and batch processes, (Dycord'95)*, pages 229–242. Oxford Elsevier Science, Helsignor, Denmark, 1995. Plenary address.

[7] E. Polak. *Optimization: Algorithms and Consistent Appoximations.* Springer Verlag New York, 1997.

[8] S. Wright. Applying new optimization algorithms to model predictive control. In *5th International Conference on Chemical Process Control*, volume CACHE, AIChe, pages 147–155, 1997.

[9] A. Bemporad, M. Morari, V. Dua, and E. Pistikopulos. The explicit linear quadratic regulator for constrained systems. Technical Report AUT99-16, Automatic Control Laboratoty, ETH-Swiss Federal Institute of Technology, 1999.

[10] J. A. De Doná and G. C. Goodwin. Elucidation of state-space regions wherein model predictive control and anti-windup strategies achieve identical control policies. *Proceedings of American Control Conference*, 2000. Chicago, Illinois, USA.

[11] M. Serón, J. A. De Doná, and G. C. Goodwin. Global analytical model predictive control with input constraints. In *Proceedings of the 39th Conference on Decision and Control, Sydney, Australia*, 2000.

[12] G.C. Goodwin, S. Graebe, and M Salgado. *Control System Design.* Prentice-Hall, Inc, 2000.

[13] R. Bitmead, M. Gevers, and V. Wertz. *Adaptive Optimal Control: the thinking's man GPC.* Prentice-Hall, Inc, 1990.

[14] J. Nocedal and S. J. Wright. *Numerical Optimization.* Springer-Verlag New York, Inc., 1999.

[15] R. E. Bellman. *Dynamic Programming.* Princeton University Press, 1957.

[16] B.D.O. Anderson and J. B. Moore. *Optimal Control. Linear Quadratic Methods.* Prentice-Hall, 1971.

[17] E. G. Gilbert and K. Tan. Linear systems with state and control cosnstraints: The theory and application of maximal output admissible sets. *IEEE Trans. on Autom. Control*, 36(9):1008–1020, 1991.

[18] M. V. Kothare, P. J. Campo, M. Morari C., and N. Nett. A unified framework for the study of anti-windup design. *Automatica*, 30(12):1869–1883, 1994.

[19] P Marquis and J. Broustail. Smoc, a bridge between state space and model predictive controllers: application to the automation of hydrotreating unit. In *Proceedings of IFAC workshop on model based process control*, pages 37–43. Oxford: Pergamon Press, 1988.

[20] Tristan Pérez and Graham C. Goodwin. Stochastic output feedback model predictive control. In Proc. of 2001 American Control Conference, Arlington Va., USA, 2001.

[21] Robert F. Stengel. *Stochastic Optimal Control: theory and application.* Wiley Interscience, 1986.

[22] Karl J. Åström. *Introduction to Stochastisc Control Theory*, volume 70 of *Mathematics in Science and Engineering*. Academic Press, 1970.

[23] D. P. Bertsekas. *Dynamic Programming and Stochastic Control*, volume 125 of *Mathematics in science and engineering*. Academic Press, New York, 1976.

[24] D. P. Bertsekas. *Dynamic Programming: Deterministic and Stochastic Models* Prentice-Hall, 1987.

[25] Tristan Pérez and Graham C. Goodwin. On output feedback constrained model predictive control. Tecnical Report EE02015. School of Electrical Eng. and Comp Science, The University of Newcastle, NSW, Australia, 2002.

[26] R. Kalman. Contributions to the theory of optimal control. *Boletín de la Sociedad Matemática Mexicana*, 5:102–119, 1960.

[27] H. Rosenbrock and P. McMorran. Good, bad, or optimal? *IEEE Trans. on Automatic Control*, pages 552–554, 1971.

[28] S. Keerthi and E. Gilbert. Optimal, infinite horizon feedback laws for a general class of constrained discrete time systems: Stability and moving-horizon approximations. *Journal of Optimization Theory and Applications*, 57:265–293, 1988.

[29] C. Chen and L. Shaw. On receeding horizon feedback control. *Automatica*, 18:349–352, 1982.

[30] J. A. De Doná, M. Serón, D. Q. Mayne, and G. C. Goodwin. Enlarged terminal sets guaranteeing stability of receiding horizon control. Technical Report EE01022, Dept. Of Elec and Comp. Engineering, The University of Newcastle, Australia, 2001. Submitted for publication.

[31] Tristan Pérez, G.C. Goodwin, and C.Y. Tzeng. Model predictive rudder roll stabilization control for ships. In *5th IFAC Conference on Maneouvering and Control of Marine Crafts. (MCMC2000), Aalborg, Denmark*, 2000.

[32] Thor I. Fossen. *Guidance and Control of Ocean Marine Vehicles*. John Wiley and Sons Ltd, New York, 1994.

[33] J. van Amerongen, P. van der Klugt, and H. van Nauta Lemke. Rudder roll stabilization for ships. *Automatica*, 26:679–690, 1990.

[34] Trygve Laudval and Thor I. Fossen. Rudder roll stabilization for ships subject to input rate saturation using a gain scheduling control law. *In: Proceeding of IFAC CAMS'98*, pages 121–126, 1998.

[35] M. Blanke and A. Christensen. Rudder roll dumping autopilot robustness to sway-yaw-roll couplings. *In: Proceedings of 10th SCSS, Ottawa, Canada*, pages 93–119, 1993.

[36] W. C. Price and R. E. D. Bishop. *Probabilistic theory of ship dynamics.* Chapman and Hall, London, 1974.

[37] G. C. Goodwin and K. S. Sin. *Adaptive Filtering, Prediction, and Control.* Prentice-Hall, 1984.

[38] G.C. Goodwin, T. Pérez, M. Serón, and C. Y. Tzeng. On fundamental limitations for rudder roll stabilization of ships. In *Proceedings of the 39th Conference on Decision and Control 2000, Sydney, Australia*, 2000.

Chapter 8

ON FEASIBILITY OF INTERPLANETARY TRAVEL: THE FLIGHT FROM EARTH TO MARS AND BACK*

Angelo Miele

Research Professor and Foyt Professor Emeritus, Rice University, Houston, Texas.

Abstract This paper deals with the major issues of round-trip Mars missions, namely: flight time, characteristic velocity, and mass ratio. Two classes of trajectories are considered: (i) slow transfer trajectories, for which the round-trip angular travel of Earth exceeds that of the spacecraft by 360 degrees; (ii) fast transfer trajectories, for which the round-trip angular travels of Earth and spacecraft are the same.

For robotic spacecraft, the best trajectory is the minimum energy trajectory, which is of type (i). For manned spacecraft, the comfort of the crew might require a trajectory of type (ii), in which a substantial decrease in flight time is achieved at the expense of a large increase in characteristic velocity and an even larger increase in mass ratio.

At this time, it appears that the best policy is to continue the exploration of Mars via robotic spacecraft, since the present state of the art is not consistent with the safe and economic exploration of Mars via manned spacecraft. Major advances have yet to be achieved in three areas: spacecraft structural factors, engine specific impulses, and life support systems.

Keywords: Flight mechanics, astrodynamics, interplanetary travel, space trajectories, optimal space trajectories, Mars trajectories.

*This paper is dedicated in friendship to Professor Y.C. (Larry) Ho for his many contributions to applied mathematics, in particular optimization techniques and applications.

1. introduction

Five years ago, we initiated at Rice University a study of the interplanetary travel from Earth to Mars for either robotic spacecraft or manned spacecraft. Our feeling was that the USA will become more and more involved with the Mars problem ([1]-[3]) in the forthcoming decades. We concentrated our initial effort on feasibility studies, rather than precision navigation. In this spirit, we adopted as basic model the restricted four-body planar model (Sun, Earth, Mars, spacecraft), which is an extension to the four-body case of the restricted three-body planar model (Earth, Moon, spacecraft) of Lagrange. The basic mission analyzed is the round-trip LEO-LMO-LEO, where LEO denotes a low Earth orbit and LMO denotes a low Mars orbit.

This paper deals with the major quantities of interest for a round-trip Mars mission, namely: flight time, characteristic velocity, and mass ratio. We summarize the major results obtained in our initial effort ([4]-[7]). In the interest of clarity, some definitions are in order.

Flight Time. The total flight time t is the sum of the component flight times, namely, the flight time of the outgoing trip τ, the flight time of the return trip τ_{out}, and the flight time associated with the stay in LMO τ_{stay} .

Characteristic Velocity. The total characteristic velocity ΔV is the sum of the moduli of the velocity changes due to propulsive actions, assumed to occur impulsively. This quantity is characteristic of the mission itself and is independent of the spacecraft structural factor and engine specific impulse.

Mass Ratio. The mass ratio η is the ratio of the departing mass to the returning mass. This quantity depends on the velocity impulses of the various stages as well as on the spacecraft structural factors and engine specific impulses of the various stages. Generally speaking, the mass ratio is a monotonically increasing function of the total characteristic velocity. Hence, minimization of the total characteristic velocity is nearly the same as minimization of the mass ratio.

Spacecraft Structural Factor. The spacecraft structural factor ϵ is the ratio of the structural mass to the sum of the structural mass and propellant mass. This quantity is a measure of the structural efficiency of the spacecraft. Engine Specific Impulse.

The engine specific impulse I_{sp} is the ratio of the thrust to the reference weight flow of propellant. In turn, the reference weight flow is the product of the mass flow and a reference acceleration, namely, the acceleration of gravity at sea level on Earth. This quantity is a measure of the engine efficiency. Its inverse is called the specific fuel consumption.

2. Trajectory Classes

In a round trip LEO-LMO-LEO, the round-trip angular travels of Earth and spacecraft either differ by an integer multiple of 360 degrees or are the same. In practice, this leads to the identification of two classes of trajectories: (i) slow transfer trajectories for which the round-trip angular travel of Earth exceeds that of the spacecraft by 360 degrees; (ii) fast transfer trajectories for which the round-trip angular travels of Earth and spacecraft are the same.

Generally speaking, for a spacecraft propelled by chemical engines, slow transfer trajectories can be generated via four velocity impulses: two accelerating velocity impulses on departure from Earth or Mars and two braking velocity impulses on arrival to Mars or Earth. On the other hand, fast transfer trajectories might require an extra velocity impulse, which might occur in either the outgoing branch or the return branch. If a fifth velocity impulse is needed, its magnitude and timewise location must be optimized.

3. Reference Systems

The study of trajectories from Earth to Mars and back requires the use of three reference systems: Sun coordinate system (SCS), Earth coordinate system (ECS), and Mars coordinate system (MCS). For each of these systems, the origin is the center of the respective celestial body. Their axes point to fixed directions in space and are defined in such a way that corresponding axes are parallel to one another.

As a consequence of the above choices, each system translates with respect to the other two, but does not rotate. Hence, the coordinate transformations leading from one system to another are particularly simple for both the position coordinates and the velocity components.

Generally speaking, the differential equations governing the motion of the spacecraft are better written in SCS Cartesian coordinates. On the other hand, the boundary conditions are better written in ECS polar coordinates for LEO and MCS polar coordinates for LMO.

In Cartesian coordinates, the position vector and the velocity vector are identified via the scalar components x, y, u, w. Their counterparts in polar coordinates are r, θ, V, η, where r is the radial distance, θ is the position phase angle with respect to the x-direction, V is the velocity modulus, and η is the velocity phase angle with respect to the x-direction.

SCS quantities are designated with a single subscript (Sun S, Earth E, Mars M, spacecraft P). ECS or MCS quantities are designated with a double subscript: the first subscript refers to the spacecraft P; the second subscript refers to the planet (Earth E, Mars M).

4. Time Notations

The special times at which the four basic velocity impulses are applied are denoted by

t_0 = departure time from LEO ($t_0 = 0$),

t_2 = arrival time to LMO,

t_3 = departure time from LMO,

t_5 = arrival time to LEO.

The supplementary times at which midcourse velocity impulses might be applied are denoted by

t_1 = time of application of a midcourse velocity impulse during the outgoing trip,

t_4 = time of application of a midcourse velocity impulse during the return trip.

The total mission time is $\tau = t_5 - t_0$, the duration of the outgoing trip is $\tau = t_2 - t_0$, the duration of the return trip is $\tau_{RET} = t_5 - t_3$, and the stay time in LMO is $\tau_{STAY} = t_3 - t_2$.

5. Optimization Problems

The optimization problems relevant to LEO-LMO-LEO transfer can be formulated as optimal control problems or mathematical programming problems ([11]-[13]). In view of the assumption that the velocity changes occur impulsively, the mathematical programming formulation is to be preferred. In all cases, the characteristic velocity is minimized.

In solving the above optimization problems, the major parameters are:

(1) radii of the terminal orbits, r_{LEO} and r_{LMO};

(2)velocity impulses at departure/arrival of the outgoing and return trips,
$\Delta V_{LEO}(t_0)$, $\Delta V_{LMO}(t_2)$, $\Delta V_{LMO}(t_3)$, $\Delta V_{LEO}(t_5)$;

(3) relative position of Mars vis--vis Earth at the departure of the outgoing and return trips,
$\Delta\theta(t_0) = \theta_M(t_0) - \theta_E(t_0)$, $\Delta\theta(t_3) = \theta_M(t_3) - \theta_E(t_3)$;

(4) relative position of the spacecraft vis--vis Earth at the departure of the outgoing trip and vis--vis Mars at the departure of the return trip, $\theta_{PE}(t_0)$ and $\theta_{PM}(t_3)$;

(5) total flight time and flight times of the outgoing trip, return trip, and stay in LMO,
τ, τ_{OUT}, τ_{RET}, τ_{STAY};

(6) for a fast transfer trajectory, the magnitude and timewise location of the fifth velocity impulse in either the outgoing trip,
$\Delta_{MID}(t_1)$ and $\tau_{MID}(t_1) = t_1 - t_0$,
or the return trip
$\Delta V_{MID}(t_4)$ and $\tau_{MID}(t_4) = t_4 - t_3$.

Some of these parameters are assumed fixed; the remaining parameters are unknowns of the optimization problem.

In solving the above optimization problem, certain constraints must be satisfied:

(7) four differential equations of the restricted four-body problem; these include two kinematic equations for the position coordinates and

two dynamic equations for the velocity components;

(8) at LEO, three initial conditions of the outgoing trip; these include a radius condition, a velocity modulus condition, and a velocity direction condition;

(9) at LMO, three final conditions of the outgoing trip; these include a radius condition, a velocity modulus condition, and a velocity direction condition;

(10) at LMO, three initial conditions of the return trip; these include a radius condition, a velocity modulus condition, and a velocity direction condition;

(11) at LEO, three final conditions of the return trip; these include a radius condition, a velocity modulus condition, and a velocity direction condition;

(12) a scalar relation tying the total flight time and the partial flight times;

(13) a scalar relation tying the stay time on LMO and the relative positions of Mars vis--vis Earth at the arrival of the outgoing trip and at the departure of the return trip.

Fortunately, computation can be organized in such a way that the constraints (7), (8), (10) are automatically satisfied for any choice of the driving parameters of the problem. As a result, we are left with the essential constraints (9), (11), (12), (13). Because some of the parameters (1)-(6) are either fixed or missing from the problem, the optimization problem is a mathematical programming problem with a limited number of degrees of freedom (3 to 5, depending on problem formulation).

With the above understanding, we can formulate two optimization problems, each involving 4 degrees of freedom, that of the minimum energy trajectory and that of the fast transfer trajectory.

(P1) Minimum Energy Trajectory (MET). Assume that: the radii of the terminal orbits (1) are fixed at the values $r_{LEO} = 6841km$, $r_{LMO} = 3597km$; four velocity impulses are employed, namely, the terminal velocity impulses (2); the total mission time τ as well as the component times τ_{OUT}, τ_{RET}, τ_{STAY} in (5) are free. Minimize the total characteristic velocity for round-trip interplanetary transfer,

(14) $\Delta V_{INT} = \Delta V_{LEO}(t_0) + \Delta V_{LMO}(t_2) + \Delta V_{LMO}(t_3) + \Delta V_{LEO}(t_5)$.

(P2) Fast Transfer Trajectory (FTT). Assume that: the radii of the terminal orbits (1) are fixed at the values $r_{LEO} = 6841km$, $r_{LMO} = 3597km$; five velocity impulses are employed, namely, the four terminal velocity impulses (2) and the midcourse velocity impulse (6) for the outgoing trip; the total mission time t and stay time in LMO are fixed at the values

(15) $t = 440$ days, $\tau_{STAY} = 30$ days,

which are much smaller than those solving Problem P1 (see Section 6). Minimize the total characteristic velocity for round-trip interplanetary transfer,

(16) $\Delta V_{INT} = \Delta V_{LEO}(t_0) + \Delta V_{MID}(t_1) + \Delta V_{LMO}(t_2) + \Delta V_{LMO}(t_3) + \Delta V_{LEO}(t_5)$.

Remark. For Problems P1 and P2, the selected LEO radius and LMO radius correspond to the following altitudes above the corresponding planet surface:
(17) $h_{LEO} = 463$km, $h_{LMO} = 200$km.

6. Optimal Solutions

Problems P1 and P2 were solved by means of the sequential gradient-restoration algorithm for mathematical programming problems ([11]). The results are shown in Tables 8.1-8.4 and Figure 8.1-8.2 for both the minimum energy trajectory (MET) and fast transfer trajectory (FTT).

Table 8.1 shows the phase angle travels of the spacecraft, Earth, and Mars expressed in loops around the Sun; to convert the angular travels from loops into degrees, one must multiply by 360 the results in Table 8.1 . Clearly, the MET is a slow transfer trajectory, since the spacecraft makes one less loop around the Sun than Earth. On the other hand, the FTT is a fast transfer trajectory, sine the number of loops around the Sun is the same for both the spacecraft and Earth. For both the MET and the FTT, the angular travel of Mars around the Sun is about 53.2 percent of the angular travel of Earth, as expected since the average angular velocity of Mars in its motion around the Sun is about 53.2 percent of the average angular velocity of Earth.

Table 8.2 shows the travel and stay times for both the MET and the FTT. For the MET, the total mission time is 970 days having the following distribution: 46.8 percent spent staying in LMO and 53.2 percent spent traveling in interplanetary space, the latter evenly divided between the outgoing trip (26.6 percent) and the return trip (26.6 percent). For the FTT, the total mission time is 440 days having the following distribution: 6.8 percent spent staying in LMO and 93.2 percent spent traveling in interplanetary space, the latter unevenly divided between the outgoing trip (59.4 percent) and the return trip (33.8 percent).

The geometry of the optimal trajectories is quite different for the MET (Figure 8.1) and the FTT (Figure 8.2). For the MET, the optimal trajectory is contained in the region bounded by the orbits of Earth and Mars around the Sun for both the outgoing and return trips; indeed, the optimal trajectory is nearly a Hohmann transfer trajectory, which is the elliptical trajectory bitangent to the orbits of Earth and Mars around the Sun. For the FTT, the optimal trajectory is contained in the region bounded by the orbits of Earth and Mars around the Sun only for the return trip, but not for the outgoing trip. For the outgoing trip of the FTT, the spacecraft undershoots the Earth orbit around the Sun and passes dangerously near the Venus orbit.

Table 8.3 shows the distribution of the velocity impulses for both the MET and the FTT. Note that the total velocity impulse is 11.30 km/s for the MET and 20.79 km/s (hence, nearly double) for the FTT. Also note that each velocity impulse of the MET is lower than the corresponding velocity impulse of the FTT.

While there is no midcourse velocity impulse for the MET (Figure 8.1), note the presence in the outgoing trip of a large velocity impulse (2.68 km/s) for the FTT. For the FTT, this large velocity impulse is applied at about the perihelion of the outgoing trajectory, namely, at the point of closest approach to the Sun (Figure 8.2).

Table 8.4 converts the velocity impulses of Table 8.3 into mass ratios under the assumption that we are in the presence of a multistage rocket with uniform structural factor e = 0.10 and uniform specific impulse Isp = 450s. Note that the mass ratio for interplanetary flight is 20.1 for the MET and 304.1 for the FTT. This means that, to return to LEO a given mass, the corresponding departing mass from LEO must be about

20 times larger for the MET and 304 times larger for the FTT. Also note that each individual mass ratio of the MET is lower than the corresponding mass ratio of the FTT, consistently with the behavior of the velocity impulses.

The fact that there is no midcourse velocity impulse for the MET translates into a mass ratio of 1.00. Correspondingly, the fact that in the outgoing trip there is a large velocity impulse for the FTT translates into a mass ratio of 2.02. Because the velocity impulses sum, while the corresponding mass ratios multiply, we can state that the fifth velocity impulse of the FTT causes the doubling of the overall mass ratio of the spacecraft.

7. Round Trip Earth-Mars-Earth

The previous results refer to a round-trip LEO-LMO-LEO. Therefore, they do not include the mass ratios necessary for ascent from Earth to LEO and subsequent descent from LEO to Earth; also, they do not include the mass ratios necessary for descent from LMO to Mars and subsequent ascent from Mars to LMO. These mass ratios are estimated to be $\eta_E = 18.0$ for Earth (15.0 for ascent and 1.2 for descent) and $\eta_M = 3.6$ for Mars (1.2 for descent and 3.0 for ascent), resulting into a planetary mass ratio $\eta_{PLAN} = 64.8$; see Table 8.5 . Upon multiplication of the interplanetary and planetary mass ratios, we see that, under the assumption of a uniform spacecraft structural factor $\epsilon = 0.1$ and uniform engine specific impulse $I_{sp} = 450$ s, the overall mass ratio for a round-trip Earth-Mars-Earth becomes $\eta_{TOT} = O(10^3)$for MET and $\eta_{TOT} = O(10^4)$ for FTT; see Table 8.6 .

8. Mass Ratio Reduction

Several techniques can be used to reduce the mass ratios, for instance:

(a) assemble the spacecraft in low Earth orbit;

(b) construct the spacecraft in modular fashion, made of three parts: Earth return module (ERM), habitation module (HBM), and Mars excursion module (MEM); note that only MEM lands on Mars and that only ERM returns to Earth;

(c) use of a cargo spacecraft to transport MEM from LEO to LMO, with subsequent rendezvous with the main spacecraft;

(d) use of aerobraking techniques for the planetary flight portions of the trip;

(e) construct a hybrid spacecraft utilizing the combination of chemical propulsion for planetary flight and electrical (ion/plasma) propulsion for interplanetary flight.

Each of these techniques has advantages and disadvantages. A detailed discussion is beyond the scopes of this paper. Technique (e) might be particularly promising in view of the fact that the specific impulses already achievable with ion propulsion (3000 sec) or projected with plasma propulsion (30000 to 60000 sec) are larger by one or two order of magnitudes than the specific impulses achievable with chemical propulsion (450s).

Concerning ion engines ([14]-[15]), it must be noted that the first space mission has already taken place. Indeed, the Deep Space One (DS1) spacecraft was launched by JPL in October 1998: first, it traveled by Asteroid Braille; then, it aimed for Comet Borelli and the encounter took place at the end of September 2001.

Concerning plasma engines ([16]-[17]), it must be noted that none has flown yet in space. Consequently, at this time, the values claimed for the specific impulses of these engines must be taken with caution.

9. Comments

The previous analyses have shown that, concerning the exploration of Mars via manned spacecraft, we are in the presence of a no win situation. If we fly a minimum energy trajectory, the flight time is too large and hence inconsistent with the well being of the astronauts. On the other hand, if we fly a fast transfer trajectory, the mass ratio becomes so large that the spacecraft might not be feasible economically, even if feasible technically.

Concerning Table 8.6 , we have ignored the fact that the spacecraft can be built in modular fashion [see (b)], and this has had a pessimistic effect on the mass ratio. On the other hand, the assumed spacecraft structural factor $\epsilon = 0.1$ might have been too small (the average structural factor of either the Space Shuttle or the Arianne rocket is closer to 0.2 than 0.1), and this has had an optimistic effect on the mass ratio.

10. Conclusions and Future Developments

At this time, it appears that the best policy is to continue the exploration of Mars via robotic spacecraft. It also appears that the present state of the art is not consistent with the safe and economic exploration of Mars via manned spacecraft. Major advances have yet to be achieved in three areas: spacecraft structural factors, engine specific impulses, and life support systems.

Because it is unlikely that a manned mission to Mars might take place with only chemical engines, consideration must be given to the hybrid combination of chemical engines for planetary flight and electrical (ion) engines for interplanetary flight. Also, the electrical engines might be powered by nuclear powerplants.

The success achieved by JPL with the DS1 spacecraft hints that the development of such hybrid spacecraft might be a promising direction in design. If this hint is correct, perhaps manned flight to Mars might become feasible in the forthcoming decades.

Table 8.1. Phase angle travel of spacecraft, Earth, and Mars(loops around Sun) for a round-trip LEO-LMO-LEO

Quantity	MET	FTT
$\frac{[\theta_P(t_5)-\theta_P(t_0)]}{2\pi}$	1.656	1.204
$\frac{[\theta_E(t_5)-\theta_E(t_0)]}{2\pi}$	2.656	1.204
$\frac{[\theta_M(t_5)-\theta_M(t_0)]}{2\pi}$	1.412	0.641

Table 8.2. Travel/stay times[days] for a round-trip LEO-LMO-LEO

	MET	FTT
τ_{OUT}	257.9	261.3
τ_{STAY}	454.3	30.0
τ_{RET}	257.8	148.7
τ	970.0	440.0

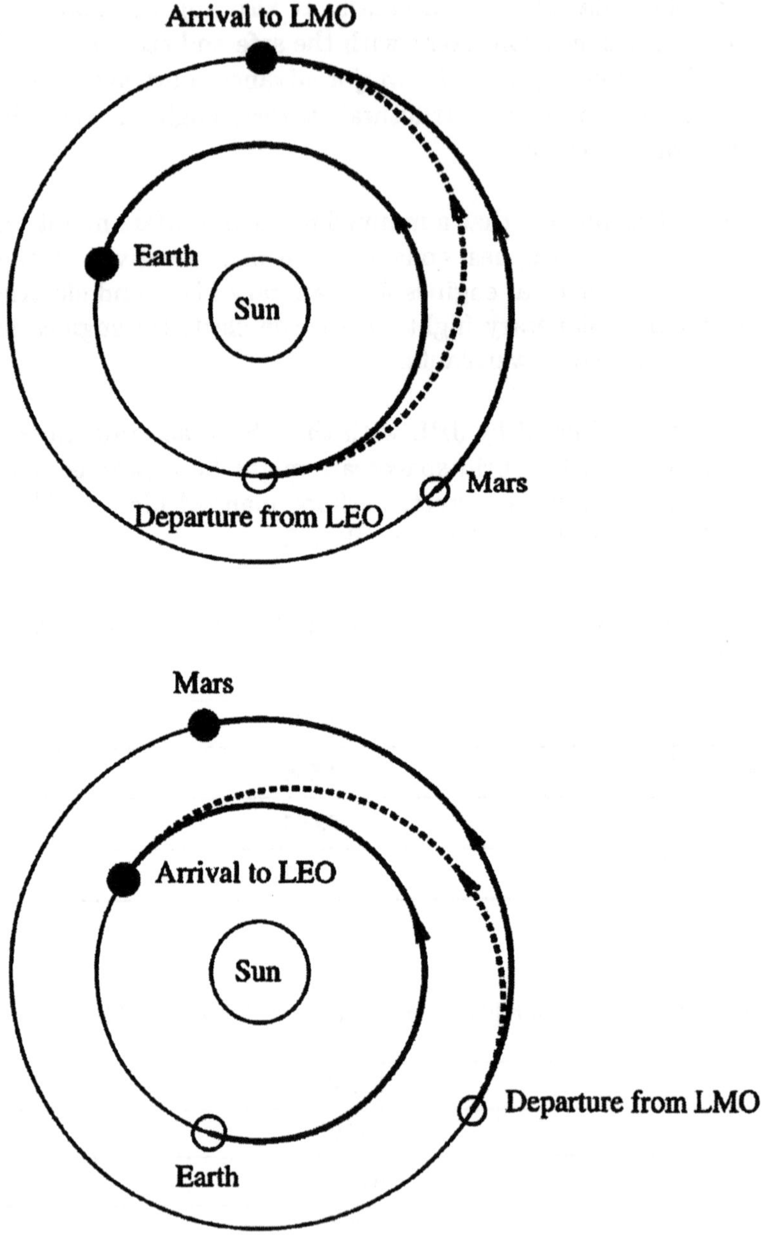

Figure 8.1. Minimum energy trajectory in interplanetary space, Sun coordinates, total time=970.0 days, stay time in LMO =454.3 days, total velocity impulse=11.30km/s, no midcourse impulse.

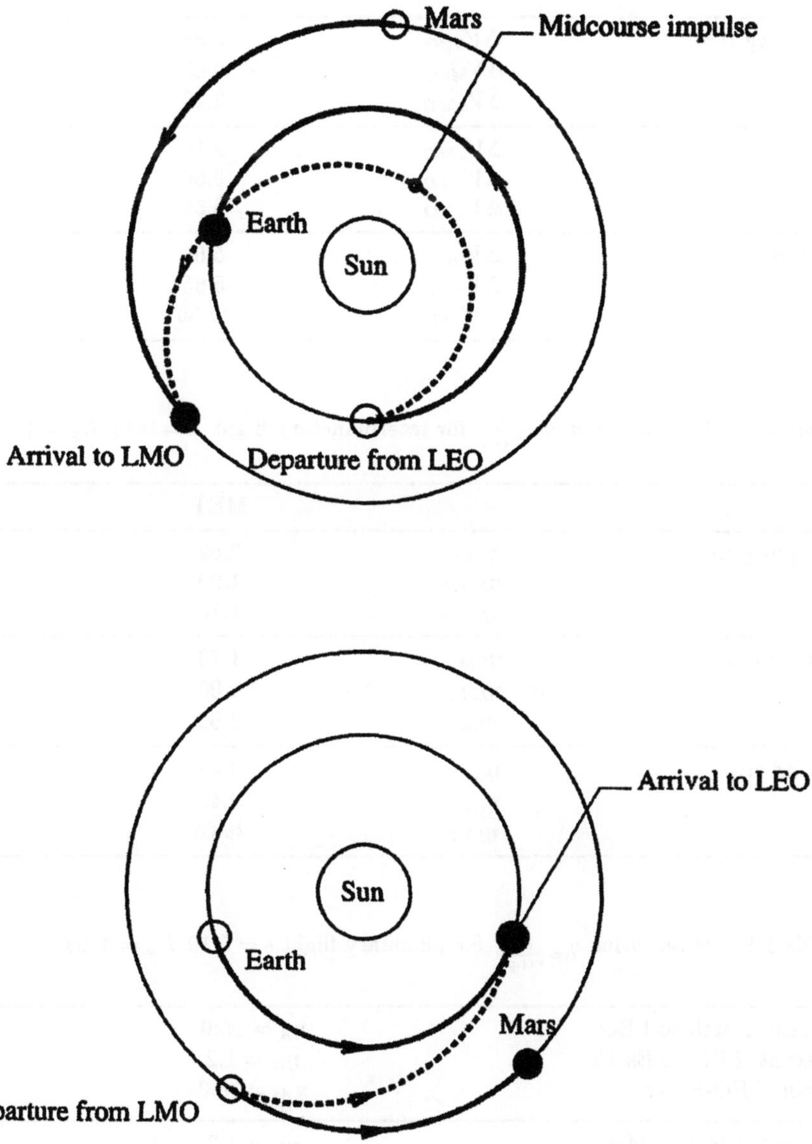

Figure 8.2. Fast trajectory in interplanetary space, Sun coordinates, total time=440.0 days, stay time in LMO =30.0 days, total velocity impulse =20.79 km/s, outgoing trip midcourse impulse

Table 8.3. Characteristic velocities[km/s] for a round-trip LEO-LMO-LEO.

	ΔV	MET	FTT
Outgoing trip	ΔV_{LEO}	3.55	5.07
	ΔV_{MID}	0.00	2.68
	ΔV_{LMO}	2.10	5.06
Return trip	ΔV_{LMO}	2.10	3.80
	ΔV_{MID}	0.00	0.00
	ΔV_{LEO}	3.55	4.18
Round trip	ΔV_{OUT}	5.65	12.81
	ΔV_{RET}	5.65	7.97
	ΔV_{INT}	11.30	20.79

Table 8.4. Mass ratios $\eta_i = \frac{m_i}{m_{i+1}}$ for interplanetary flight, $\epsilon = 0.10, I_{sp} = 450s$

		MET	FTT
Outgoing trip	η_{LEO}	2.59	4.15
	η_{MID}	1.00	2.02
	η_{LMO}	1.73	4.14
Return trip	η_{LMO}	1.73	2.79
	η_{MID}	1.00	1.00
	η_{LEO}	2.59	3.13
Round trip	η_{OUT}	4.48	34.81
	η_{RET}	4.48	8.74
	η_{INT}	20.06	304.14

Table 8.5. Mass ratios $\eta_i = \frac{m_i}{m_{i+1}}$ for planetary flight, $\epsilon = 0.10, I_{sp} = 450s$

AScent, Earth to LEO	$\eta_A = 15.0$
Descent, LEO to Earth	$\eta_D = 1.2$
Earth -LEO-Earth	$\eta_{\frac{A}{D}} = 18.0$
Descent, LMO to Mars	$\eta_D = 1.2$
Ascent, Mars to LMO	$\eta_A = 3.0$
LMO-Mars-LMO	$\eta_{\frac{D}{A}} = 3.6$
Round trip	$\eta_{PLAN} = 64.8$

References

Table 8.6. **Cumulative mass ratios for round-trip Earth-Mars-Earth, $\epsilon = 0.10, I_{sp} = 450s$.**

	MET	FTT
Interplanetary flight	$\eta_{INT} = 20.1$	$\eta_{INT} = 304.1$
planetary flight	$\eta_{PLAN} = 64.8$	$\eta_{PLAN} = 64.8$
Round-trip Earth-Mars-Earth	$\eta_{TOT} = 1302.5$	$\eta_{TOT} = 19705.7$

[1] G. Walberg, How Shall We Go to Mars? A Review of Mission Scenarios, Journal of Spacecraft and Rockets, Vol. 30, No. 2, pp. 129-139, 1993.

[2] P.F. Wercinski, Mars Sample Return: A Direct and Minimum Risk Design, Journal of Spacecraft and Rockets, Vol. 33, No. 3, pp. 381-385, 1996.

[3] W.J. O'Neil and C. Cazaux, The Mars Sample Return Project, Acta Astronautica, Vol.47, Nos. 2-9, pp. 453-465, 2000.

[4] A. Miele and T. Wang, Optimal Trajectories for Earth-to-Mars Flight, Journal of Optimization Theory and Applications, Vol. 95, No. 3, pp. 467-499, 1997.

[5] A. Miele and S. Mancuso, Optimal Ascent Trajectories and Feasibility of Next-Generation Orbital Spacecraft, Journal of Optimization Theory and Applications, Vol. 97, No. 1, pp. 519-550, 1998.

[6] A. Miele and T. Wang, Optimal Transfers from an Earth Orbit to a Mars Orbit, Acta Astronautica, Vol. 45, No. 3, pp. 119-133, 1999.

[7] A. Miele and T. Wang, Optimal Trajectories and Mirror Properties for Round-Trip Mars Missions, Acta Astronautica, Vol. 45, No. 11, pp. 655-668, 1999.

[8] A. Miele and S. Mancuso, Design Feasibility via Ascent Optimality for Next-Generation Spacecraft, Acta Astronautica, Vol. 45, No. 12, pp. 705-715, 1999.

[9] A. Miele, T. Wang, and S. Mancuso, Assessment of Launch Vehicle Advances to Enable Human Mars Excursions, Acta Astronuatica, Vol. 49, No. 11, pp. 563-580, 2001.

[10] A. Miele, T. Wang, and S. Mancuso, Optimal Free-Return Trajectories for Moon Missions and Mars Missions, Journal of the Astronautical Sciences, Vol. 48, Nos. 2/3, pp. 183-206, 2000.

[11] A. Miele, H.Y. Huang, and J.C. Heideman, Sequential Gradient-Restoration Algorithm for the Minimization of Constrained Func-

tions: Ordinary and Conjugate Gradient Versions, Journal of Optimization Theory and Applications, Vol. 4, No. 4, pp. 213-243, 1969.

[12] A. Miele, R. E. Pritchard, and I.N. Damoulakis, Sequential Gradient-Restoration Algorithm for Optimal Control Problems, Journal of Optimization Theory and Applications, Vol. 5, No. 4, pp. 235-282, 1970.

[13] A. Miele, T. Wang, and V.K. Basapur, Primal and Dual Formulations of Sequential Gradient-Restoration Algorithms for Trajectory Optimization Problems, Acta Astronautica, Vol. 13, No. 8, pp. 491-505, 1986.

[14] G. B. Ganapathi and C.S. Engelbrecht, Performance of the Xenon Feed System on Deep Space One, Journal of Spacecraft and Rockets, Vol. 37, No. 3, pp. 392-398, 2000.

[15] S.N. William and V. Coverstone-Carroll, Mars Missions Using Solar Electric Propulsion, Journal of Spacecraft and Rockets, Vol. 37, No. 1, pp.71-77, 2000.

[16] F.R. Chang-Diaz, J.P. Squire, R. D. Bengston, B.N. Breizman, F.W. Baity, and M.D. Carter, The Physics and Engineering of the VASIMR Engine, Paper AIAA 2000 3756, 36th AIAA/ASME/SAE/ASEE Joint Propulsion Conference, Huntsville, Alabama, 2000.

[17] A.G. Accettura, C. Bruno, S. Casotto, and F. Marzari, Mission to Mars Using Integrated Propulsion Concepts: Considerations, Opportunities, and Strategies, Paper IAF-01-R.4.08, 52nd IAF Congress, Toulouse, France, 2001.

Chapter 9

LINEAR QUADRATIC CONTROL REVISITED: A VIEW THROUGH SEMIDEFINITE PROGRAMMING*

David D. Yao*
IEOR Department, 302 Mudd Building
Columbia University, New York, NY 10027
yao@ieor.columbia.edu

Shuzhong Zhang†
Dept. of Systems Engineering and Eng Mgmt
The Chinese University, Shatin, Hong Kong
zhang@se.cuhk.edu.hk

Xun Yu Zhou‡
Dept. of Systems Engineering and Eng Mgmt
The Chinese University, Shatin, Hong Kong
xyzhou@se.cuhk.edu.hk

*Research supported by Hong Kong RGC Grant CUHK4175/00E.
*Research undertaken while on leave at Department of Systems Engineering and Engineering Management, The Chinese University of Hong Kong, Shatin, Hong Kong; supported in part by NSF under Grant ECS-97-05392.
†Research supported in part by RGC Earmarked Grant CUHK4181/00E.
‡Research supported in part by RGC Earmarked Grants CUHK 4054/98E and CUHK 4435/99E.

Abstract

We present a unified approach to both deterministic and stochastic linear-quadratic (LQ) control via the duality theory of semi-definite programming (SDP). This new framework allows the control cost matrix to be singular or even indefinite (in the stochastic setting), a useful feature in applications such as the optimal portfolio selection of financial assets. We show that the complementary duality condition of the SDP is necessary and sufficient for the existence of an optimal LQ control under certain stability conditions. When the complementary duality does hold, an optimal state feedback control is constructed explicitly in terms of the solution to the primal SDP. Furthermore, if the *strict* complementarity holds, then a new optimal feedback control, which is always stabilizing, is generated via the dual SDP. On the other hand, for cases where the complementary duality fails and the LQ problem has no attainable optimal solution, we develop an ϵ-approximation scheme that achieves asymptotic optimality.

1. Introduction

It is a great pleasure and distinct privilege for us to contribute an article to this volume, honoring Larry Ho's life-long achievements in advancing the state of knowledge in control and systems theory and related applications.

The plan we set for this article is to revisit what is perhaps the most extensively studied control problem, the linear quadratic (LQ) control, but from a new perspective, namely, that of *semidefinite programming.* LQ control has a long history that can be traced to the pioneering works of Kalman [21] (for deterministic systems) and Wonham [44] (for stochastic systems). Also refer to [4, 5, 6, 9, 16] and the references therein. The stochastic version of the problem can be stated as follows:

$$\text{(SLQ)} \quad \min \quad J(x_0, u(\cdot)) := \mathsf{E} \int_0^{+\infty} [x(t)^T Q x(t) + u(t)^T R u(t)] dt \tag{1}$$

$$\text{s.t.} \quad dx(t) = [Ax(t) + Bu(t)]dt + [Cx(t) + Du(t)]dW(t) \tag{2}$$

$$x(0) = x_0 \in \Re^n.$$

Here A, B, C, D and Q, R are constant matrices, with Q and R being symmetric matrices; the superscript T denotes the transpose of matrices and vectors; $W(\cdot)$ is a one-dimensional standard Brownian motion (with $t \in [0, +\infty)$ and $W(0) = 0$), defined on a filtered probability space $(\Omega, \mathcal{F}, \mathcal{F}_t, P)$; and $u(\cdot)$ denotes the (open-loop) control, which belongs to $L^2_{\mathcal{F}}(\Re^m)$, the space of all $\Re^m$-valued, $\mathcal{F}_t$-adapted measurable processes satisfying:

$$\mathsf{E}\int_0^{+\infty} \|u(t)\|^2 dt < +\infty.$$

(Note that allowing multi-dimensional Brownian motion will render no additional difficulty to the results below.)

The dynamics in (2) involve multiplicative noise in both the state and the control. This is a useful feature in many applications. For instance, the trade of a large block of an asset may affect its volatility; production control may affect not only the mean but also the standard deviation of (future) demands, as the forecast error (measured by the coefficient of variation – the ratio of standard deviation to mean) stays constant. See also an example below of tracking a targeted growth rate through the control of a portfolio.

A classical tool in solving LQ control problems is the following (stochastic algebraic) Riccati equation, with the symmetric matrix P being the unknown:

$$\begin{aligned} &A^TP + PA + Q + C^TPC \\ &- \quad (PB + C^TPD)(R + D^TPD)^{-1}(B^TP + D^TPC) = 0. \end{aligned} \tag{3}$$

Suppose P^* is a solution to the above equation with $R + D^TP^*D \succ 0$ (positive definite). Then,

$$u^*(t) = -(R + D^TP^*D)^{-1}(B^TP^* + D^TP^*C)x^*(t), \tag{4}$$

assuming that it is mean-square stabilizing (see Section 6 for definition), is an optimal state feedback control for (SLQ).

The classical theory has an apparent limitation. There is no easy way to solve the Riccati equation (2). As the matrix inverse term involved in the equation also involves the unknown P, there is no guarantee for $R + D^TP^*D \succ 0$, except when $R \succ 0$. Recent studies (e.g., [12, 15, 24, 13, 14, 2, 1, 45]) have, however, made a case for meaningful applications, especially financial applications ([50, 22, 25, 23]), in which R is singular or even indefinite. (Also see the application example below.) Indeed, thanks to the D^TP^*D term, even if the control cost R is zero or negative, we can still have a meaningful problem, with $R + D^TP^*D \succeq 0$ (positive semidefinite).

Our objective here is to present semidefinite programming (SDP) as a unifying approach to solve the LQ control problem. In particular, we focus on the connection between the LQ control problem and the duality theory of SDP.

The rest of the paper is organized as follows. In the rest of this introductory section, §1.1 presents an application example that motivates

LQ control with singular cost matrices Q and R, and §1.2 overviews the related literature. In §2, we collect preliminary materials, including the generalized Riccati equation, and the SDP primal-dual problems that correspond to the generalized Riccati equation. Our main results are presented in §3, where we demonstrate that the complementary duality is the key linkage between the SDP and LQ control. The results there are divided into two parts: the first part starts with $Q \succ 0$, so that the resulting control is guaranteed to be stable; the second part addresses the more general case of $Q \succeq 0$, where stability becomes a key issue, and hence the dual SDP plays a central role. This section is followed by a detailed comparison in §4 between our primal-dual SDP approach and two major approaches to singular LQ problems in the existing literature: dimension reduction and distributional control. In §5 we propose an ϵ-approximation scheme that achieves asymptotic optimality when complementary duality fails (and hence the LQ control problem has no attainable solution). To get the main ideas across, in sections §2 through §5, we shall focus on the deterministic version of the (SLQ) problem, i.e., with $C = D = 0$. (Much of §3.1 is adapted from [48]; other sections are new.) In §6, we revisit the (SLQ) problem, and discuss the main results in the stochastic setting, pointing out any necessary modifications. Concluding remarks are summarized in §7.

1.1 A Motivating Example

Consider a market where there is one bond and one stock, with price dynamics governed, respectively, by

$$dP_0(t) = rP_0(t)dt, \quad P_0(0) = p_0,$$

and

$$dP_1(t) = P_1(t)\left[bdt + \sigma dW(t)\right], \quad P_1(0) = p_1;$$

where $W(\cdot)$ denotes the one-dimensional standard Brownian motion. Suppose an agent, with an initial endowment z_0, wants to track a (stochastic) wealth trajectory (e.g., that of an index fund) $I(t)$ determined by the following equation:

$$dI(t) = I(t)[b_I dt + \sigma_I(t)dW(t)], \quad I(0) = i_0.$$

At any time $t \geq 0$ the total wealth of the agent is denoted by $z(t)$, of which the market value of the stock is denoted by $\pi(t)$. If we assume that the stock is traded continuously, and that there is no transaction cost, dividend payment and withdrawal for consumption, then $z(t)$ must satisfy the following (see, e.g., [49, Chapter 2, Section 3.2]):

$$dz(t) = [rz(t) + (b - r)\pi(t)]dt + \sigma\pi(t)dW(t), \quad z(0) = z_0.$$

The objective of the agent is to choose $\pi(\cdot)$ so as to minimize the following objective:

$$J(z(0), \pi(\cdot)) = \mathsf{E} \int_0^{+\infty} e^{-\rho t} |z(t) - I(t)|^2 dt,$$

where ρ is the discount rate.

To transform the above into the formulation in (1) and (2), define the state and control variables as follows:

$$(x(t), y(t)) = e^{-\frac{1}{2}\rho t}(z(t), I(t)), \qquad u(t) = e^{-\frac{1}{2}\rho t}\pi(t).$$

Then, the state dynamics become:

$$\begin{aligned} dx(t) &= [(r - \frac{1}{2}\rho)x(t) + (b - r)u(t)]dt + \sigma u(t) dW(t), \;\; x(0) = z_0, \\ dy(t) &= (b_I - \frac{1}{2}\rho)y(t) + \sigma_I y(t) dW(t), \;\; y(0) = i_0. \end{aligned}$$

And, the objective function can be rewritten as follows:

$$\begin{aligned} & J(x(0), y(0), u(\cdot)) \\ &= \mathsf{E} \int_0^{+\infty} |x(t) - y(t)|^2 dt \\ &= \mathsf{E} \int_0^{+\infty} \left[\, x(t), \;\; y(t) \,\right] \begin{bmatrix} 1, & -1 \\ -1, & 1 \end{bmatrix} \begin{bmatrix} x(t) \\ y(t) \end{bmatrix} dt. \end{aligned}$$

This way, we are in the framework of (1) and (2), with the control cost $R \equiv 0$, and the state cost $Q = \begin{bmatrix} 1, & -1 \\ -1, & 1 \end{bmatrix}$ being a singular matrix.

1.2 Related Literature

A traditional method to solve the Riccati equation (in the case of $D = 0$) is to consider the so-called associated Hamiltonian matrix ([8]). In this case it is known that the Riccati equation has a solution if and only if the associated Hamiltonian matrix admits no pure imaginary eigenvalues, a condition that can be verified using a Routh-Hurwitz type test on a set of polynomial inequalities involving the given matrices. If the associated Hamiltonian matrix passes this test, then the solution to the Riccati equation can be constructed using the eigenvectors of the associated Hamiltonian matrix. This procedure, however, does not apply when R is indefinite, as the matrix inverse in (2), and hence the Riccati equation and the associated Hamiltonian matrix themselves, may not be well-defined.

We next comment on the related literature under the heading of singular (deterministic) LQ control, i.e., with the control cost matrix R being singular ; refer to, e.g., [10, 11, 19, 34, 37, 43]. In [19, 43] the existence of optimal controls for singular LQ problems is studied; but the class of controls is defined in the sense of impulsive distributions, which allow controls that can *instantaneously* steer the system dynamics from one state to another state so as to reduce the singular problem to a regular one. (One such case is shown in Example 14 below.) In addition to technical complications (such as Dirac measure and its derivatives), this kind of controls is usually impossible to implement in practice. More importantly, there are singular LQ problems that do possess regular (i.e., conventional) controls that are optimal, and our approach can identify such cases; see Examples 10 and 6 below. For these problems, it is clearly preferable that the optimal controls are conventional ones that do not involve impulse distributions.

Another technique, as elegantly elaborated in [10], is to reduce the problem, through an orthonormal diagonalization of R, to one with a lower dimension in control and with a non-singular control cost; and then solve the reduced problem via the classical Riccati approach. However, this approach is premised upon the existence of an optimal solution to the original LQ problem; otherwise, the orthonormal transformation of R becomes quite meaningless, and, as prescribed in [10], one has to resort to the distributional controls to reduce the dimension of both the state and control. Our results, however, has led to new insights to the approach: we demonstrate that the dimension reduction approach is feasible if and only if the SDP complementary duality holds.

Yet another approach to singular LQ control is based on the connection between the Riccati equation and a certain linear matrix inequality (LMI). The LMI approach dates back to Yakubovich [46] and Willems [42]. It works for the singular case because LMI, in contrast to the Riccati equation, does not involve any matrix inverse; hence, in particular, it is not restricted to a non-singular R. Based on this idea, in [11] a necessary condition and a (different) sufficient condition are derived for the well-posedness of a singular LQ problem over a finite-time horizon; while in [34] it is shown that the maximal solution to the LMI provides the optimal cost value of the singular LQ problem via a transfer matrix technique. A similar LMI-based approach has been also applied to singular H_2 control problems; see [35, 33]. However, we shall demonstrate in Example 6 below, that in the case of the infinite-time horizon the optimal feedback control constructed via LMI may not be stabilizing when Q and R are both singular.

In comparison, our result via SDP is at once stronger (well-posedness is a considerably weaker condition than the existence of optimal solutions), more complete (ours is an "if and only if" condition), and computationally more viable (standard SDP algorithms are primal-dual based). Moreover, the dual SDP leads to a new class of feedback control, which is stabilizing and optimal under a mild and easily verifiable condition. Finally, our result applies to both regular and singular LQ control problems in a unified framework.

SDP has emerged in recent years as an important computational tool to solve Riccati equations and LMI's; refer to [40], and also to [17, 18, 27, 29, 38, 39]. In addition, two recent works, [7, 41], have shown that the optimal value of the LQ control lies within the duality gap of the corresponding SDP. However, the results are limited to non-singular R. Indeed, one result in [41], which points out when the duality gap will close and the optimal solution becomes attainable, is very much in the same spirit as our Lemma 6 below.

In contrast, our focus here is to develop a direct connection between the LQ control problem and the duality theory of SDP. In particular, we demonstrate that to extend beyond the confines of the classical LQ theory (with positive definite cost matrices) the central issue is ***stability***; and stability is intimately related to the *complementary duality* of the SDP associated with the LQ problem. We establish several equivalence relations between the stability/optimality of the LQ problem and the duality of the SDP, and demonstrate that a new class of optimal controls can be constructed based on the dual SDP. Furthermore, exploiting the primal–dual structure of the SDP also leads to powerful and efficient computational means, based on the newly developed primal–dual interior-point technologies (refer to, e.g., [36]), to solving the LQ problem. In short, while the LMI approach is a primal-only method, our primal–dual SDP approach applies to a more general class of problems, generates new theoretical results, and leads to practical computational algorithms.

2. Generalized Riccati Equation and SDP

As mentioned in the Introduction, to facilitate exposition, we shall present our main results focusing on the deterministic version of the (SLQ) problem, i.e., by setting the matrices $C = D = 0$ in (SLQ). This will be the setup for this section and the following three sections. In §6, we present the extension of the results to the stochastic setting.

Consider the following deterministic linear-quadratic (LQ) control problem:

$$\text{(LQ)}\quad \min \quad J(x_0, u(\cdot)) := \int_0^\infty [x(t)^T Q x(t) + u(t)^T R u(t)] dt \tag{5}$$

$$\text{s.t.} \quad \begin{cases} \dot{x}(t) = Ax(t) + Bu(t), \\ x(0) = x_0 \in \Re^n. \end{cases} \tag{6}$$

Here, the control $u(\cdot)$ is an element of $L^2(\Re^m)$, the set of all $\Re^m$-valued, measurable functions satisfying

$$\int_0^{+\infty} \|u(t)\|^2 dt < +\infty,$$

where $\|u(t)\| := [\sum_i u_i(t)^2]^{1/2}$. To account for the infinite-horizon nature of the problem, we further require the control to be stabilizing such that the corresponding state trajectory converges to zero as time goes to infinity.

In the classical setting, with $R \succ 0$ and $Q \succ 0$, the optimal feedback control is

$$u^*(t) = -R^{-1} B^T P x^*(t),$$

where P is the solution to the Riccati equation:

$$Q + A^T P + PA - PBR^{-1}B^T P = 0. \tag{7}$$

Here, we want to study a more general (deterministic) LQ problem — in particular, with $R \succeq 0$ and $Q \succeq 0$. Note that $R \succeq 0$ is *necessary* for the LQ problem to be well-posed; see for example [49, Chapter 6, Proposition 2.4]. In the stochastic setting, the analogous condition is $R + D^T P^* D \succeq 0$, where we can allow R to be non-definite, as explained earlier.

First, we present some regularity conditions concerning the problem (LQ).

(i) An open-loop control $u(\cdot)$ is called *admissible* (w.r.t. x_0), if it is (asymptotically) *stabilizing* (w.r.t. x_0), namely, if the state process under the control, $x(\cdot)$ of (6), with initial state x_0, satisfies $\lim_{t\to+\infty} x(t) = 0$. The set of all admissible controls w.r.t. x_0 is denoted as U^{x_0}.

(ii) A feedback control $u(t) = Kx(t)$, where K is a constant matrix, is called (asymptotically) *stabilizing*, if for every initial state x_0, we have $\lim_{t\to+\infty} x(t) = 0$, where $x(\cdot)$ is the solution to (6), with $u(t) = Kx(t)$.

(iii) Accordingly, the system in (6) is called (asymptotically) *stabilizable*, if there exists a stabilizing feedback control of the form $u(t) = Kx(t)$.

(iv) (LQ) is called *well-posed* (w.r.t. x_0), if its cost objective has a finite infimum:

$$-\infty < \inf_{u(\cdot)\in U^{x_0}} J(x_0, u(\cdot)) < +\infty.$$

(v) (LQ) is called *attainable* (w.r.t. x_0), if it is well-posed and if there exists a control that attains the infimum $\inf_{u(\cdot)\in U^{x_0}} J(x_0, u(\cdot))$, in which case the control is called *optimal.*

Since we allow R to be singular, the classical Riccati equation is no longer defined. A natural extension is to consider the following *generalized Riccati equation*:

$$F(P) := A^T P + PA + Q - PBR^+ B^T P = 0 \tag{8}$$

where R^+ stands for the pseudo-inverse of R. Note that R^+ satisfies the following properties (refer to [30], and note the symmetry of R):

$$R^+ \succeq 0, \quad (R^+)^T = R^+, \quad R^+ R = RR^+;$$
$$RR^+ R = R, \qquad R^+ RR^+ = R^+.$$

Next, we introduce an affine transformation of the matrix P,

$$\mathcal{L}(P) := \begin{bmatrix} R, & B^T P \\ PB, & Q + A^T P + PA \end{bmatrix}. \tag{9}$$

The following lemma ([3]) shows that $F(P)$ and $\mathcal{L}(P)$ are closely related.

Lemma 4 *$\mathcal{L}(P) \succeq 0$ if and only if $F(P) \succeq 0$ and $(I - RR^+)B^T P = 0$.*

Consider the following SDP:

$$\begin{array}{lll} \text{(P)} & \max & \langle I, P\rangle \\ & \text{s.t.} & \mathcal{L}(P) \succeq 0 \\ & & P \in \mathcal{S}^{n\times n}. \end{array}$$

Here and below, $\mathcal{S}^{n\times n}$ denotes the set of $n \times n$ symmetric matrices, and $\langle X, Y\rangle := \sum_{i,j} X_{ij} Y_{ij}$ denotes the matrix inner-product. In particular, $\langle I, P\rangle$ (with I being the identity matrix) is equal to the trace of the matrix P. Note that Lemma 4 implies that any feasible solution of (P) is necessarily constrained by the equality $(I - RR^+)B^T P = 0$ (which becomes redundant should R be non-singular).

In the classical setting when $R \succ 0$ and $Q \succ 0$, it is known that the solution to the Riccati equation $F(P) = 0$ can be obtained by solving the SDP problem (P); refer to [8, 40]. In the more general setting here (allowing $R \succeq 0$), the SDP is still a well defined problem; in particular, it does not impose any restrictions on the definiteness of R. Hence, a viable approach to (LQ) is to solve the SDP first, and then study the relationship between the SDP solution and the solution to (LQ).

To this end, consider the dual of (P), which is also an SDP. Let

$$Z := \begin{bmatrix} Z_b, & Z_u \\ Z_u^T, & Z_n \end{bmatrix}$$

denote the dual variable associated with the primal constraint $\mathcal{L}(P) \succeq 0$, with Z_b, Z_u and Z_n being a block partitioning of Z with appropriate dimensions. (Here we borrowed a similar notion from linear programming: The index set b denotes the "basis" part of the partition, and n denotes the "non-basis" part of the partition; see e.g. [32]. Note that in linear programming, the so-called strict complementarity holds, and hence the non-basis part of optimal dual variables are strictly positive.)

The dual of (P) is

$$\begin{array}{lll} \text{(D)} & \min & \langle R, Z_b \rangle + \langle Q, Z_n \rangle \\ & \text{s.t.} & I + Z_u^T B^T + B Z_u + Z_n A^T + A Z_n = 0 \\ & & Z \succeq 0. \end{array}$$

The semidefinite programs are known to be special forms of *conic optimization problems*, for which there exists a well-developed duality theory; see, e.g. [29, 40, 26] and the references therein. Key points of the theory can be highlighted as follows:

- The weak duality always holds, i.e. any feasible solution to the primal (maximization) problem always possesses an objective value that is no greater than the (dual) objective value of any dual feasible solution (the dual being a minimization problem).

- In contrast, the strong duality — that the optimal values of the primal and dual problems coincide — needs not always hold (unlike the case of linear programming).

- A sufficient condition for the strong duality is that there exists a pair of *complementary* optimal solutions, i.e., both the primal and dual SDP problems have attainable optimal solutions, and that these solutions are complementary to each other (i.e., no duality gap). For (P) and (D) above, this means that the primal optimal solution P^* and the dual optimal solution Z^* both exist and satisfy $\mathcal{L}(P^*)Z^* = 0$.

- If both (P) and (D) satisfy the *strict feasibility* (also known as *Slater's condition*), namely, there exist primal and dual feasible solutions P^0 and Z^0 such that $\mathcal{L}(P^0) \succ 0$ and $Z^0 \succ 0$, then the complementary solutions exist.

A mild regularity condition, which is assumed throughout the paper, is that the system in (6) is stabilizable as defined at the end of §1. In terms of SDP, this is equivalent to (D) satisfying Slater's condition. Refer to the lemma below.

Lemma 5 The following conditions are equivalent.

(i) The system in (6) is stabilizable.

(ii) Problem (D) satisfies Slater's condition.

Proof. First assume that the system in (6) is stabilizable by some feedback control $u(t) = Kx(t)$. Then all the eigenvalues of the matrix $A+BK$ have negative real parts. Consequently, the following Lyapunov equation has a positive solution $Z_n \succ 0$:

$$(A + BK)Z_n + Z_n(A + BK)^T = -I.$$

Set $Z_u = KZ_n$. The above relation can be rewritten as

$$I + Z_u^T B^T + BZ_u + Z_n A^T + AZ_n = 0.$$

Now choose $Z_b = \epsilon I + Z_u(Z_n)^{-1} Z_u^T$. Then by Schur's lemma,

$$Z = \begin{bmatrix} Z_b, & Z_u \\ Z_u^T, & Z_n \end{bmatrix}$$

is strictly feasible to (D), namely, Problem (D) satisfies Slater's condition.

Conversely, if Z is strictly feasible to (D), then $Z_n \succ 0$ by Schur's lemma. Putting $K = Z_u(Z_n)^{-1}$, then Z satisfying the equality constraint of (D) yields

$$(A + BK)Z_n + Z_n(A + BK)^T = -I.$$

By constructing a quadratic Lyapunov function $x^T Z_n x$, it is easily verified that the system in (6) is stabilizable. □

The above lemma suggests that, in view of the duality of the semidefinite programming, the stability can be regarded as a *dual* concept of the optimality.

As to the primal problem (P), in the non-singular setting (i.e., when $Q \succ 0, R \succ 0$), we recover the following classical result.

Lemma 6 Suppose $Q \succ 0, R \succ 0$. Then, there exists a *maximal* solution $P^* \succ 0$ to the Riccati equation $F(P^*) = 0$ (i.e., $P^* - P \succeq 0$ for any symmetric P that satisfies $F(P) \succeq 0$). And, P^* is also the unique optimal solution to the SDP problem (P). In this case, (LQ) has an optimal feedback control $u(t) = -R^{-1}B^T P^* x(t)$, with an optimal value $x_0^T P^* x_0$ for any initial state x_0.

Proof. It is known that the classical Riccati equation $F(P) = 0$ has a maximal solution $P^* \succ 0$ when $Q \succ 0$ and $R \succ 0$ (cf. [31]). From Schur's lemma and the maximality of P^*, it is clear that P^* is an optimal solution to (P).

To show that it is the unique optimal solution, let $\bar{P}$ be any optimal solution to (P). Then $\langle I, P^* - \bar{P}\rangle = 0$. However, $P^* - \bar{P} \succeq 0$ due to the maximality of P^*. Hence $P^* = \bar{P}$. The rest of the lemma is well known. □

Note that the non-singularity of Q is *necessary* for the constructed control to be stabilizing. For example, take $n = 1, A = 0, B = 1, Q = 0$ and $R = 1$. In this case the Riccati equation is $P^2 = 0$ having the only solution $P = 0$. The corresponding control suggested by the above lemma is $u^*(t) = 0$, which is not stabilizing if the initial state is non-zero.

Underlying the elegant simplicity of the results summarized in Lemma 6 is the fact that in the non-singular setting both primal and dual SDP's satisfy Slater's condition. To see this, note that $P^0 = 0$ is strictly feasible for the primal problem (as evident from $\mathcal{L}(P) \succ 0$ following (9), taking into account $Q \succ 0$ and $R \succ 0$), while the dual is strictly feasible by virtue of the system in (6) being stabilizable as discussed earlier. Hence complementary duality holds automatically in the non-singular setting. This leads to a constructive way of solving the Riccati equation through solving the SDP, for which efficient interior point codes are available (e.g., [38]). However, with the possible singularity of R, the situation becomes more complicated, as the primal problem may no longer satisfy Slater's condition, and consequently, complementary duality may fail. (Refer to Examples 14 and 15 below.)

In contrast to Lemma 6, we have the following result, which will also be used later.

Proposition 3 Suppose $Q \succeq 0, R \succeq 0$. Then (P) has a unique optimal solution $P^* \succeq 0$.

Proof. Consider the following perturbed problem of (P) along with its dual, where $\epsilon > 0$:

$$(\mathrm{P}_\epsilon) \quad \max \quad \langle I, P \rangle$$
$$\text{s.t.} \quad \begin{bmatrix} R+\epsilon I, & B^T P \\ PB, & Q+\epsilon I + A^T P + PA \end{bmatrix} \succeq 0$$

and

$$(\mathrm{D}_\epsilon) \quad \min \quad \langle R+\epsilon I, Z_b \rangle + \langle Q+\epsilon I, Z_n \rangle$$
$$\text{s.t.} \quad I + Z_u^T B^T + B Z_u + Z_n A^T + A Z_n = 0$$
$$Z := \begin{bmatrix} Z_b, & Z_u \\ Z_u^T, & Z_n \end{bmatrix} \succeq 0.$$

Both problems satisfy Slater's condition, and therefore complementary optimal solutions exist. Observe that the feasible set of (D_ϵ) is independent of ϵ. Take any dual feasible solution Z^0. By weak duality, we have

$$\langle I, P \rangle \leq \langle R+\epsilon I, Z_b^0 \rangle + \langle Q+\epsilon I, Z_n^0 \rangle. \tag{10}$$

Further, by Lemma 6, the unique optimal solution for (P_ϵ), denoted P_ϵ^*, is positive definite: $P_\epsilon^* \succ 0$. This, together with (10), implies in particular that P_ϵ^* is contained in a compact set, where $0 \leq \epsilon \leq \epsilon_0$, and $\epsilon_0 > 0$ is a pre-determined constant. Now, take a convergent subsequence such that

$$\lim_{i \to \infty} P_{\epsilon_i}^* = P_0^* \succeq 0,$$

with $\epsilon_i \to 0$ as $i \to \infty$. Clearly, P_0^* is a feasible solution of (P) since the feasible region of (P_ϵ) monotonically shrinks as $\epsilon \downarrow 0$. Now it suffices to show that $P_0^* = P^*$. Indeed, since P^* is feasible for (P_ϵ), and by Lemma 6 P_ϵ^* is the maximal solution to the corresponding Riccati equation, we have $P_\epsilon^* \succeq P^*$, resulting in $P_0^* \succeq P^*$. But P^* is optimal; hence, $\langle I, P^* \rangle \geq \langle I, P_0^* \rangle$. Therefore, we have $P^* = P_0^* \succeq 0$. The uniqueness is evident from the above argument. □

3. A Unified LQ Theory via SDP Duality

Throughout, we shall assume $R \succeq 0$, in particular, allowing the matrix to be singular. A possibly singular Q, however, creates complications in stability. So we shall first focus on the special case of $Q \succeq 0$ in §3.1 below, where stability is guaranteed, as long as complimentary duality (of the SDP) holds. In §3.2, we consider the general case of $Q \succeq 0$, where stability requires the *strict* complementarity, or Slater condition.

3.1 The Case of $Q \succ 0$

Theorem 1 Suppose $Q \succ 0$. The following three statements are equivalent:

(A) (LQ) has an attainable optimal feedback control

(B) (P) has an optimal solution P^* that satisfies the generalized Riccati equation: $F(P^*) = 0$.

(C) (P) and (D) have complementary optimal solutions P^* and Z^*.

If either **(B)** or **(C)** holds, then the optimal feedback control is

$$u^*(t) = -R^+ B^T P^* x^*(t), \tag{11}$$

where P^* is the optimal solution to (P).

As discussed earlier, in the non-singular setting when $Q \succ 0$ and $R \succ 0$, complementary duality, and hence **(C)** is automatically satisfied. Therefore, Theorem 1 reduces to Lemma 6.

Among the three equivalent conditions in the above theorem, **(A)** concerns the original (LQ) problem, whereas **(B)** and **(C)** are easy to verify numerically — via SDP. (Notice that **(B)** does not require solving the generalized Riccati equation, which could be a difficult task; rather, it only verifies if an optimal solution to (P) satisfies the Riccati equation.)

From another angle, **(A)** and **(C)** equate verifying complementary duality of the SDP to solving the original (LQ) problem, whereas **(B)** plays the role of an intermediary between the two, with the generalized Riccati equation substituting for the dual SDP. Note that computationally **(B)** is not needed, as most SDP codes are primal-dual based, which directly verifies **(C)**.

Below, we break the proof of Theorem 1 into four parts, each with a different implication result. The first part asserts that complementary duality is the key for the solvability of the generalized Riccati equation.

Theorem 2 [**(C)**$\Rightarrow$**(B)**] If (P) and (D) have complementary optimal solutions P^* and Z^*, respectively, then P^* must satisfy the generalized Riccati equation: $F(P^*) = 0$.

Proof. By Lemma 4, we have $(I - RR^+)B^T P^* = 0$. Thus, the following Schur decomposition holds true:

$$\mathcal{L}(P^*) = \begin{bmatrix} I, & 0 \\ P^* B R^+, & I \end{bmatrix} \begin{bmatrix} R, & 0 \\ 0, & F(P^*) \end{bmatrix} \begin{bmatrix} I, & R^+ B^T P^* \\ 0, & I \end{bmatrix}. \tag{12}$$

From the relation $\mathcal{L}(P^*)Z^* = 0$, it follows that

$$\begin{aligned}
&\begin{bmatrix} R, & 0 \\ 0, & F(P^*) \end{bmatrix} \begin{bmatrix} I, & R^+B^TP^* \\ 0, & I \end{bmatrix} \begin{bmatrix} Z_b^*, & Z_u^* \\ (Z_u^*)^T, & Z_n^* \end{bmatrix} \\
= &\begin{bmatrix} R(Z_b^* + R^+B^TP^*(Z_u^*)^T), & R(Z_u^* + R^+B^TP^*Z_n^*) \\ F(P^*)(Z_u^*)^T, & F(P^*)Z_n^* \end{bmatrix} \\
= &\begin{bmatrix} 0, & 0 \\ 0, & 0 \end{bmatrix}.
\end{aligned}$$

Therefore

$$F(P^*)(Z_u^*)^T = 0 \quad \text{and} \quad F(P^*)Z_n^* = 0,$$

and hence,

$$Z_u^*F(P^*) = 0 \quad \text{and} \quad Z_n^*F(P^*) = 0.$$

Since Z^* is dual feasible, we have

$$I + (Z_u^*)^TB^T + BZ_u^* + Z_n^*A^T + AZ_n^* = 0.$$

Multiplying $F(P^*)$ on both sides above yields

$$\begin{aligned}
0 &= F(P^*)(I + (Z_u^*)^TB^T + BZ_u^* + Z_n^*A^T + AZ_n^*)F(P^*) \\
&= F(P^*)^2,
\end{aligned}$$

which implies $F(P^*) = 0$. □

Theorem 3 [(B)⇒(A)] If (P) has an optimal solution P^* satisfying $F(P^*) = 0$, then (LQ) has an attainable optimal feedback control as determined by (11).

Proof. To start with, consider any primal feasible solution P, and any admissible (therefore stabilizing) control $u(\cdot) \in U^{x_0}$. We have,

$$\begin{aligned}
&\frac{d}{dt}(x(t)^TPx(t)) \\
= \; &(Ax(t) + Bu(t))^TPx(t) + x(t)^TP(Ax(t) + Bu(t)) \\
= \; &x(t)^T(A^TP + PA)x(t) + 2u(t)^TB^TPx(t). \qquad (13)
\end{aligned}$$

Integrating (13) over $[0, \infty)$ and making use of $x(t)^TPx(t) \to 0$ as $t \to \infty$, we have

$$0 = x_0^TPx_0 + \int_0^\infty \Big[x(t)^T(A^TP + PA)x(t) + 2u(t)^TB^TPx(t)\Big]dt.$$

Therefore,

$$
\begin{aligned}
& J(x_0, u(\cdot)) \\
= & \int_0^\infty [x(t)^T Qx(t) + u(t)^T Ru(t)]dt \\
= & x_0^T P x_0 + \int_0^\infty \Big[x(t)^T (A^T P + PA + Q)x(t) \\
& +2u(t)^T B^T P x(t) + u(t)^T Ru(t) \Big] dt \\
= & x_0^T P x_0 + \int_0^\infty \Big[x(t)^T F(P) x(t) \\
& +(u(t) + R^+ B^T P x(t))^T R(u(t) + R^+ B^T P x(t)) \Big] dt. \qquad (14)
\end{aligned}
$$

Since P is feasible, we have $F(P) \succeq 0$. This means

$$
J(x_0, u(\cdot)) \geq x_0^T P x_0, \qquad (15)
$$

for any P feasible to (P) and for any admissible control $u(\cdot) \in U^{x_0}$. On the other hand, under the feedback control $u^*(t) = -R^+ B^T P^* x(t)$, taking into account $P^* \succeq 0$ (Proposition 3), we have

$$
\begin{aligned}
0 \leq & J(x_0, u^*(\cdot)) \\
= & \int_0^\infty [x(t)^T Qx(t) + u^*(t)^T Ru^*(t)]dt \\
= & \lim_{t\to\infty} \int_0^t [x(\tau)^T Qx(\tau) + u^*(\tau)^T Ru^*(\tau)]d\tau \\
= & \lim_{t\to\infty} \{x_0^T P^* x_0 - x(t)^T P^* x(t) + \int_0^t \Big[x(\tau)^T (A^T P^* + P^* A + Q)x(\tau) \\
& +2u(\tau)^T B^T P^* x(\tau) + u^*(\tau)^T Ru^*(\tau) \Big] d\tau\} \\
\leq & x_0^T P^* x_0 + \lim_{t\to\infty} \int_0^t \Big[x(\tau)^T F(P^*) x(\tau) \\
& +(u^*(\tau) + R^+ B^T P^* x(\tau))^T R(u^*(\tau) + R^+ B^T P^* x(\tau)) \Big] d\tau \\
= & x_0^T P^* x_0. \qquad (16)
\end{aligned}
$$

The above shows that if the feedback control $u^*(\cdot)$ incurs a finite cost (w.r.t. any initial state x_0), then it must be stabilizing (and hence admissible). This is because a finite cost in (5) implies $\lim_{t\to+\infty} x^*(t)^T Qx^*(t) = 0$, where $x^*(\cdot)$ is the corresponding state trajectory; and since $Q \succ 0$, we must have $\lim_{t\to+\infty} x^*(t) = 0$. On the other hand, (16) yields

$J(x_0, u^*(\cdot)) \le x_0^T P^* x_0$. Thus, in view of (15) we conclude that $u^*(\cdot)$ is an optimal control. □

The last piece in establishing the equivalence relations in Theorem 1 is to show **(A)⇒(C)**. To do so, we need to first establish another result, which is useful in its own right. We want to show that **(C)** is, in fact, implied by (a weaker version of) **(B)**. That is, complementary duality is actually *necessary* for any positive semidefinite and feasible (as opposed to optimal) solution of (P) to satisfy the generalized Riccati equation.

Theorem 4 [**(B)⇒(C)**] If (P) has a feasible solution P^* satisfying $P^* \succeq 0$ and $F(P^*) = 0$, then there exist complementary optimal solutions to (P) and (D); and in particular, P^* is optimal to (P).

Proof. Denote $K := -R^+ B^T P^*$. First we show that the feedback control given by $u(t) = Kx(t)$ must be stabilizing. Indeed, going through the same calculation as (16) and noting the assumption that $P^* \succeq 0$, we conclude that $u(\cdot)$ incurs a finite cost with respect to any initial state. Hence it must be stabilizing (as shown in the proof of Theorem 3). It follows that the following Lyapunov equation

$$(A + BK)Y + Y(A + BK)^T + I = 0 \tag{17}$$

has a positive solution; let it be $Y^* \succ 0$. Let

$$Z_n^* = Y^*, \qquad Z_u^* = KY^*, \qquad Z_b^* = KY^*K^T.$$

By this construction, we can easily verify the following:

$$\begin{bmatrix} Z_b^*, & Z_u^* \\ (Z_u^*)^T, & Z_n^* \end{bmatrix} = \begin{bmatrix} I, & K \\ 0, & I \end{bmatrix} \begin{bmatrix} 0, & 0 \\ 0, & Z_n^* \end{bmatrix} \begin{bmatrix} I, & 0 \\ K^T, & I \end{bmatrix} \succeq 0,$$

and

$$I + (Z_u^*)^T B^T + BZ_u^* + Z_n^* A^T + AZ_n^* = 0.$$

Therefore, Z^* is a feasible solution of (D). Moreover,

$$\begin{aligned}
&\mathcal{L}(P^*) \begin{bmatrix} Z_b^*, & Z_u^* \\ (Z_u^*)^T, & Z_n^* \end{bmatrix} \\
&= \begin{bmatrix} I, & 0 \\ -K^T, & I \end{bmatrix} \begin{bmatrix} R, & 0 \\ 0, & F(P^*) \end{bmatrix} \begin{bmatrix} I, & -K \\ 0, & I \end{bmatrix} \begin{bmatrix} Z_b^*, & Z_u^* \\ (Z_u^*)^T, & Z_n^* \end{bmatrix} \\
&= \begin{bmatrix} I, & 0 \\ -K^T, & I \end{bmatrix} \begin{bmatrix} R(Z_b^* - K(Z_u^*)^T), & R(Z_u^* - KZ_n^*) \\ F(P^*)(Z_u^*)^T, & F(P^*)Z_n^* \end{bmatrix} \\
&= \begin{bmatrix} 0, & 0 \\ 0, & 0 \end{bmatrix}.
\end{aligned}$$

This means that P^* and Z^* are complementary solutions. In particular, P^* is optimal to (P). □

Theorem 5 **[(A)⇒(C)]** If (LQ) has an attainable optimal control w.r.t. any initial condition x_0, then (P) and (D) must have complementary optimal solutions.

Proof. Since (LQ) has an attainable optimal control w.r.t. any initial condition x_0, it is known ([4, p.21]) that there exists $M \succeq 0$ such that

$$\inf_{u(\cdot)\in U^{x_0}} J(x_0, u(\cdot)) = x_0^T M x_0.$$

For the time being, suppose the matrix M is a feasible solution to (P). Fix an initial x_0 and let $u^*(\cdot)$ be the optimal control w.r.t. x_0. Since M is feasible to (P), using (14), we obtain the following identity:

$$J(x_0, u^*(\cdot)) = x_0^T M x_0 + \int_0^\infty \Big[x(t)^T F(M) x(t)$$
$$+(u^*(t) + R^+ B^T M x(t))^T R (u^*(t) + R^+ B^T M x(t))\Big] dt.$$

Since $J(x_0, u^*(\cdot)) = x_0^T M x_0$, we have

$$\int_0^\infty \Big[(u^*(t) + R^+ B^T M x(t))^T R (u^*(t) + R^+ B^T M x(t))$$
$$+x(t)^T F(M) x(t)\Big] dt = 0. \tag{18}$$

Thus, $x(t)^T F(M) x(t) = 0$ for all $t \in [0, \infty)$. Since x_0 can be chosen arbitrarily, we conclude that $F(M) = 0$. The desired result then follows from Theorem 4. Furthermore, we know M must be optimal to (P).

What remains is to show the primal feasibility of M. To this end we consider the perturbed problem (P_ϵ) and its dual (D_ϵ) introduced in the proof of Proposition 3. For the optimal solution of (P_ϵ), denoted P_ϵ^*, there is a convergent subsequence such that

$$\lim_{i\to\infty} P_{\epsilon_i}^* = P_0^*,$$

with $\epsilon_i \to 0$ as $i \to \infty$. Clearly, P_0^* is a feasible solution of (P), since the feasible region of (P_ϵ) shrinks as $\epsilon \to 0$. We now show that $P_0^* = M$.

First, it follows from Lemma 6 that

$$\inf_{u(\cdot)\in U^{x_0}} J_{\epsilon_i}(x_0, u(\cdot)) = x_0^T P_{\epsilon_i}^* x_0$$

for all i, where

$$J_\epsilon(x_0, u(\cdot)) = \int_0^\infty \left[x(t)^T(Q+\epsilon I)x(t) + u(t)^T(R+\epsilon I)u(t)\right]dt.$$

Let $\bar{u}(\cdot)$ be the optimal control of (LQ) w.r.t. x_0 and $\bar{x}(\cdot)$ be the corresponding state. Then, we have

$$\begin{aligned} 0 &\le \inf_{u(\cdot)\in U^{x_0}} J_\epsilon(x_0, u(\cdot)) - \inf_{u(\cdot)\in U^{x_0}} J(x_0, u(\cdot)) \\ &= \inf_{u(\cdot)\in U^{x_0}} J_\epsilon(x_0, u(\cdot)) - J(x_0, \bar{u}(\cdot)) \\ &\le J_\epsilon(x_0, \bar{u}(\cdot)) - J(x_0, \bar{u}(\cdot)) \\ &= \epsilon \int_0^\infty (\|\bar{x}(t)\|^2 + \|\bar{u}(t)\|^2)dt. \end{aligned}$$

It then follows that

$$\begin{aligned} \lim_{i\to\infty} x_0^T P_{\epsilon_i}^* x_0 &\equiv \lim_{i\to\infty} \inf_{u(\cdot)\in U^{x_0}} J_{\epsilon_i}(x_0, u(\cdot)) \\ &= \inf_{u(\cdot)\in U^{x_0}} J(x_0, u(\cdot)) \equiv x_0^T M x_0. \end{aligned}$$

The above yields

$$x_0^T P_0^* x_0 = x_0^T M x_0$$

for all x_0, implying $M = P_0^*$. This shows that M is indeed a primal feasible solution, and consequently (P) and (D) have complementary optimal solutions as we discussed before. □

3.2 The General Case: $Q \succeq 0$

The key advantage with a non-singular Q, as we have observed above, is that for any primal feasible solution P^* satisfying $P^* \succeq 0$ and $F(P^*) = 0$, the control in (11) is automatically stabilizing (as evident from the proof of Theorem 3). This stability is no longer guaranteed when Q is possibly singular, as illustrated in the example below.

Example 6 Let

$$A = \begin{bmatrix} 0, & 1 \\ 1, & 0 \end{bmatrix}, \qquad B = \begin{bmatrix} 1, & 1 \\ 1, & 1 \end{bmatrix},$$

and

$$Q = \begin{bmatrix} 2, & -2 \\ -2, & 2 \end{bmatrix} \succeq 0, \qquad R = \begin{bmatrix} 1, & 0 \\ 0, & 0 \end{bmatrix} \succeq 0.$$

This system is easily seen to be stabilizable as $A - B = -I$. Moreover, one can verify that

$$P^* = \begin{bmatrix} 1, & -1 \\ -1, & 1 \end{bmatrix} \succeq 0$$

is feasible to the primal SDP and satisfies the generalized Riccati equation. On the other hand, one easily checks that

$$Z_b = \begin{bmatrix} 0, & 0 \\ 0, & \alpha \end{bmatrix}, \quad Z_u = \begin{bmatrix} 0, & 0 \\ -\beta, & -\beta \end{bmatrix}, \quad Z_n = \begin{bmatrix} \beta, & \beta - 1/2 \\ \beta - 1/2, & \beta \end{bmatrix}$$

form a feasible solution to the dual SDP, where $\beta \geq 1/4$ and $\alpha > 0$ are such that

$$\begin{bmatrix} Z_b, & Z_u \\ Z_u^T, & Z_n \end{bmatrix} \succeq 0.$$

Moreover, complementary duality holds, as the dual solution achieves the same objective value as the primal solution. However,

$$A - BR^+B^TP^* \equiv A,$$

which has an eigenvalue 1. Hence the control $u^*(t) = -R^+B^TP^*x^*(t)$ is *not* stabilizing.

Clearly, Theorem 1, let alone the classical LQ theory or the LMI-based approach, does not have much to say about the above simple example. Let us now take a closer look at this example: It can be presented as follows:

$$\begin{array}{ll} \min & 2\int_0^\infty [x_1(t) - x_2(t)]^2 dt + \int_0^\infty u_1(t)^2 dt \\ \text{s.t.} & \dot{x}_1(t) = x_2(t) + u_1(t) + u_2(t) \\ & \dot{x}_2(t) = x_1(t) + u_1(t) + u_2(t) \\ & x_1(0) = x_{01} \\ & x_2(0) = x_{02}. \end{array}$$

From the first two state equations we obtain $\dot{x}_1(t) - \dot{x}_2(t) = -(x_1(t) - x_2(t))$. Hence, regardless of the control, we always have

$$x_1(t) - x_2(t) = (x_{01} - x_{02})e^{-t}.$$

In other words, any control with $u_1(t) \equiv 0$ and $u_2(t)$ stabilizing the system

$$\begin{cases} \dot{x}_1(t) = x_2(t) + u_2(t) \\ \dot{x}_2(t) = x_1(t) + u_2(t) \\ x_1(0) = x_{01} \\ x_2(0) = x_{02} \end{cases}$$

will be optimal (e.g., $u_1(t) = 0$, $u_2(t) = -x_1(t) - x_2(t)$). Of course, identifying the optimal control via inspection is possible only due to the simplicity of this particular example. The general approach to this type of problems is based on the *dual* SDP, as detailed in the following:

Theorem 7 Suppose that (P) and (D) have complementary optimal solutions, with a primal optimal solution P^* and a dual optimal solution

$$Z^* = \begin{bmatrix} Z_b^*, & Z_u^* \\ (Z_u^*)^T, & Z_n^* \end{bmatrix} \succeq 0.$$

If $Z_n^* \succ 0$ then the following state feedback control

$$u^*(t) = Z_u^*(Z_n^*)^{-1}x^*(t)$$

is stabilizing and optimal to the control problem (LQ).

Proof. First of all, we note that if $Z_n^* \succ 0$ then the feedback control

$$u^*(t) = Z_u^*(Z_n^*)^{-1}x^*(t)$$

is stabilizing; see the proof of Lemma 5.

Second, from the proof of Theorem 2, we know $F(P^*) = 0$. Moreover, equation (14) holds for any stabilizing controls, including the above one. This yields

$$\begin{aligned} J(x_0, u^*(\cdot)) &= x_0^T P^* x_0 \\ &+ \int_0^\infty \Big[(u^*(t) + R^+B^TP^*x^*(t))^T R(u^*(t) + R^+B^TP^*x^*(t))\Big] dt. \end{aligned}$$

Obviously the optimal value of (LQ) can never be lower than $x_0^TP^*x_0$. Therefore to prove the optimality of $u^*(\cdot)$ it suffices to show

$$(u^*(t) + R^+B^TP^*x^*(t))^T R(u^*(t) + R^+B^TP^*x^*(t)) = 0. \qquad (19)$$

Using $u^*(t) = Z_u^*(Z_n^*)^{-1}x^*(t)$ we have

$$\begin{aligned} & (u^*(t) + R^+B^TP^*x^*(t))^T R(u^*(t) + R^+B^TP^*x^*(t)) \\ = \; & (x^*(t))^T \Big[(Z_u^*(Z_n^*)^{-1} + R^+B^TP^*)^T R(Z_u^*(Z_n^*)^{-1} + R^+B^TP^*)\Big] x^*(t) \\ = \; & (x^*(t))^T \Big[(Z_n^*)^{-1}(Z_u^*)^T R Z_u^*(Z_n^*)^{-1} + (Z_n^*)^{-1}(Z_u^*)^T RR^+B^TP^* \\ & + P^*BR^+RZ_u^*(Z_n^*)^{-1} + P^*BR^+B^TP^*\Big] x^*(t). \end{aligned} \qquad (20)$$

Now that the complementary duality holds, i.e., $\mathcal{L}(P^*)Z^* = 0$, it follows that

$$RZ_u^* + B^TP^*Z_n^* = 0. \qquad (21)$$

Therefore,

$$RZ_u^*(Z_n^*)^{-1} + B^T P^* = 0$$

which leads to

$$P^* BR^+ RZ_u^*(Z_n^*)^{-1} + P^* BR^+ B^T P^* = 0 \tag{22}$$

Similarly, it follows from (21) that

$$RR^+ B^T P^* Z_n^* = -RR^+ RZ_u^* = -RZ_u^*$$

and

$$RR^+ B^T P^* = -RZ_u^*(Z_n^*)^{-1}.$$

Multiplying $(Z_n^*)^{-1}(Z_u^*)^T$ on the left side of the above equation yields

$$(Z_n^*)^{-1}(Z_u^*)^T RZ_u^*(Z_n^*)^{-1} + (Z_n^*)^{-1} Z_u^* RR^+ B^T P^* = 0. \tag{23}$$

Using (22), (23) and (20) we obtain

$$(u^*(t) + R^+ B^T P^* x^*(t))^T R(u^*(t) + R^+ B^T P^* x^*(t)) = 0$$

as desired. The optimality of $u^*(t) = Z_u^*(Z_n^*)^{-1} x^*(t)$ thus follows. □

A sufficient condition to guarantee $Z_n^* \succ 0$ is the so called *strict complementarity* which plays an important role in the analysis of primal-dual interior point methods for SDP problems (see, e.g., [28]).

Theorem 8 Suppose that (P) and (D) have strictly complementary optimal solutions P^* and Z^* respectively, i.e. $\mathcal{L}(P^*)Z^* = 0$ and $\mathcal{L}(P^*) + Z^* \succ 0$. Then, $Z_N^* \succ 0$.

Proof. Using (12) we have

$$\mathcal{L}(P^*) = (H^{-1})^T \text{diag}(R, 0) H^{-1}$$

where

$$H = \begin{bmatrix} I, & K \\ 0, & I \end{bmatrix} \text{ and } K = -R^+ B^T P^*.$$

Let $\bar{Z}^* := H^{-1} Z^* (H^{-1})^T$.

It is readily seen that $\bar{Z}_N^* = Z_N^*$. Moreover, diag $(R, 0) = H^T \mathcal{L}(P^*) H$ and $\bar{Z}^* = H^{-1} Z^* (H^{-1})^T$ are nonnegative definite and they are complementary to each other. We shall further show that they stay strictly complementary. To see this we first note that the range space of $\mathcal{L}(P^*)$, range$(\mathcal{L}(P^*))$, and the range space of Z^*, range(Z^*), form an orthogonal decomposition of the whole space. Clearly, range$(H^T \mathcal{L}(P^*) H)$ has the same dimension as that of range$(\mathcal{L}(P^*))$; and range$(\bar{Z}^*)$ has the same dimension as that of range(Z^*). Finally, range($H^T\mathcal{L}(P^*)H$) and range($\bar{Z}^*$) remain orthogonal to each other. Hence they span the whole space too. Therefore,

$$H^T\mathcal{L}(P^*)H + \bar{Z}^* \succ 0.$$

As we noted before, the last diagonal block of the above matrix is Z_N^*. Hence, $Z_N^* \succ 0$, as the proposition stipulates. □

If P^* and Z^* are strictly complementary optimal solutions with $R \succ 0$ and $Z_N^* \succ 0$, then the feedback control $u^*(t) = Z_u^*(Z_n^*)^{-1}x^*(t)$ coincides with the control $u(t) = -R^{-1}B^TP^*x(t)$ due to (21). In general, however, they can be quite different, as shown below.

Example 6 (continued) By virtue of Theorem 7 we can now claim that

$$\begin{bmatrix} u_1^*(t) \\ u_2^*(t) \end{bmatrix} = Z_u^*(Z_n^*)^{-1}x^*(t)$$

$$= \begin{bmatrix} 0, & 0 \\ -\beta, & -\beta \end{bmatrix} \begin{bmatrix} \frac{\beta}{\beta-1/4}, & -\frac{\beta-1/2}{\beta-1/4} \\ -\frac{\beta-1/2}{\beta-1/4}, & \frac{\beta}{\beta-1/4} \end{bmatrix} \begin{bmatrix} x_1^*(t) \\ x_2^*(t) \end{bmatrix}$$

$$= \begin{bmatrix} 0 \\ -\frac{\beta}{2\beta-1/2}(x_1^*(t) + x_2^*(t)) \end{bmatrix}$$

is optimal feedback control, where $\beta > 1/4$, which can be directly verified to be correct. In particular, taking $\beta = 1/2$, we recover the control mentioned earlier by direct inspection.

We remark that in SDP parlance, the dual (D), rather than the primal (P), is in the so-called standard form; and hence for most numerical solvers, it is much easier to solve (D) instead of (P). If one uses a primal-dual interior point code, then the primal variable P^* is reported in the 'dual' part of the output (dual of the dual is primal). Therefore, in implementation it is advisable to concentrate on (D) instead of (P).

Theorem 9 Suppose $Q \succeq 0$. Consider the following three statements:

(A) (LQ) has an attainable optimal feedback control.

(B) (P) has an optimal solution P^* such that

(i) it satisfies the generalized Riccati equation: $F(P^*) = 0$;

(ii) the feedback control $u^*(t) = -R^+B^TP^*x^*(t)$ is stabilizing.

(C) (P) and (D) have complementary optimal solutions P^* and Z^* with $Z_n^* \succ 0$ (as in Theorem 7).

We have the following implications:

- **(A)** $\Rightarrow$ **(B(i))**;
- **(B)** $\Rightarrow$ **(A)**, with the control $u^*(t) = -R^+B^TP^*x^*(t)$ being optimal;
- **(B)** $\Rightarrow$ **(C)**;
- **(C)** $\Rightarrow$ **(A)**, with the control $u^*(t) = Z_u^*(Z_n^*)^{-1}x^*(t)$ being optimal.

Proof. **(A)** $\Rightarrow$ **(B(i))**: The proof of Theorem 5 implies that there is a primal feasible solution M with

$$\inf_{u(\cdot)\in U^{x_0}} J(x_0, u(\cdot)) = x_0^T M x_0, \text{ and } F(M) = 0.$$

On the other hand, in view of (15), we have

$$x_0^T M x_0 \equiv \inf_{u(\cdot)\in U^{x_0}} J(x_0, u(\cdot)) \geq x_0^T P x_0,$$

for any P feasible to (P). Hence M is optimal for (P), and set $P^* = M$.
(B) $\Rightarrow$ **(A)**: Follows from the proof of Theorem 3, where the control in (11) is stabilizing due to Q being non-singular. Here, the stability of the control is assumed.
(B) $\Rightarrow$ **(C)**: Follows from the proof of Theorem 4 (where the stabilizing property of $u(t) = -R^+B^TP^*x(t)$ is guaranteed by $Q \succ 0$). That $Z_n^* \succ 0$ follows from $Z_n^* = Y^* \succ 0$, which is the solution to the Lyapunov equation (17).
(C) $\Rightarrow$ **(A)**: Follows from Theorem 7. □

We illustrate the above results by another example.

Example 10 Let

$$A = \begin{bmatrix} 1, & 1 \\ 4, & 1 \end{bmatrix}, \quad B = \begin{bmatrix} 1, & 1 \\ 1, & -2 \end{bmatrix},$$

and

$$Q = \begin{bmatrix} 12, & 6 \\ 6, & 3 \end{bmatrix} \succeq 0, \quad R = \begin{bmatrix} 1, & 0 \\ 0, & 0 \end{bmatrix} \succeq 0.$$

This system is easily seen to be stabilizable. To identify a positive semidefinite feasible solution P^* to the primal SDP that satisfies the generalized Riccati equation $F(P^*) = 0$, first consider the constraint $(I - RR^+)B^TP^* = 0$ as stipulated by Lemma 4. This gives rise to

$$P^* = \begin{bmatrix} 2p, & p \\ p, & \frac{p}{2} \end{bmatrix}$$

for some p. Substituting the above into the generalized Riccati equation yields

$$A^T P^* + P^* A + Q - P^* B R^+ B^T P^*$$
$$\equiv \begin{bmatrix} -9p^2 + 12p + 12, & -\frac{9}{2}p^2 + 6p + 6 \\ -\frac{9}{2}p^2 + 6p + 6, & -\frac{9}{4}p^2 + 3p + 3 \end{bmatrix} = 0.$$

Solving for p leads to $p = 2$. Thus,

$$P^* = \begin{bmatrix} 4, & 2 \\ 2, & 1 \end{bmatrix} \succeq 0$$

is a primal feasible solution that satisfies $F(P^*) = 0$. On the other hand, in this case

$$A - BR^+B^T P^* = \begin{bmatrix} -5, & -2 \\ -2, & -2 \end{bmatrix},$$

which has eigenvalues -1 and -6. Hence $u^*(t) = -R^+B^TP^*x^*(t)$ is stabilizing. By Theorem 9, the control must be optimal as well.

4. Connections to Singular LQ Control

Here we discuss how our approach complements existing approaches to singular LQ control as outlined in the introductory section. We focus on two types of existing approaches: dimension reduction and distributional controls.

Dimension reduction refers to the approach that replaces the original singular problem by a regular problem with a lower control dimension, via an orthonormal transformation of control coordinates. This approach is well documented in [10], although in the context of a finite-horizon LQ problem (hence the issue of stability can be avoided). Below, we bring out the connection of this approach to our primal-dual SDP approach. To sharpen the comparison, we shall assume that $Q \succ 0$ (so as to avoid any stability related complications), whereas $R \succeq 0$ is singular. Notice that $R \in \mathcal{S}^{m\times m}$ can be diagonalized by an orthonormal transformation as follows:

$$R = W^T \Lambda_r W$$

where $W^TW = I$, r denotes the rank of R, and Λ_r denotes a diagonal matrix with the first r diagonal components being positive and the remaining $m - r$ components all zeros.

Theorem 11 Suppose $Q \succ 0$; and suppose (P) and (D) have complementary optimal solutions. Suppose $R \in \mathcal{S}^{m\times m}$, and rank($R$)=$r$; and write $R = W^T\Lambda_r W$ as an orthonormal transformation. Then, the last $m - r$ columns of the matrix BW^T must be zero vectors.

Proof. By Theorem 1, the existence of a pair of complementary optimal solutions for (P) and (D), P^* and Z^*, implies that

$$F(P^*) = Q + A^T P^* + P^* A - P^* B R^+ B^T P^* = 0.$$

We now claim that P^* cannot be singular. To see this, consider any ξ satisfying $P^*\xi = 0$. Pre- and post-multiplying ξ^T and ξ on both sides of the above equation yields $\xi^T Q \xi = 0$, and consequently $\xi = 0$, due to $Q \succ 0$. This shows that P^* is non-singular.

Once again observe $F(P^*) = 0$ where P^* is primal optimal. It follows from (12) that $\text{rank}(\mathcal{L}(P^*)) = \text{rank}(R) = r$. Notice that

$$\begin{bmatrix} W, & 0 \\ 0, & I \end{bmatrix} \mathcal{L}(P^*) \begin{bmatrix} W^T, & 0 \\ 0, & I \end{bmatrix} = \begin{bmatrix} \Lambda_r, & WB^T P^* \\ P^* B W^T, & Q + A^T P^* + P^* A \end{bmatrix}.$$

Therefore,

$$\text{rank}([\Lambda_r, WB^T P^*]) = r,$$

and hence the last $m-r$ rows of $WB^T P^*$ must be zero vectors. Because P^* is non-singular as argued above, we conclude that the last $m-r$ rows of WB^T must all be zero vectors too. □

Theorem 11 implies that in the case when R is singular and when the SDP identifies complementary dual solutions, through a transformation on the control variables, $\bar{u}(t) := W^T u(t)$, the last $m-r$ control variables, $\bar{u}_{r+1}(t), \cdots, \bar{u}_m(t)$, will vanish from *both* the cost objective and the system dynamics, and hence can be eliminated from the problem. The new LQ problem becomes one with a non-singular cost matrix that is the non-zero diagonal block of Λ_r. In summary, complementary duality guarantees the viability of dimension reduction. The next result asserts that complementary duality is, in fact, also *necessary* to carry out dimension reduction.

Theorem 12 Suppose $Q \succ 0$ and $R \succeq 0$ is singular. Then a dimension reduction can be carried out if and only if (P) and (D) have complementary optimal solutions.

Proof. It suffices to prove the "only if" part. Suppose after a control dimension reduction one has a non-singular LQ problem with a non-singular control cost matrix. By Lemma 6 the reduced problem has attainable optimal control w.r.t. any initial state which, in turn, implies that the original LQ problem has attainable optimal control. Hence it follows from Theorem 5 that (P) and (D) must have complementary optimal solutions. □

The following example illustrates how dimension reduction connects to SDP.

Example 13 Suppose

$$A = \begin{bmatrix} 0, & 1 \\ 1, & 0 \end{bmatrix}, \qquad B = \begin{bmatrix} 2, & 2 \\ 2, & 2 \end{bmatrix},$$

and

$$Q = \begin{bmatrix} 1, & 0 \\ 0, & 1 \end{bmatrix} \succ 0, \qquad R = \begin{bmatrix} 1, & 1 \\ 1, & 1 \end{bmatrix} \succeq 0.$$

Similar to Example 6, this problem is stabilizable. Solving the primal SDP problem (P) yields

$$P^* = \begin{bmatrix} 1/2, & 0 \\ 0, & 1/2 \end{bmatrix}$$

which satisfies the generalized Riccati equation $F(P) = 0$. Hence Theorem 4 indicates that the complementary duality holds, and consequently the problem is reducible in control dimension. On the other hand, by Theorem 3, the following is the optimal feedback control:

$$u^*(t) = -R^+B^TP^*x^*(t) = -\frac{1}{2}\begin{bmatrix} 1, & 1 \\ 1, & 1 \end{bmatrix} x^*(t),$$

or,

$$u_1^*(t) = -\frac{1}{2}x_1^*(t) - \frac{1}{2}x_2^*(t), \qquad u_2^*(t) = -\frac{1}{2}x_1^*(t) - \frac{1}{2}x_2^*(t).$$

Clearly, by a variable transformation $\bar{u}(t) := \frac{\sqrt{2}}{2}(u_1(t) + u_2(t))$ the original problem will be reduced as Theorem 11 asserts.

As prescribed in [10], if dimension reduction cannot be carried out, then distributional controls have to be used to iteratively reduce both control and state dimensions until one of them reaches zero. The distributional control approach was studied in depth in [19, 43]. The essential idea of this approach can be illustrated in the following simple example.

Example 14

$$\begin{aligned} \min \quad & \int_0^\infty x^2(t)dt \\ \text{s.t.} \quad & \begin{cases} \dot{x}(t) = -x(t) + u(t), \\ x(0) = x_0 \in \Re^1. \end{cases} \end{aligned}$$

A quick inspection suggests that the infimum of the objective is zero, which is however not achievable when $x_0 \neq 0$. This can be confirmed by checking the corresponding SDP's. Indeed, the primal SDP reads:

$$
\begin{array}{lll}
(\mathrm{P}_1) & \max & p \\
 & \text{s.t.} & \begin{bmatrix} 0, & p \\ p, & 1-2p \end{bmatrix} \succeq 0.
\end{array}
$$

This problem has a unique feasible solution $p = 0$, which is hence optimal. The generalized Riccati equation in this case is $F(p) = 1 - 2p = 0$. So the optimal solution does not satisfy the generalized Riccati equation, and consequently by Theorem 1 the complementary duality fails and the LQ problem has no attainable optimal control.

The distributional control approach "solves" this problem by applying an impulsive control to bring the non-zero initial state x_0 to 0 instantaneously at time zero. Afterwards, the state trajectory will stay in the reduced state space $x = 0$ if the control is chosen to be constantly 0. The key here is that the impulse control does not incur any cost due to the singularity of the original problem. Hence, the overall control strategy is "optimal", albeit within a class of controls that is broader than the conventional admissible class.

In [43], a key reference for singular LQ control, the approach is to apply a geometric transformation technique leading to distributional optimal controls to all LQ problems with a singular R. Our Examples 10 and 6 suggest, however, that such LQ problems may not necessarily exclude *regular* optimal controls. In fact, as illustrated in those examples, we have indeed identified regular optimal controls based on dual SDP solutions. In the following example which motivates the analysis in [43], we shall further demonstrate that our approach can identify a singular LQ problem as being "hopelessly" singular, in the sense that no attainable regular control exists, and thereby justifies the use of distributional controls as a last resort.

Example 15

$$
\begin{array}{ll}
\min & \int_0^\infty (x_1(t)^2 + x_2(t)^2)dt \\
\text{s.t.} & \dot{x_1}(t) = A_{11}x_1(t) + A_{12}x_2(t) + u(t) \\
 & \dot{x_2}(t) = A_{21}x_1(t) + A_{12}x_2(t) \\
 & x_1(0) = x_{01} \\
 & x_2(0) = x_{02}.
\end{array}
$$

Assume $A_{21} \neq 0$, so that the system is controllable. The associated primal and dual SDP problems are as follows:

$$\begin{array}{lll} \text{(P)} & \max & p_{11} + p_{22} \\ & \text{s.t.} & \begin{bmatrix} 0 & p_{11} & p_{12} \\ p_{11} & 1 & A_{21}p_{22} \\ p_{12} & A_{21}p_{22} & 1 + 2A_{22}p_{22} \end{bmatrix} \succeq 0 \end{array}$$

and

$$\begin{array}{lll} \text{(D)} & \min & z_{11} + z_{22} \\ & \text{s.t.} & 1 + 2x + 2(A_{11}z_{11} + A_{12}z_{21}) = 0 \\ & & y + A_{11}z_{12} + A_{21}z_{22} + A_{21}z_{11} + A_{22}z_{21} = 0 \\ & & 1 + 2(A_{21}z_{12} + A_{22}z_{22}) = 0 \\ & & \begin{bmatrix} z_b & x & y \\ x & z_{11} & z_{12} \\ y & z_{21} & z_{22} \end{bmatrix} \succeq 0. \end{array}$$

Simple calculation shows that (P) has an optimal solution $p_{11}^* = 0$, $p_{12}^* = 0$ and $p_{22}^* = \frac{A_{22}+\sqrt{A_{22}^2+A_{21}^2}}{A_{21}^2}$.

However, (D) does not have any attainable optimal solutions. This is because its optimal value could only be attained when

$$\begin{aligned} z_{11} &= \frac{(A_{22} + \sqrt{A_{21}^2 + A_{22}^2})^2}{2A_{21}^2\sqrt{A_{21}^2 + A_{22}^2}} \\ z_{12} &= -\frac{A_{22} + \sqrt{A_{21}^2 + A_{22}^2}}{2A_{21}^2\sqrt{A_{21}^2 + A_{22}^2}} \\ z_{22} &= \frac{1}{2\sqrt{A_{21}^2 + A_{22}^2}} \end{aligned}$$

in which case the positive semidefiniteness constraint

$$\begin{bmatrix} z_b & x & y \\ x & z_{11} & z_{12} \\ y & z_{21} & z_{22} \end{bmatrix} \succeq 0$$

becomes impossible to satisfy. This implies (by Theorem 5) that there exists no attainable regular control to the original problem, making it necessary to consider distributional controls. (For a distributional control solution to this particular example, see [43, p. 325].)

5. Asymptotic Optimality

In this section, we consider the case when complementary duality fails. As an alternative to applying distributional controls, we develop an approximation scheme to obtain asymptotically optimal (regular) controls. We note that this regularization idea was originally developed by Jacobson, Gershwin and Lele in [20]. It appears interesting, however, to recast the idea in light of our SDP framework.

If Theorem 1 or Theorem 7 indicates that there is no optimal solution to the original LQ problem, then we consider the perturbed problem, (LQ_ϵ), obtained by keeping all the data A and B unchanged, and letting $Q_\epsilon = Q+\epsilon I$ (this transformation is not necessary when Q is non-singular) and $R_\epsilon = R+\epsilon I$ with $\epsilon > 0$. The corresponding perturbed SDP's have already appeared in the proofs of Proposition 3 and Theorem 5.

By Lemma 6, (LQ_ϵ) has an optimal feedback control for each $\epsilon > 0$; and both (P_ϵ) and (D_ϵ) satisfy Slater's condition.

Theorem 16 Let $J_\epsilon^*(x_0)$ and $J^*(x_0)$ be the optimal values of (LQ_ϵ) and (LQ), respectively. Then,

$$\lim_{\epsilon\downarrow 0} J_\epsilon^*(x_0) = J^*(x_0).$$

Proof. Let the optimal solution of (P_ϵ) be P_ϵ^*. Following Lemma 6, we know that

$$u^\epsilon(t) = -(R+\epsilon I)^{-1}B^T P_\epsilon^* x^\epsilon(t) \tag{24}$$

is optimal for (LQ_ϵ), with the corresponding optimal objective value equal to $J_\epsilon^*(x_0) = x_0^T P_\epsilon^* x_0$.

Following the same argument as in the proof of Proposition 3, we know that P_ϵ^* is contained in a compact set, with $0 < \epsilon \le 1$. Moreover, since by definition $J_\epsilon^*(x_0)$ decreases monotonically as $\epsilon \downarrow 0$, so does P_ϵ^*. Therefore, P_ϵ^* itself also converges as $\epsilon \downarrow 0$.

What remains is to show that $x_0^T P_0^* x_0$ is equal to the true infimum of (LQ), now denoted as $J^*(x_0)$. To this end, first note that

$$x_0^T P_\epsilon^* x_0 = J_\epsilon^*(x_0) \ge J^*(x_0),$$

where the inequality is due to the positive perturbation in (P_ϵ). Letting $\epsilon \to 0$, we obtain

$$x_0^T P_0^* x_0 \ge J^*(x_0).$$

On the other hand, since P_0^* is feasible to (P) (see the proof of Proposition 3), it follows from (15) that

$$J^*(x_0) = \inf_{u(\cdot)\in U^{x_0}} J(x_0, u(\cdot)) \ge x_0^T P_0^* x_0.$$

This completes the proof. □

The above theorem says that the optimal values of the original and the perturbed problems are very close to each other or, more precisely, they are the same asymptotically. The next result is concerned with the asymptotically optimal feedback control of the original LQ problem.

Theorem 17 The feedback control $u^\epsilon(\cdot)$ constructed by (24) is asymptotically optimal for (LQ), namely,

$$\lim_{\epsilon \downarrow 0} J(x_0, u^\epsilon(\cdot)) = J^*(x_0).$$

Proof. Denote by $J_\epsilon(x_0, u(\cdot))$ the cost of the perturbed problem (LQ_ϵ) under an admissible control $u(\cdot) \in U^{x_0}$ w.r.t. the initial state x_0. Then for any $\eta > 0$, there is an ϵ_0 such that when $0 < \epsilon < \epsilon_0$:

$$\begin{aligned} J^*(x_0) &\leq J(x_0, u^\epsilon(\cdot)) \\ &\leq J_\epsilon(x_0, u^\epsilon(\cdot)) \\ &= J_\epsilon^*(x_0) \\ &\leq J^*(x_0) + \eta, \end{aligned}$$

where the last inequality is due to Theorem 16. This proves our claim. □

To understand the approximation scheme in our framework consider Example 14 again.

Example 14 (continued) With perturbation, the primal SDP becomes

$$\begin{array}{ll} (\text{P}_\epsilon) & \max\ p \\ & \text{s.t.} \begin{bmatrix} \epsilon, & p \\ p, & 1-2p \end{bmatrix} \succeq 0. \end{array}$$

The optimal solution is:

$$p_\epsilon^* = \sqrt{\epsilon^2 + \epsilon} - \epsilon.$$

Clearly, $p_\epsilon^* = O(\sqrt{\epsilon})$, and hence the optimal value, $p_\epsilon^* x_0^2$, converges to 0 as $\epsilon \downarrow 0$. Further, to obtain an asymptotic optimal feedback control for the original problem, we apply the *same* feedback law

$$u^\epsilon(t) = -(\sqrt{1 + \frac{1}{\epsilon}} - 1)x^\epsilon(t),$$

which is optimal to the perturbed problem, to the original problem. It is straightforward to check that the corresponding cost under this control is

$$J(x_0, u^\epsilon(\cdot)) = \frac{\sqrt{\epsilon}}{2\sqrt{1+\epsilon}} x_0^2,$$

which converges to 0 as $\epsilon \downarrow 0$.

6. Extensions to the Stochastic Setting

We now return to the stochastic LQ problem (SLQ) formulated at the beginning of §1. A key difference between stochastic and deterministic systems, is reflected in their respective Riccati equations (2) and (7): $R + D^T PD$ now takes place of R. Hence, the analogy of allowing $R \succeq 0$ is to allow $R + D^T PD \succeq 0$ in the stochastic system, which, in turn, allows R to be possibly indefinite (in particular, negative definite). Note that in stochastic systems, the possible singularity of $R + D^T PD$ is an even more challenging issue than the singularity of R in the deterministic setting, as R is given and hence its singularity readily verifiable, whereas $R + D^T PD$ involves the unknown P.

In what follows, we present the main results in parallel to or in contrast with those for the deterministic systems. We shall omit all proofs, which can be found in [47].

First of all, all the definitions on admissible control, well-posednedss, etc. for deterministic systems, as listed in Section 2, can be extended naturally to stochastic systems. The only exception is for the stability. In the stochastic case, an open-loop control $u(\cdot)$ is called *(mean-square) stabilizing* at x_0, if the corresponding state $x(\cdot)$ of (2) with the initial state x_0 satisfies $\lim_{t\to+\infty} \mathsf{E}[x(t)^T x(t)] = 0$. The stabilizability can then be defined accordingly. Throughout this section we assume that the system under consideration is mean-square stabilizable.

The generalized Ricatti equation now takes the following form:

$$\begin{aligned} F(P) &:= A^T P + PA + Q + C^T PC \\ &\quad -(PB + C^T PD)(R + D^T PD)^+(B^T P + D^T PC) \\ &= 0. \end{aligned} \tag{25}$$

The corresponding primal and dual SDP's are:

$$\begin{array}{lll} \text{(P)} & \max & \langle I, P \rangle \\ & \text{s.t.} & \mathcal{L}(P) \succeq 0, \end{array}$$

where

$$\mathcal{L}(P) := \begin{bmatrix} R + D^T PD, & B^T P + D^T PC \\ PB + C^T PD, & Q + C^T PC + A^T P + PA \end{bmatrix}; \tag{26}$$

and

$$\begin{array}{lll} \text{(D)} & \min & \langle R, Z_B\rangle + \langle Q, Z_N\rangle \\ & \text{s.t.} & I + Z_U^T B^T + B Z_U + Z_N A^T + A Z_N \\ & & \quad + C Z_N C^T + D Z_U C^T + C Z_U^T D^T + D Z_B D^T = 0 \\ & & Z := \begin{bmatrix} Z_B, & Z_U \\ Z_U^T, & Z_N \end{bmatrix} \succeq 0. \end{array}$$

With the above, the main results in §3 can be extended to (SLQ). The only change is the optimal feedback control based on the primal SDP solution; it now takes the following form:

$$u^*(t) = -(R + D^T P^* D)^+ (B^T P^* + D^T P^* C) x^*(t). \tag{27}$$

The control based on the dual SDP remains in the same form:

$$u^*(t) = Z_U^* (Z_N^*)^{-1} x^*(t). \tag{28}$$

Theorem 18 All implication relations in Theorem 9 hold for (SLQ) as well. Furthermore, if the feedback control in (27) is stabilizing, then the equivalence statements in Theorem 1 also hold for (SLQ).

We next present an example to illustrate the SDP approach in the presence of the singularity of $R + D^T P^* D$. It also shows that optimal stabilizing controls can be obtained by both primal and dual SDP's, leading to different optimal controls.

Example 19 Let $m = n = 1$; $A = C = -1$, $B = D = 1$; $Q = 1$ and $R = -1$. Namely, the problem is this:

$$\begin{array}{ll} \min & \mathsf{E} \int_0^\infty [x(t)^2 - u(t)^2] dt \\ \text{s.t.} & dx(t) = [-x(t) + u(t)]dt + [-x(t) + u(t)]dW(t) \\ & x(0) = x_0. \end{array}$$

This system is mean-square stabilizable, as $u(t) = \alpha x(t)$ is stabilizing for any α with $|\alpha| < 1$. To see this, applying Itô's formula to the system (1) under the above feedback control, we obtain

$$d\mathsf{E}[x(t)^2] = (\alpha^2 - 1)\mathsf{E}[x(t)^2]dt, \quad \mathsf{E}[x(0)^2] = x_0^2.$$

Hence

$$\mathsf{E}[x(t)^2] = e^{(\alpha^2-1)t} x_0^2, \tag{29}$$

which converges to 0 as $t \to +\infty$.

Now, the primal SDP is

$$\begin{array}{ll} \max & p \\ \text{s.t.} & \begin{bmatrix} -1+p, & 0 \\ 0, & 1-p \end{bmatrix} \succeq 0. \end{array}$$

The above has an optimal solution $p^* = 1$ (the only feasible solution), which also satisfies the generalized Riccati equation $F(p) = 1 - p = 0$. (Note that singularity occurs in this solution). The feedback control given by (27) reduces to $u^*(t) = 0$, which is stabilizing as shown above ($\alpha = 0 < 1$). In fact, $u^*(t) = 0$ is *one* optimal control of the LQ problem. Moreover, the corresponding objective value is

$$\mathsf{E} \int_0^\infty [x^*(t)]^2 dt = x_0^2 \int_0^\infty e^{-t} dt = x_0^2, \tag{30}$$

where the first equality is due to (29) with $\alpha = 0$.

Next, we can obtain additional — in fact, infinitely many more — optimal controls based on the dual SDP, which in this case is:

$$\begin{array}{ll} \min & -z_b + z_n \\ \text{s.t.} & 1 + z_b - z_n = 0, \\ & z := \begin{bmatrix} z_b, & z_u \\ z_u, & z_n \end{bmatrix} \succeq 0. \end{array}$$

It can be directly verified that the above has multiple optimal solutions:

$$(z_b, z_u, z_n) = (z_b, z_u, 1 + z_b),$$

parameterized by (z_u, z_b) with

$$z_b \geq 0, \quad z_u^2 \leq z_b(1 + z_b). \tag{31}$$

In particular, note that the above ensures $z_n = 1 + z_b > 0$. Furthermore, these (parameterized) solutions are all complementary to the primal optimal solution $p^* = 0$. Hence, the test in Step 5 of the numerical procedure is passed, which gives rise to (multiple) optimal controls

$$u^*(t) = z_u z_n^{-1} x^*(t) \equiv \alpha x^*(t).$$

Notice that the feedback gain α satisfies

$$|\alpha| = \frac{z_u}{1 + z_b} \leq \frac{\sqrt{z_b(1 + z_b)}}{1 + z_b} = \sqrt{\frac{z_b}{1 + z_b}} < 1$$

where the first inequality follows from (31). Therefore, these controls are indeed stabilizing by (29). Finally, the optimal cost corresponding to these controls is:

$$\begin{aligned} J = & \ \mathsf{E} \int_0^\infty (1-\alpha^2)[x(t)]^2 dt \\ = & \ x_0^2(1-\alpha^2) \int_0^\infty e^{(\alpha^2-1)t} dt \\ = & \ x_0^2, \end{aligned}$$

which coincides with (30).

As in the deterministic case, when (SLQ) does not have any attainable optimal solution (but is otherwise well posed), we will consider (SLQ$_\epsilon$), obtained by keeping all the data A, B, C and D in (SLQ) unchanged, and letting $R_\epsilon = R + \epsilon I$ and $Q_\epsilon = Q + \epsilon I$ with $\epsilon > 0$. The associated SDP's for (SLQ$_\epsilon$) are

$$\begin{array}{lll} (\mathrm{P}_\epsilon) & \max & \langle I, P \rangle \\ & \text{s.t.} & \begin{bmatrix} R + \epsilon I + D^T P D, & B^T P + D^T P C \\ PB + C^T P D, & Q + \epsilon I + C^T P C + A^T P + P A \end{bmatrix} \succeq 0 \\ & & P \in \mathcal{S}^{n \times n} \end{array}$$

and

$$\begin{array}{lll} (\mathrm{D}_\epsilon) & \min & \langle R + \epsilon I, Z_B \rangle + \langle Q + \epsilon I, Z_N \rangle \\ & \text{s.t.} & I + Z_U^T B^T + B Z_U + Z_N A^T + A Z_N \\ & & +C Z_N C^T + D Z_U C^T + C Z_U^T D^T + D Z_B D^T = 0 \\ & & \begin{bmatrix} Z_B, & Z_U \\ Z_U^T, & Z_N \end{bmatrix} \succeq 0. \end{array}$$

Theorem 20 Let $J_\epsilon^*(x_0)$ and $J^*(x_0)$ be the optimal values of (SLQ$_\epsilon$) and (SLQ), respectively. Then,

$$\lim_{\epsilon \downarrow 0} J_\epsilon^*(x_0) = J^*(x_0).$$

Let the optimal solution of (P$_\epsilon$) be P_ϵ^*. Construct the following feedback control

$$u^\epsilon(t) = -(R_\epsilon + D^T P_\epsilon^* D)^+ (B^T P_\epsilon^* + D^T P_\epsilon C) x^\epsilon(t). \tag{32}$$

Theorem 21 The feedback control in (32) is asymptotically optimal for (SLQ), namely,

$$\lim_{\epsilon \downarrow 0} J(x_0, u^\epsilon(\cdot)) = J^*(x_0).$$

The following example illustrates a situation where the approximation is necessary.

Example 22 Suppose $m = n = 1$; $A = C = 0$, $B = D = 1$; $Q = 4$ and $R = -1$. The control system is as follows:

$$\begin{array}{ll} \min & \mathsf{E}\int_0^\infty [4x(t)^2 - u(t)^2]dt \\ \text{s.t.} & dx(t) = u(t)dt + u(t)dW(t) \\ & x(0) = x_0. \end{array}$$

Consider a feedback control $u(t) = -kx(t)$. Applying Itô's lemma yields

$$d[x(t)]^2 = (k^2 - 2k)[x(t)]^2 - 2k[x(t)]^2 dW(t).$$

Clearly, such a feedback control is stabilizing if and only if $k^2 - 2k < 0$, or, $0 < k < 2$. In particular, this implies that $u(t) = -x(t)$ is stabilizing while $u(t) = -2x(t)$ is not.

For any $0 < k < 2$, it follows that $\mathsf{E}\int_0^\infty [x(t)]^2 dt = x_0^2/(2k - k^2)$. Therefore, the control $u(t) = -kx(t)$ has a cost

$$(4 - k^2)\mathsf{E}\int_0^\infty [x(t)]^2 dt = (1 + \frac{2}{k})x_0^2.$$

As $k \uparrow 2$ we see that the cost can be arbitrarily close to $2x_0^2$. Nevertheless, this optimum is not attainable when $x_0 \neq 0$.

In terms of the corresponding SDP's, the primal reads

$$\begin{array}{ll} \max & p \\ \text{s.t.} & \begin{bmatrix} -1 + p, & p \\ p, & 4 \end{bmatrix} \succeq 0. \end{array}$$

This problem has only one feasible solution $p^* = 2$, which is necessarily the optimal solution. It clearly satisfies the Riccati equation:

$$4 + \frac{p^2}{1 - p} = 0.$$

The control in (27) is hence:

$$u^*(t) = -\frac{p^*}{-1 + p^*}x^*(t) = -2x^*(t),$$

which is *not* stabilizing as we discussed before.

Since **(B(ii))** does not hold, we expect **(C)** to fail as well, as **(C)** and **(C)** are equivalent under $R + D^T P^* D \succ 0$. So, let us now examine the dual:

$$\begin{array}{ll} \min & 4z_n - z_b \\ \text{s.t.} & 1 + 2z_u + z_b = 0, \\ & z := \begin{bmatrix} z_b, & z_u \\ z_u, & z_n \end{bmatrix} \succeq 0. \end{array}$$

This problem is strictly feasible, since the original LQ problem is stabilizable. For instance, $(z_b, z_u, z_n) = (1, -1, 2)$ is a strictly feasible solution. Hence ([40, Theorem 3.1]), the infimum of the dual objective value must coincide with the supremum of the primal objective value, which is 2. This means, should the dual optimal solution z^* exist, it must satisfy $4z_n^* - z_b^* = 2$, or $z_n^* = (2 + z_b^*)/4$. This, along with $z_u^* = -(1 + z_b^*)/2$, leads to

$$z_b^* z_n^* = z_b^* \cdot \frac{2 + z_b^*}{4} < \frac{(1 + z_b^*)^2}{4} = (z_u^*)^2,$$

which violates $z \succeq 0$. Consequently, the dual does not have an attainable optimal solution, and the complementary duality fails.

Now, we are going to use the approximation approach to solve the LQ problem. With perturbation, the corresponding primal SDP becomes

$$\begin{array}{ll} \max & p \\ \text{s.t.} & \begin{bmatrix} -1 + \epsilon + p, & p \\ p, & 4 + \epsilon \end{bmatrix} \succeq 0. \end{array}$$

Solving this problem yields

$$p_\epsilon^* = \frac{4 + \epsilon + \sqrt{(4+\epsilon)^2 + 4(4+\epsilon)(-1+\epsilon)}}{2}.$$

Clearly, $p_\epsilon^* = 2 + O(\sqrt{\epsilon})$, and hence the optimal value of (P_ϵ), $p_\epsilon^* x_0^2$, converges to $2x_0^2$ as $\epsilon \downarrow 0$.

7. Concluding Remarks

We have developed a unified theory, based on primal-dual SDP, to LQ control allowing the cost matrices (R in particular) to be singular, or even indefinite in the stochastic setting. Specifically, we examine a pair of primal-dual SDP problems; if complementary solutions exist, then either of the two feedback controls (based on the primal and dual SDP solutions) is optimal. If complementary duality fails, then either the LQ control problem cannot be solved by the SDP approach (and neither can it be solved by any other known approaches); or we can conclude that there exists no attainable optimal control for the LQ problem (e.g., when $Q \succ 0$ in the deterministic setting, since **(A)** $\Rightarrow$ **(C)**). In this case, we have developed a convergent approximation scheme that achieves asymptotic optimality.

Our results also complement two main approaches to deterministic singular LQ control problems, dimension reduction and distributional controls, in the following sense: We show that dimension reduction is feasible if and only if the SDP complementary duality holds; and that the non-existence of attainable (regular) controls confirmed by the SDP approach justifies the distributional controls as a last resort to problems that are identified as "hopelessly" singular.

References

[1] M. Ait Rami, X. Chen, J.B. Moore, and X.Y. Zhou, "Solvability and asymptotic behavior of generalized Riccati equations arising in indefinite stochastic LQ controls", *IEEE Trans. Autom. Contr.*, vol. AC-46, pp. 428-440, 2001.

[2] M. Ait Rami and X. Y. Zhou, "Linear matrix inequalities, Riccati equations, and indefinite stochastic linear quadratic controls", *IEEE Trans. Aut. Contr.*, vol. AC-45, pp. 1131-1143, 2000.

[3] A. Albert, "Conditions for positive and nonnegative definiteness in terms of pseudo-inverses," *SIAM J. Appl. Math.*, vol. 17, pp. 434-440, 1969.

[4] B.D.O. Anderson and J.B. Moore, *Optimal Control - Linear Quadratic Methods.* New Jersey: Prentice-Hall, 1989.

[5] M. Athens, *Special issues on linear–quadratic–Gaussian problem*, IEEE Trans. Auto. Contr., vol. AC-16, pp. 527-869, 1971.

[6] A. Bensoussan, "Lecture on stochastic control, part I", *Lecture Notes in Math.*, vol. 972, pp. 1-39, 1983.

[7] V. Balakrishnan and L. Vandenberghe, "Connections between duality in control theory and convex optimization", *Proc. Amer. Contr. Conf.*, pp. 4030-4034, 1995.

[8] S. Boyd, L. El Ghaoui, E. Feron and V. Balakrishnan, *Linear Matrix Inequality in Systems and Control Theory.* Philadelphia: SIAM, 1994.

[9] A.E. Bryson, Jr. and Y.C. Ho, *Applied Optimal Control: Optimization, Estimation, and Control.* New York: John Wiley & Sons, 1975.

[10] D.J. Clements and B.D.O. Anderson, *Singular Optimal Control: The Linear–Quadratic Problem.* Berlin Heidelberg: Springer-Verlag, 1978.

[11] D.J. Clements, B.D.O. Anderson and P.J. Moylan, "Matrix inequality solution to linear-quadratic singular control problems", *IEEE Trans. Auto. Contr.*, vol. AC-22, pp. 55-57, 1977.

[12] S. Chen, X. Li and X.Y. Zhou, "Stochastic linear quadratic regulators with indefinite control weight costs", *SIAM J. Contr. Optim.*, vol. 36, pp. 1685-1702, 1998.

[13] S. Chen and J. Yong, "Stochastic linear quadratic optimal control problems with random coefficients", *Chinese Ann. Math. Ser. B*, vol. 21, pp. 323-338, 2000.

[14] S. Chen and J. Yong, "Stochastic linear quadratic optimal control problems", *Appl. Math. Optim.*, vol. 43, pp. 21-45, 2001.

[15] S. Chen and X.Y. Zhou, "Stochastic linear quadratic regulators with indefi nite control weight costs. II", *SIAM Journal on Control and Optimization*, Vol. 39, pp. 1065-1081, 2000.

[16] M.H.A. Davis, Linear Estimation and Stochastic Control. Chapman and Hall, London, 1977.

[17] L. El Ghaoui, R. Nikoukhah and F. Delebecque, "`LMITOOL`: A front-end for LMI optimization, user's guide", 1995. Available via anonymous ftp to `ftp.ensta.fr`, under `/pub/elghaoui/lmitool`.

[18] K. Fujisawa, M. Fukuda, M. Kojima and K. Nakata, "Numerical evaluation of SDPA", to appear in *Proceedings of HPOPT97*, The Netherlands, 1999.

[19] M.L.J. Hautus and L.M. Silverman, "System structure and singular control", *Lin. Algebra Appl.*, vol. 50, pp. 369-402, 1983.

[20] D.H. Jacobson, S.B. Gershwin and M.M. Lele, "Computation of optimal singular controls", *IEEE Trans. Aut. Contr.*, vol. AC-15, pp. 67-73, 1970.

[21] R. E. Kalman, "Contributions to the theory of optimal control", *Bol. Soc. Math. Mexicana*, vol. 5, pp. 102-119, 1960.

[22] M. Kohlmann and X.Y. Zhou, "Relationship between backward stochastic differential equations and stochastic controls: A linear-quadratic approach", *SIAM J. Contr. Optim.*, vol. 38, pp. 1392-1407, 2000.

[23] X. Li, X.Y. Zhou and A.E.B. Lim, "Dynamic mean–variance portfolio selection with no-shorting constraints", to appear in *SIAM J. Contr. Optim.*

[24] A.E.B. Lim and X.Y. Zhou, "Optimal stochastic LQR control with integral and indefinite control weights", *IEEE Trans. Autom. Contr.*, Vol. AC-44, pp. 1359-1369, 1999.

[25] A.E.B. Lim and X.Y. Zhou, "Mean–variance portfolio selection with random parameters", to appear in *Math. Oper. Res.*

[26] Z.Q. Luo, J.F. Sturm and S. Zhang, "Duality results for conic convex programming", Report 9719/A, Econometric Institute, Erasmus University Rotterdam, 1997.

[27] Z.Q. Luo, J.F. Sturm and S. Zhang, "Conic convex programming and self-dual embedding", *Optim. Meth. Software*, vol. 14, pp. 169-218, 2000.

[28] Z.Q. Luo, J.F. Sturm and S. Zhang, "Superlinear convergence of a symmetric primal-dual path following algorithms for semidefinite programming", *SIAM J. Optim.*, vol. 8, pp. 59-81, 1998.

[29] Yu. Nesterov and A. Nemirovski, *Interior Point Polynomial Methods in Convex Programming.* Philadelphia: SIAM, 1994.

[30] R. Penrose, "A generalized inverse of matrices", *Proc. Cambridge Philos. Soc.*, vol. 52, pp. 17-19, 1955.

[31] A. Ran and R. Vreugdenhil, "Existence and comparison theorems for algebraic Riccati equations for continuous and discrete time systems", *Linear Algebra Appl.*, vol. 99, pp. 63-83, 1988.

[32] C. Roos, T. Terlaky and J.-Ph. Vial, *Theory and Algorithms for Linear Optimization.* New York: John Wiley & Sons, 1997.

[33] M.A. Rotea, "The generalized H_2 control problem", *Automatica*, vol. 29, pp. 373-385, 1993.

[34] J.M. Schumacher, "The role of the dissipation matrix in singular optimal control", *Syst. Contr. Lett.*, vol. 2, pp. 262-266, 1983.

[35] A.A. Stoorvogel, "The singular H_2 control problem", *Automatica*, vol. 28, pp. 627-631, 1992.

[36] J.F. Sturm, *Primal-Dual Interior Point Approach to Semidefinite Programming*, Ph.D thesis, Tinbergen Institute Series 156, Erasmus University Rotterdam, 1997.

[37] R.E. Skelton, T. Iwasaki and K. Grigoriadis, *A Unified Algebraic Approach to Linear Control Design.* London: Taylor & Francis, 1998.

[38] J.F. Sturm, "Using SeDuMi 1.02, a MATLAB toolbox for optimization over symmetric cones", *Optim. Meth. Software*, vol. 11-12, pp. 625-653, 1999.

[39] K.C. Toh, M.J. Todd and R.H. Tütüncü, "SDPT3 — A MATLAB software package for semidefinite programming", *Optim. Meth. Software*, vol. 11-12, pp. 545-581, 1999.

[40] L. Vandenberghe and S. Boyd, "Semidefinite Programming', *SIAM Rev.*, vol. 38, pp. 49-95, 1996.

[41] L. Vandenberghe and V. Balakrishnan, "Semidefinite programming duality in linear system theory: connections and implications for computation", *Proc. IEEE Conf. Decision and Control*, pp. 989-994, 1999.

[42] J.C. Willems, "Least square stationary optimal control and the algebraic Riccati equation", *IEEE Trans. Aut. Contr.*, vol. AC-16, pp. 621-634, 1971.

[43] J.C. Willems, A. Kitapci and L.M. Silverman, "Singular optimal control: a geometric approach", *SIAM J. Contr. Optim.*, vol. 24, pp. 323-337, 1986.

[44] W. M. Wonham, "On a matrix Riccati equation of stochastic control", *SIAM J. Contr.*, vol. 6, pp. 312-326, 1968.

[45] H. Wu and X.Y. Zhou, "Stochastic frequency characteristics", *SIAM J. Contr. Optim.*, vol. 40, pp. 557-576, 2001.

[46] V.A. Yakubovich, "The solution of certain linear matrix inequalities in automatic control theory", *Soviet Math. Dokl.*, vol. 5, pp. 620-623, 1964.

[47] D.D. Yao, S. Zhang and X. Y. Zhou, "Stochastic LQ control via semidefinite programming", *SIAM J. Contr. Optim.*, **40** (2001), 801-823.

[48] D.D. Yao, S. Zhang and X. Y. Zhou, "A primal-dual semidefinite programming approach to linear quadratic control," *IEEE Trans. Aut. Contr.*, **46** (2001), 1442-1447.

[49] J. Yong and X. Y. Zhou, *Stochastic Controls: Hamiltonian Systems and HJB Equations*, New York: Springer, 1999.

[50] X.Y. Zhou and D. Li, "Continuous-time mean-variance portfolio selection: A stochastic LQ framework", *Appl. Math. Optim.*, vol. 42, pp. 19-33, 2000.

Chapter 10

DISCRETE EVENTS: TIMETABLES, CAPACITY QUESTIONS, AND PLANNING ISSUES FOR RAILWAY SYSTEMS *

Geert Jan Olsder
g.j.olsder@math.tudelft.nl

Antoine F. de Kort
Faculty of Information Technology and Systems
Delft University of Technology
P.O.Box 5031, 2600 GA Delft, the Netherlands

1. Introduction and motivation

The theory of Discrete Event Systems (DES's) is a research area of current vitality. The development of this theory is largely stimulated by discovering general principles which are (or are hoped to be) useful to a wide range of application domains. In particular, technological and/or 'man-made' manufacturing systems, communication networks, transportation systems, and logistic systems, all fall within the class of DES's. One of the key features that characterize these systems is that their dynamics are *event-driven* as opposed to *time-driven*, i.e., the behavior of a DES is governed only by occurrences of different types of events over time rather than by ticks of a clock.

In the corresponding literature, one can distinguish three main fields of activities, all claiming and using the phrase 'discrete event (dynamic) system'. They are:

- The logical approach, where DES's are modeled as formal languages. One of the seminal papers is [17].

*Presented at the Symposium in honour of Larry Ho, Harvard University, June 23, 24 , 2001.

- Perturbation analysis on time series resulting from DES's. A seminal paper here is [13]; see also [12].
- Modeling in terms of the max-plus algebra. There are clear links with timed Petri nets. A seminal contribution for this direction is [7].

For an overview article which discusses these three approaches, see [6] or [5]. In the current paper two applications of the third approach within the Dutch railway system will be given. In addition, yet another application with respect to the synchronization of traffic lights will briefly be mentioned. The underlying theory, viz. that of the max-plus algebra, will also be briefly introduced. For detailed treatment of this theory, with various extensions, see [1], [5] and [11].

1.1 Design of timetables for trains

The first application is concerned with a closed network of railway connections and studies questions related to the design of timetables according to which the trains run within this network. Specifically, we will study the intercity network of the Netherlands. This network, given in figure 10.1, consists of 11 lines (routes along which the trains run in both directions). An example of such a line is the line with number 10 which starts (respectively terminates) at Amsterdam and terminates (respectively starts) at Vlissingen. There are no circular lines, but inclusion of circular lines into the model would not constitute any difficulties. We will not consider international connections: it is assumed in our model that trains arriving at the border return direction and continue in opposite direction (this is what actually happens with some trains at some border stations). Hence the name 'closed' network. On each line, both directions, there is a fixed number of trains which return at both endpoints, in the opposite direction. Thus each line can be considered as a closed circuit (or as a circular line). The travelling times between stations is given and assumed to be deterministic.

Suppose there is no timetable. A very simple policy for trains to run would be to drop and pick up passengers at stations and then directly continue. This is not very realistic since one would like trains to remain at a station for some time such as to wait for other arriving trains and allow change overs. Now suppose that realistic constraints are included which state which trains have to wait for each other, at each station, such as to allow change overs before the trains can continue on their respective lines.

Without an apriori timetable, and with the constraints of trains waiting for each other, how fast can the system operate? The actual timetable

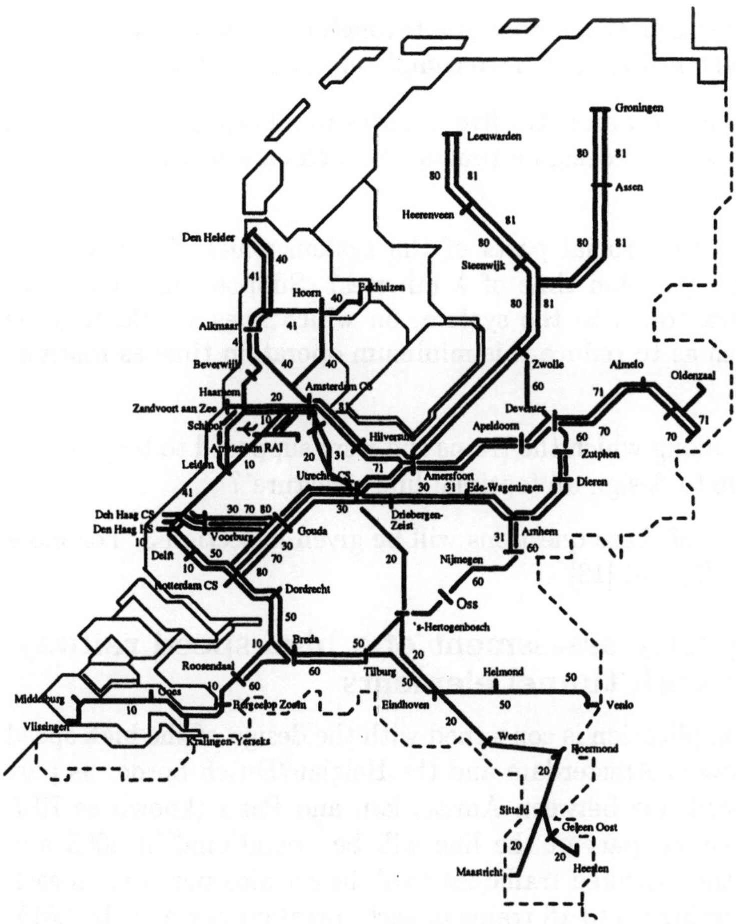

Figure 10.1. The Dutch intercity network

is based on a half hour schedule and hence from this fact of real life we know that without timetable the system must be 'operable' in a time of at most half an hour (this time will be indicated by λ which turns out to be equal to about 27 minutes). The difference between the actual half hour and the theoretical λ minutes is the flexibility in the system which causes propagation of possible delays to disappear in finite time. Various questions can be posed now, some of which are:

- How do perturbations progagate through the system and how long does it take before they have completely disappeared?
- If one would add an extra five minutes to all changeover times, is it still possible to design a timetable based on a half hour service? (answer: yes).
- What are the crucial parts of the system which determine the minimum operation time of λ minutes? Suppose one could add some extra trains to the system, on which lines should they be set in such as to reduce this minimum operation time as much as possible?
- The lines along which the trains run were supposed to be given. Is it possible to design an 'optimal line structure'?

Answers to most of these questions will be given in section 3. For more background see [3] and [18]

1.2 Capacity assessment of a high speed railway line with tunnel elements

The second application is concerned with the design of the high speed connection between Amsterdam and the Belgian/Dutch border as part of the high speed line between Amsterdam and Paris (known as HSL South). The Dutch part of the line will be operational in 2005 and from then on, the expected traffic load will be 8 trains per hour in each direction, increasing up to 16 trains in each direction per hour by 2015. The line will be used for transporting passengers only (no cargo).

The Dutch part of the line includes three special tunnels, each with separated tunnel tubes for both running directions. Figure 10.2 shows the line schematically, with the tunnels under consideration represented as boxes. From north to south (left to right), the tunnels are called "Groene Hart" tunnel, "Oude Maas" tunnel and "Dordtsche Kil" tunnel, respectively. The nodes are fictitious points distinguishing the respective track parts for which we provide time values later on in section 4. The corresponding track distances are depicted in kilometers. There will

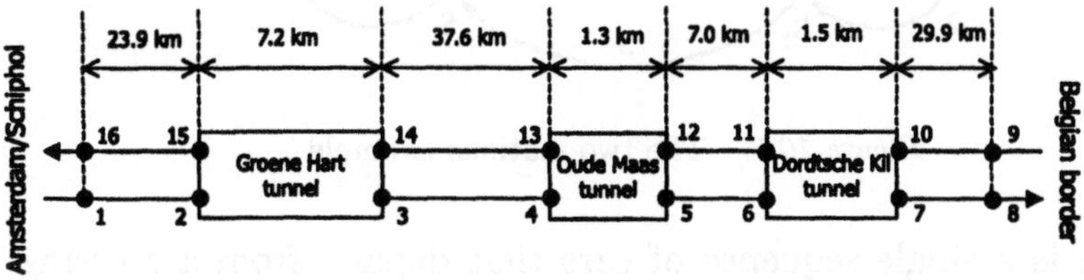

Figure 10.2. Distances along HSL South and its three tunnels

be no emergency exits to ground level, nor will these three tunnels be equipped with service channels. Instead, each tunnel will be provided with cross–sections connecting both tunnel tubes. This implies that, in case of an emergency (e.g. a train getting stuck in the tunnel while being on fire), passengers can only be evacuated via the opposite tunnel tube. Evidently, for this escape route to be safe, the opposite tunnel tube must be free of trains at the time of evacuation.

A special traffic regime has been proposed to guarantee passengers at any time a safe escape route from each of these tunnels. This regime permits the presence of at most one train at a time in each tunnel (thus both tubes). In other words, if a train occupies one of the tunnel tubes, the opposite tube is immediately blocked for other trains as well. Consequently, even though HSL South is an entirely double track line, all three tunnels behave like single track parts and as such, the capacity of the entire line may be restricted too much to establish the intended train amounts as indicated above. An alternative would then be to build an extra tube (e.g. a service channel). However, drilling a railway–tunnel tube costs about $50,000 per running meter (including installation of equipment), which explains the importance of carefully examining the capacity of the line. This problem will be considered in more detail in section 4.

1.3 The control of traffic lights

In this paragraph yet another application will be mentioned, viz. a Traffic Route System (TRS) for a (part of a) city plan, see [9] for details. Given a set of interacting traffic routes, the problem is to determine the Signal Timing Plan (STP) that minimizes the time duration that cars have to wait for green lights while following their predetermined routes (along the route no different choices can be made). Compared to the timetable design problem for a railroad network, the railroad stations have become crossings in the city plan and trains have become platoons

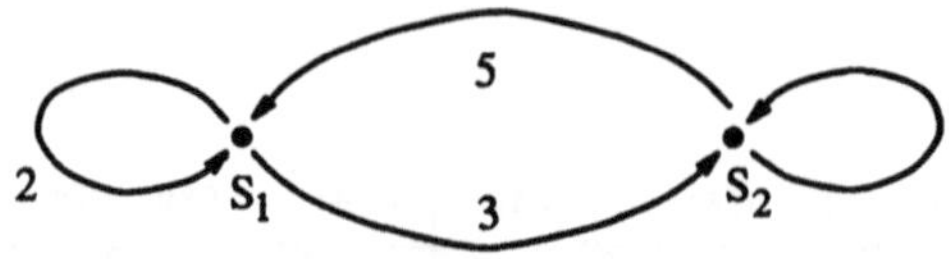

Figure 10.3. The two stations example

(a platoon is a single sequence of cars that departs from a crossing in the same direction during a green phase).

2. The max-plus algebra approach to DES

This section starts with the modelling of an example, in terms of an extremely simple railway network, in subsection 2.1, and subsequently this modelling is given a more abstract setting in subsections 2.2 and 2.3.

2.1 The stage for the max-plus approach

In a metropolitan area there are two railway stations, S_1 and S_2, which are interconnected by a railway system as indicated in figure 10.3. This railway system consists of an innercircle and of two outer circles. The trains on these outer circles deliver and pick up passengers in the suburbs. The stations in the suburbs have not been drawn since they do not play any role in the problem to be formulated.

Suppose there are four trains (two at each station) and they leave the stations at time 0, one along each track. They reach the other (or the same) station after a certain time which is indicated in the figure. The arriving trains at a station have to wait for each other such as to allow the passengers to change trains.

Suppose that there is no timetable and that the trains leave directly after the change over of the passengers at the stations and that the time needed for change overs has been incorporated in the travelling time. If this process of departing and arriving trains is continued, the departure time $x_i(k+1)$ for the $(k+1)^{\text{st}}$ departure at station S_i satisfies

$$\begin{aligned} x_1(k+1) &= \max(x_1(k)+2,\ x_2(k)+5), \\ x_2(k+1) &= \max(x_1(k)+3,\ x_2(k)+3), \end{aligned} \tag{1}$$

for $k = 0, 1, 2, \ldots$.

With $x_1 = 0$, $x_2 = 0$, the evolution of equations (1) becomes

$$\underset{x(0)}{\begin{pmatrix} 0 \\ 0 \end{pmatrix}} \rightarrow \underset{x(1)}{\begin{pmatrix} 5 \\ 3 \end{pmatrix}} \rightarrow \underset{x(2)}{\begin{pmatrix} 8 \\ 8 \end{pmatrix}} \rightarrow \underset{x(3)}{\begin{pmatrix} 13 \\ 11 \end{pmatrix}} \rightarrow \underset{x(4)}{\begin{pmatrix} 16 \\ 16 \end{pmatrix}} \rightarrow \cdots$$

This pattern of departure times shows a periodic solution superimposed on a linear drift; the 'period' equals 2 and the average time between two subsequent departures is 4. From a timetable point of view (timetables must be as 'regular' as possible), it is better to start with the initial departures $x_1 = 1$, $x_2 = 0$, since then the evolution becomes

$$\underset{x(0)}{\begin{pmatrix} 1 \\ 0 \end{pmatrix}} \rightarrow \underset{x(1)}{\begin{pmatrix} 5 \\ 4 \end{pmatrix}} \rightarrow \underset{x(2)}{\begin{pmatrix} 9 \\ 8 \end{pmatrix}} \rightarrow \underset{x(3)}{\begin{pmatrix} 13 \\ 12 \end{pmatrix}} \rightarrow \cdots$$

where the interdeparture time is now exactly 4 at each station and thus the departure times are very regular (they have 'period' 1). It turns out that, whatever the initial condition, after possibly a short transient period of time, a periodic behavior of either period 1 or 2 is obtained with (average) interdeparture times 4. A solution with an (average) departure time smaller than 4 is not possible, since for a train to go around in the innercircle costs $3 + 5 = 8$ time units. There are two trains on the innercircle and therefore the (average) interdeparture time is limited from below by $8/2 = 4$.

With the above sketched railway system and trains it is not possible to design a timetable with interdeparture times smaller than 4. If one wants a faster timetable, one must change the problem. To this end, let us add a train on the innercircle, such that three trains will run along this circle all the time. Suppose that initially this extra train is situated at station S_1. The equations for the departure times now become

$$\begin{aligned} x_1(k+1) &= \max(x_1(k) + 2,\ x_2(k) + 5), && (2)\\ x_2(k+1) &= \max(x_1(k-1) + 3,\ x_2(k) + 3), && (3) \end{aligned}$$

which can be rewritten as a set of first order equations as

$$\begin{aligned} x_1(k+1) &= \max(x_1(k) + 2,\ x_2(k) + 5),\\ x_2(k+1) &= \max(x_3(k) + 3,\ x_2(k) + 3),\\ x_3(k+1) &= x_1(k). \end{aligned} \qquad (4)$$

With the initial condition $x_1 = 0$, $x_2 = 0$, $x_3 = 0$ the evolution of the latter set of equations becomes

$$\underset{x(0)}{\begin{pmatrix} 0 \\ 0 \\ 0 \end{pmatrix}} \rightarrow \underset{x(1)}{\begin{pmatrix} 5 \\ 3 \\ 0 \end{pmatrix}} \rightarrow \underset{x(2)}{\begin{pmatrix} 8 \\ 6 \\ 5 \end{pmatrix}} \rightarrow \underset{x(3)}{\begin{pmatrix} 11 \\ 9 \\ 8 \end{pmatrix}} \rightarrow \underset{x(4)}{\begin{pmatrix} 14 \\ 12 \\ 11 \end{pmatrix}} \rightarrow \cdots$$

which shows, after a transient part, a regular behavior of 'period' 1 with interdeparture times 3. This interdeparture time is caused by the outerloop at station S_2; on this loop there is one train which needs 3 time units to travel around. The innerloop is not the bottleneck anymore: this innerloop itself would lead to a lower limit of 8/3 (travelling time of the loop divided by the number of trains on this loop).

2.2 Formalization

The basic form of the systems under consideration is

$$x_i(k+1) = \max_j(a_{ij} + x_j(k)), \quad i = 1, \ldots, n \ . \tag{5}$$

It is common practice to change the notation somewhat. Addition + will be written as ⊗ and max will be written as ⊕. This change of notation makes the resemblance with conventional linear difference systems visible:

$$x_i k + 1) = \bigoplus_j (a_{ij} \otimes x_j(k)), \quad i = 1, \ldots, n \ , \tag{6}$$

which in vector notation will be written as

$$x(k+1) = A \otimes x(k). \tag{7}$$

Of the latter equation one speaks as a linear (difference) equation in the max-plus algebra, this in clear analogy with linear difference equations in the conventional, 'plus-times', algebra. If it is clear from the context that the underlying algebra is the max-plus one, one even writes $x(k+1) = Ax(k)$ for (7). If the initial condition for (7) is $x(0) = x_0$, then

$$x(1) = A \otimes x_0 \ ,$$

$$x(2) = A \otimes x(1) = A \otimes (A \otimes x_0) = (A \otimes A) \otimes x_0 = A^2 \otimes x_0 \ .$$

It can be shown that indeed $A \otimes (A \otimes x_0) = (A \otimes A) \otimes x_0$. Instead of $A \otimes A$ we simply write A^2. We get, for the general case,

$$x(k) = \underbrace{(A \otimes A \otimes \cdots \otimes A)}_{\text{k times}} \otimes x_0 = A^k \otimes x_0 \ .$$

The matrices A^2, A^3,..., can be calculated directly. Let us consider the A-matrix of (1),

$$A = \begin{pmatrix} 2 & 5 \\ 3 & 3 \end{pmatrix}$$

then

$$A^2 = \begin{pmatrix} \max(2+2, 5+3) & \max(2+5, 5+3) \\ \max(3+2, 3+3) & \max(3+5, 3+3) \end{pmatrix} = \begin{pmatrix} 8 & 8 \\ 6 & 8 \end{pmatrix} .$$

In general

$$(A^2)_{ij} = \bigoplus_l a_{il} \otimes a_{lj} = \max_l(a_{il} + a_{lj}) \ . \tag{8}$$

In terms of the railway example, the quantity $(A^2)_{ij}$ can be interpreted as the maximum (with respect to l) of all connections from station S_j via station S_l to station S_i. One speaks of paths of length two between the stations S_j and S_i. In graph-theory terminology, the stations are called nodes and the tracks between stations are called arcs. More generally, $(A^k)_{ij}$ denotes the maximum of all paths of length k, starting at node j and ending at node i.

In many networks such as a railway net there will not be an arc from each node to each other node. If there is no arc from node S_j to node S_i then the behaviour of node S_i is not directly influenced by that of node S_j. In such a situation it is useful to consider the element a_{ij} to be equal to $-\infty$. In (5) a term $-\infty + x_j(k)$ does not influence $x_i(k+1)$ as long as $x_j(k)$ is finite. The number $-\infty$ will occur frequently in the sequel and it will be indicated by ε.

A seeming generalization of (7) is

$$x(k+1) = A_0 x(k+1) \oplus A_1 x(k) \oplus \cdots \oplus A_{l+1} x(k-l), \tag{9}$$

which is implicit in $x(k+1)$ and which has extra delays. By repeated substitution of the whole right-hand side of (9) for the term $x(k+1)$ in this right-hand side, one gets

$$x(k+1) = A_0^* A_1 x(k) \oplus \cdots \oplus A_0^* A_{l+1} x(k-l) \tag{10}$$

where

$$A_0^* := I \oplus A_0 \oplus A_0^2 \oplus A_0^3 \oplus \cdots$$

(Later on we will also need the notation A_0^+, which is, by definition,

$$A_0^+ := A_0 \oplus A_0^2 \oplus A_0^3 \oplus \cdots)$$

The notation I refers to the identity matrix in the max-plus algebra: it has zeros on the main diagonal and ε's elsewhere. Equation (10) only makes sense if A_0^* is well defined (its elements are finite or ε). This is for instance the case if the precedence graph of A, see subsection 2.3, does not contain circuits, because then $A^k = \varepsilon$ for $k \geq n$. Equation (10) can be rewritten as a first order difference equation by augmenting the state space. This is a standard trick in system theory and has already been used in subsection 2.1.

2.3 Periodic behaviour

Given a square matrix A, we consider the problem of existence of eigenvalues and eigenvectors in the max-plus algebra, that is, the existence of λ and $v \neq \varepsilon$ such that:

$$Av = \lambda v \ . \tag{11}$$

This equation has to be interpreted in the max-plus algebra sense; the expression λv means that one adds λ to each component of v. We already have seen examples of eigenvalues and eigenvectors in subsection 2.1; v corresponds to an initial state resulting in a solution with 'period' 1 and λ is the interdeparture time.

Before formulating a theorem about eigenvalues, some graph theory must be recapitulated. In the following definition the starting point is a square matrix, the entries of which may again assume the 'value' ε.

Definition 2 (Precedence graph) *The precedence graph of an $n \times n$ matrix A is a weighted digraph with n nodes and an arc (j, i) if $a_{ij} \neq \varepsilon$, in which case the weight of this arc receives the numerical value of a_{ij}. The precedence graph is denoted $\mathcal{G}(A)$.*

It is not difficult to see that any weighted digraph $\mathcal{G} := (\mathcal{V}, \mathcal{E})$, with $\mathcal{V}$ being the set of nodes and $\mathcal{E}$ being the set of arcs, is the precedence graph of an appropriately defined square matrix. The weight a_{ij} of the arc from node j to node i is defined as the ij^{th} entry of a matrix A. If an arc does not exist, the corresponding entry of A becomes ε. The matrix A thus defined has $\mathcal{G}$ as its precedence graph.

As we have seen before, the element (i, j) of $A^k = A \otimes \cdots \otimes A$, considered within the max-plus algebra denotes the maximum weight with respect to all paths of length k which go from node j to node i. If no such path exists, then $(A^k)_{ij} = \varepsilon$. The weight of a path ρ is denoted $|\rho|_{\mathrm{w}}$ and its length is denoted $|\rho|_{\mathrm{l}}$.

Definition 3 *The* mean weight of a path *is defined as the sum of the weights of the individual arcs of this path, divided by the length of this path. If the path is denoted by ρ, then the mean weight equals $|\rho|_{\mathrm{w}}/|\rho|_{\mathrm{l}}$. If such a path is a circuit one talks about the mean weight of the circuit, or simply the* cycle mean.

We are interested in the maximum of these cycle means, where the maximum is taken over all circuits in the graph. This number will be called the *maximum cycle mean.* If the cycle mean of a circuit equals the maximum cycle mean, then the circuit is called critical. The graph consisting of all critical circuits (if there happen to be more than one) is called the *critical graph* and denoted by $\mathcal{G}^c$.

Theorem 1 *We are given a square matrix A. If $\mathcal{G}(A)$ is strongly connected, then there exists one and only one eigenvalue and at least one eigenvector. The eigenvalue is equal to the maximum cycle mean of the graph:*

$$\lambda = \max_{\zeta} \frac{|\zeta|_w}{|\zeta|_1},$$

where ζ ranges over the set of circuits of $\mathcal{G}(A)$.

Definition 4 (Cyclicity of a graph) *Given a strongly connected graph, its cyclicity equals the greatest common divisor of the lengths of all its circuits. The cyclicity of an arbitrary graph (which may consist of several strongly connected subgraphs) equals the least common multiple of the cyclicities of all its maximal strongly connected subgraphs.*

Definition 5 (Cyclicity of a matrix) *A matrix A is said to be cyclic if there exist scalars M, λ and d such that $\forall m \geq M$, $A^{m+d} = \lambda^d A^m$. The least such d is called the cyclicity of A. The quantity λ equals the maximum cycle mean of A.*

The expression $A^{m+d} = \lambda^d A^m$ in the definition above must be interpreted in the max-plus algebra sense of course. Thus λ^d in the max-plus algebra means $d\lambda$ in the conventional algebra and $\lambda^d A^m$ refers to the addition of λ^d to each element of A^m.

Theorem 2 *Any irreducible matrix is cyclic. The cyclicity of the irreducible matrix A equals the cyclicity of $\mathcal{G}^c(A)$, being the critical graph corresponding to matrix A.*

If we go back to system (7) then the eigenvalue λ of the system matrix A can be interpreted as the "speed" of the system. Similarly the speed λ of the generalised system (9) is defined by

$$v = \mathcal{A}(\lambda^{-1})v, \text{ where } \mathcal{A}(\lambda^{-1}) := \bigoplus_{k=0}^{l+1} A_k \lambda^{-k}, \tag{12}$$

and where v is a "generalized eigenvector".

2.4 Extensions

Various extensions are known nowadays. We will only mention some of them:

- Stochastic max-plus algebra; the travelling times in the railway application are stochastic.
- Numerical procedures for the calculation of the eigenvalue, especially important when the dimension of the state is large (say $n > 1000$).
- The γ-transform as the max-plus equivalent of the z−transform in automatic control. Transfer matrices for input/output systems in the max-plus algebra.
- Relation to Petri net theory.
- Min-max-plus systems and their imbedding in nonexpansive mappings.
- Continuous flow systems with synchronization constraints.

For detailed treatments the reader is referred to the literature.

3. Railway network

3.1 Intercity trains

The problem description has been given in section 1. With the theory just given, answers to the various questions posed can be given now. The model which follows directly from the problem description is of the form (9), with the vector x being 53-dimensional. (It has been described explicitly in [3].) After rewriting it in the form of (7), a model with state vector of dimension 79 resulted. The matrix A will not be given here: it is made up of the different travelling times and standard change over times as set by the railway company. The answer to question 3 turns out to be $\lambda = 27\frac{1}{12}$. The critical circuit (this is not to be confused with a line of the net) turns out to be from Venlo to Eindhoven to Utrecht to Amsterdam to Zandvoort aan Zee and back. Hence if one wants a faster timetable, one should add extra trains to a line (or lines) which forms part of this critical circuit (i.e. the lines numbered 20 and 50). If one would add $5\frac{2}{3}$ minutes to all change over times (question 2), to be incorporated into the model by adjusting the A matrix, then it turns out that eigenvalue $\lambda = 30$ minutes. Of course this eigenvalue is monotone with respect to the duration of the change over times. The critical circuit in this case with maximum change over time which allows a half hour timetable is Venlo to Eindhoven to Utrecht to Amsterdam to Haarlem to Den Haag HS to Breda to Eindhoven to Venlo. If one would consider a very unfriendly way of handling passengers by not waiting for them, i.e. trains would simply stop at stations, deliver passengers and pick up passengers which happen to be there and depart again, then we can design a faster timetable. It turns out that in this case $\lambda = 26\frac{2}{11}$. The critical circuit is now Venlo to Eindhoven to Breda to Den Haag CS and back.

The question of the design of an 'optimal line structure' cannot be answered directly by an application of the theory. One can of course compare different designs by calculating their critical circuits and minimum interdeparture times. A FIFO-rule at a crossing (i.e. the train which arrives first can pass first) can be modelled if in addition to addition and maximization one also introduces the operation minimization. Thus one enters the realm of the theory of min-max-plus algebra, see [15], [11] and [18].

3.2 All trains

In this subsection some results (see [18] for detals) will be given by considering all passengers trains, i.e. intercity trains, express trains and local trains, together. In addition to the 11 lines for intercity trains, we now have 50 more lines for the two types of slower trains (one physical track on which both intercity and express trains move is counted as two). The corresponding model (7) now is 441-dimensional. By some reduction techniques this dimension could be brought back to 214. Since some local and express trains have a hourly pattern in the existing timetable (in contrast to intercity trains which depart every half hour), this model is based on one full hour of operation (during which all intercity connections appear twice). One thus would expect the calculated eigenvalue ($\lambda = 53.4$) to be at least twice as high as for the half hour model of the intercity trains ($\lambda = 27\frac{1}{12}$). The reason for this not being true is that somewhat different data for travelling times and required connections have been used (different years). A conclusion that can be drawn anyhow is that the addition of the slower trains does not really slow down the system.

A question also considered in [18] is the optimal allocation of trains. How should one distribute the extisting trains over all lines, while guaranteeing all connections of change overs, such that the corresponding λ becomes as small as possible? A lower bound for this quantity can easily be given if one allows noninteger-valued numbers of trains by making all cycle means (see Definition 3) equal. In that case one can talk about the train-density (number of trains per km). This lower bound turned out to be 43.3.

3.3 Stabibity and the propagation of delays

A few notions with respect to the concepts of stability and delays will be given. Consider

$$x(k+1) = \bigoplus_{p=0}^{l+1} A_p \otimes x(k-p+1) \oplus d(k+1), \tag{13}$$

where $d(k) = (d_1(k), \ldots, d_n(k))'$, the symbol $'$ denoting transposed, is the vector of the scheduled k^{th} departure times. For a periodic timetable with period length (or cycle time) T one has

$$d(k+1) = T^k d(1), \quad k = 1, 2, \ldots.$$

A necessary condition for the timetable to be feasible is $d(k+1) \geq \bigoplus_{p=0}^{l+1} A_p d(k-p+1)$.

Definition 6 (Stability margin) *The stability margin Δ of the system (13) is the minimium time increase of all travelling times such that no longer a stable timetable exists with period T.*

It can be shown that $\Delta = -\mu$, where μ is the solution of $\mathcal{A}(T^{-1})v = \mu v$.

Definition 7 (Recovery matrix) *Consider the system (13). The* recovery matrix R *is the matrix of which the entry $(R)_{ij}$ equals the maximum delay of $x_j(k)$ such that $x_i(m)$ is not delayed, for all $m \geq k$.*

Theorem 3 *Consider the max-plus system (13). The recovery matrix R is given as*

$$(R)_{ij} = d_i - d_j - (\mathcal{A}(T^{-1}))^+_{ij},$$

where $\mathcal{A}$ was defined in (12)

4. Dimensioning tunnels for the HSL south

4.1 The generic building block

Figure 10.4 shows the layout of the generic building block (which coincides with a tunnel element in figure 10.2). This layout resembles that of an elementary bottleneck in any railway network. The generic building block contains two parts of double track (i.e. separate tracks for opposite running directions), linked together by a single track. Each track lies between two adjacent (fictive) nodes, numbered from 1 to 10, representing the boundaries of each track. Altogether, the building block thus represents a network of seven tracks, numbered $\tau_1 \ldots, \tau_7$ as shown in 10.4. The geographical location is shown in figure 10.5.

We let an infinite number of trains reside in *reservoirs* in front of the isolated building block. For our model, we require two reservoirs, one for each direction, represented by nodes 1 and 6, respectively. Hence, trains are generated at either reservoir and move to the downstream node (either node 2 or 7) where they enter the building block. From these points, trains run on consecutive tracks according to the arrows. So, trains running from left to right will visit tracks τ_1, τ_3, τ_4 and τ_6 (in consecutive order), whereas trains from right to left will do so via tracks τ_7, τ_5, τ_4 and τ_2. Trains leave the building block at nodes 5 and 10, respectively. Accordingly, nodes 5 and 10 represent the *sinks* of the building block. Observe that τ_4 is used by all trains from both directions. Consequently, this track should only be occupied by one train at a time to prevent deadlock and train collisions. Because of this, τ_4 acts as the bottleneck of our building block.

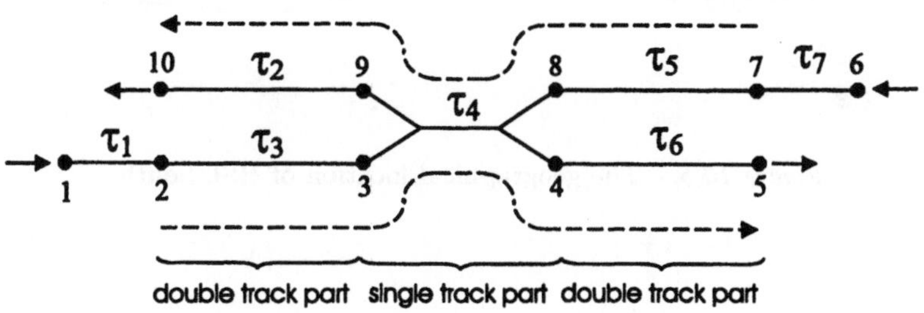

Figure 10.4. Layout of the generic building block

In what follows, we provide conditions for departure times of a train from each node in the generic building block, as a function of departure times of previous trains. To this end, we define $x_i(k)$ as the departure time of the kth train from node i, for $i = 1, \ldots, 10$.

For the internal nodes of the system (i.e. all nodes except the reservoirs and sinks), the departure time of the kth train depends on two conditions: 1. the train must be ready to leave the node; 2. the safety system should have authorized the train to enter the downstream track.

Clearly, condition 1 means that the departure time of the kth train from node i can not be earlier than its arrival time at node i which is equal to the sum of its departure time from the node upstream of i (according to its running direction), say $\pi(i)$, and the travel time on the track between $\pi(i)$ and i. Let $a_{i\,\pi(i)}(k)$ represent the required travel time between $\pi(i)$ and i for the kth train (including all relevant time components for running, dwelling, etcetera). Then, the first condition

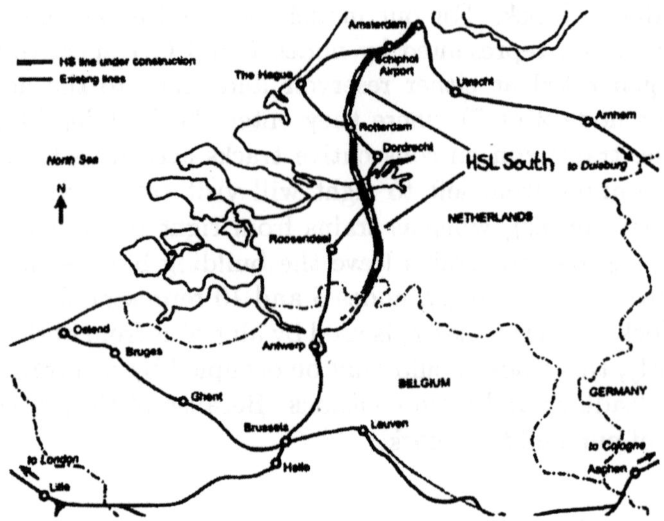

Figure 10.5. The geographical location of HSL South

becomes

$$x_i(k) \geq x_{\pi(i)}(k) + a_{i\,\pi(i)}(k), \quad \text{for } k \geq 1\,, \tag{14}$$

where $a_{i\,j} \geq 0$ if $j = \pi(i)$ and $a_{i\,j} = -\infty$ otherwise for all $1 \leq i \leq 10$ and $k = 1, 2, \ldots$.

Condition 2 is satisfied if the *rear end* of the $(k-1)^{\text{st}}$ train has reached a safe position in front of the kth train. For simplicity, assume that this 'track release' is achieved some time after the front of the $(k-1)^{\text{st}}$ train has departed from the node downstream of i, say $\sigma(i)$. Then, the 'track access condition' reads

$$x_i(k) \geq x_{\sigma(i)}(k-1) + r_{i\,\sigma(i)}(k-1), \quad \text{for } k \geq 2\,, \tag{15}$$

where $r_{j\,i}(k)$ denotes the time elapsing before the track between i and j is released by the kth train (emphasizing that train length must not be neglected). We call this type of time variable the *release time* of the associated track throughout this section. Furthermore, for all $1 \leq i \leq 10$ and $k \geq 1$ we have $r_{i\,j}(k) \geq 0$ if $j = \sigma(i)$ and $r_{i\,j}(k) = -\infty$ otherwise (as explained in the remark below).

Remark. *Equation (15) represents a wide variety of safety regimes. More precisely, we can make the condition suitable for any specific safety principle by choosing the appropriate value for the release time. For* *example, setting* $r_{ji}(k) \geq 0$, *no more than one train will occupy the track between nodes* i *and* j *at any time; whereas setting* $r_{ji}(k) < 0$ *enables track occupation by several trains at the same time (e.g. with a moving block safety system). Theoretically, we may even observe* $r_{i\,\pi(i)}(k) = -\infty$ *on long distance tracks, implying the immediate access of the* k^{th} *train, regardless of the probable presence of other trains on the track. Still, the trains should follow each other at least at absolute braking distance to guarantee safe operation. This explains why the number of trains present on the same track will be finite at any time in practice. See [2] for a detailed overview of principles and properties of safety and signaling systems in Europe.*

At track τ_4, we actually have two track access conditions, namely one condition for consecutive trains running in the same direction (see above) and another condition for trains running in the opposite direction. In the latter case, we have to deal with the order of succession in which trains enter track τ_4. For our analysis, we assume that trains from either direction visit track τ_4 *alternately*, starting with a train running from left to right. This agrees with common railway practice where dispatchers will only deviate from this alternate passing of trains on single track parts of infrastructure to prevent large disturbances.

As for our formal analysis, the fixed order assumption allows us to model the generic building block with the (max,+) semi–ring.

The above fixed order assumption implies that the kth train running from right to left will occupy τ_4 after the kth train in the opposite direction has done so. Accordingly, the kth departure time at node 8 depends on the kth departure from node 4 according to

$$x_8(k) \geq r_{8\,4}(k) + x_4(k)\,,\; k \geq 1\,,$$

where $r_{8\,4}(k)$ is again a release time, like $r_{3\,4}(k)$. However, here $r_{8\,4}(k)$ refers to the release time between *reverse* train movements, whereas $r_{3\,4}(k)$ applies to trains running in the same direction. Furthermore, $r_{8\,4}(k) \geq 0\,,\; k \geq 1$ (indeed, setting $r_{8\,4}(k) < 0$, would imply that track τ_4 is also a double track!).

Conversely, the kth departure from node 3 depends on the $(k-1)^{\text{st}}$ departure from node 9. Hence,

$$x_3(k) \geq r_{3\,9}(k-1) + x_9(k-1)\,,\; k \geq 2\,.$$

Analogously to $r_{8\,4}(k)$, $r_{3\,9}(k) \geq 0$ for all $k \geq 1$.

The departure conditions for the reservoirs are

$$x_1(k) \geq z_1(k) + x_1(k-1)\,,\;\; k \geq 2$$

and

$$x_6(k) \geq z_6(k) + x_6(k-1)\,, \quad k \geq 2\,,$$

where $z_i(k)$ denotes the time between generation of the $(k-1)^{\text{st}}$ train and the kth train.

In addition, trains cannot leave either reservoir unless the respective downstream track is released. Thus, track access conditions given in (15) should also be fulfilled at the reservoirs.

Finally, trains leave the system immediately, once they have arrived at the sinks. Consequently, train departure times from the sinks only depend on the required travel time along the respective upstream tracks, that is, track access conditions can be omitted in this case. Thus, the departure conditions from the sinks are expressed as

$$\begin{aligned} x_5(k) \geq &\ \ x_4(k) + a_{5\,4}(k), \quad & k \geq 1, \\ x_{10}(k) \geq &\ \ x_9(k) + a_{10\,9}(k), \quad & k \geq 1. \end{aligned}$$

Next, we assume that each train departure from the respective nodes takes place immediately after all conditions are satisfied. We then obtain the following difference equations

$$\left\{ \begin{array}{rcl} x_1(k) & = & \max(z_1(k) + x_1(k-1), r_{1\,2}(k-1) + x_2(k-1)) \\ x_2(k) & = & \max(a_{2\,1}(k) + x_1(k), r_{2\,3}(k-1) + x_3(k-1)) \\ x_3(k) & = & \max(a_{3\,2}(k) + x_2(k), r_{3\,4}(k-1) + x_4(k-1), \\ & & r_{3\,9}(k-1) + x_9(k-1)) \\ x_4(k) & = & \max(a_{4\,3}(k) + x_3(k), r_{4\,5}(k-1) + x_5(k-1)) \\ x_5(k) & = & a_{5\,4}(k) + x_4(k) \\ x_6(k) & = & \max(z_6(k) + x_6(k-1), r_{6\,7}(k-1) + x_7(k-1)) \\ x_7(k) & = & \max(a_{7\,6}(k) + x_6(k), r_{7\,8}(k-1) + x_8(k-1)) \\ x_8(k) & = & \max(r_{8\,4}(k) + x_4(k), a_{8\,7}(k) + x_7(k), \\ & & r_{8\,9}(k-1) + x_9(k-1)) \\ x_9(k) & = & \max(a_{9\,8}(k) + x_8(k), r_{9\,10}(k-1) + x_{10}(k-1)) \\ x_{10}(k) & = & a_{10\,9}(k) + x_9(k)\,, \end{array} \right. \tag{16}$$

for $k = 2, 3, \ldots$, with the initial condition

$$x_1(1) = x_6(1) = 0 \text{ and } x_j(1) = -\infty \text{ for } j \neq 1, 6,$$

indicating that the first train from either reservoir can depart immediately (as the system is initially empty). As a result of our fixed train order assumption, we can simplify the above expressions for $x_3(k)$ and $x_8(k)$ as follows. We have assumed that, independent of the release times r_{ij}, the kth train running from left to right always passes track τ_4 before the kth train from right to left. Consequently, the $(k-1)^{\text{th}}$ departure time from node 9 dominates over $(k-1)^{\text{th}}$ departure time from node 4 in the release of τ_4, that is

$$r_{34}(k-1) \otimes x_4(k-1) \leq r_{39}(k-1) \otimes x_9(k-1).$$

Hence, we conclude that the term $r_{34}(k-1) \otimes x_4(k-1)$ is superfluous in the expression for $x_3(k)$ and

$$x_3(k) = (a_{32}(k) \otimes x_2(k)) \oplus (r_{39}(k-1) \otimes x_9(k-1)).$$

Likewise, we have

$$r_{89}(k-1) + x_9(k-1) \leq r_{84}(k) + x_4(k).$$

Consequently, we may leave out the term $r_{89}(k-1) + x_9(k-1)$ in the expression for $x_8(k)$, which in turn results in

$$x_8(k) = \max(r_{84}(k) + x_4(k)\,,\, a_{87}(k) + x_7(k)).$$

In order to determine whether the two–tube tunnel layout for HSL South offers sufficient capacity to process the intended hourly train amounts, we proceed as follows. Since each tunnel behaves as a single track part, we may split up the line into three tunnel elements, each of which can be considered as a generic building block as defined above. We obtain for each of the three tunnel elements a (max,+) model $x^{(i)}(k+1) \; = \; A^{(i)}(k) \otimes x^{(i)}(k)$, $i = 1, 2, 3$.

Capacity in isolation is consecutively determined by means of the deterministic and probabilistic (max,+) approach. In this paper only the deterministic case will be dealth with. For the stochastic approach, see, as for more extensive discussions on the deterministic case, [14]. (In the stochastic situation the travelling times have a stochastic component on top of a deterministic part; the stochastic part is described by a combination of a uniform distribution on a fixed small time interval and an exponential distribution.) By using both methods, we gain insight to what extent incorporation of stochastic time variables affects the outcomes. Below, we describe how the deterministic method is applied to HSL South.

For deterministic time variables, we let the intended number of trains to be processed, denoted by N_{intended}, vary from 6 to 20 per hour in both directions. The corresponding average inter–arrival time at either source thus becomes

$$z_1 = z_6 = T_{\text{interval}} = \frac{60}{N_{\text{intended}}}.$$

As a result, we get $3 \leq T_{\text{interval}} \leq 10$ minutes. For each value of T_{interval}, we evaluate the (max, +) eigenvalue, $\tau^{(i)}(T_{\text{interval}})$, of the corresponding matrix $A^{(i)}$. Next, we check if tunnel element i can process the intended number of trains without waiting. This is only possible if

$$\tau^{(i)}(T_{\text{interval}}) \le T_{\text{interval}} \,.$$

Subsequently, we take

$$\tau(T_{\text{interval}}) = \max\{\, \tau^{(i)}(T_{\text{interval}}) : 1 \le i \le 3 \,\}$$

as an estimator for the asymptotic inter–departure time of the entire line. In fact, $\tau(T_{\text{interval}})$ is a lower bound for the true asymptotic inter–departure time, since we ignore dependencies between the tunnel elements.

Another approach to consider capacity is to combine the three separate models into one large model for all three tunnel elements together. The reason that that has not (yet) been done is that the distances between the three tunnel elements are rather large such that trains can brake and come to a stillstand in between.

4.2 Technical data

Below we discuss the points of departure for determining travel times and release times as presented in the next subsection. These assumptions concern the characteristics of the HSL South track layout, the underlying safety system and deployed rolling stock.

HSL South has a total length of approximately 100 kilometers. The maximum permitted speed is 300 km/hour which is only achieved nearby and inside the Groene Hart tunnel, and to the south of the Dordtsche Kil tunnel, towards the Belgian border. Anywhere else, more restrictive speed indications apply, in particular at the other two tunnels under consideration (to guarantee that trains will not accelerate too much while descending) and near curves (to prevent derailments). Table 10.1 shows the maximum permitted speeds for TGV trains (to be shortly introduced) at the entrances and exits of the three tunnels. All running times are based on an accurate description of the HSL South track layout, including locations and values for curves, slopes and (rolling stock dependent) local speed restrictions.

The line is assumed to be equipped with an ERTMS/ETCS level 3 moving block safety system. The interested reader could refer to [8] for further details about this generation safety system. With ERTMS/ETCS level 3, trains may theoretically follow each other at absolute breaking distance, that is the distance required by the following train to break down from actual speed to stand still. Meanwhile, track access to a train is given via a so–called Movement Authority (MA) comprising a part of track that is exclusively reserved for a particular train.

Table 10.1. **Speed restrictions at HSL South tunnels for TGV trains**

tunnel	direction	maximum speed at entrance [km/h]	maximum speed at exit [km/h]
Groene	North–South	300	300
Hart	South–North	300	300
Oude	North–South	200	220
Maas	South–North	250	250
Dordtsche	North–South	280	280
Kil	South–North	300	300

All time variables are determined for one specific type of rolling stock, namely TGV2 (i.e. two coupled TGV trains). Table 10.2 contains the main characteristics of this train type. Of all types of rolling stock that will actually run on HSL South, TGV2 trains require the smallest travel times. Consequently, the results in practice will be worse than the ones obtained if only TGV2 trains were deployed.

Table 10.2. **Main characteristics for TGV2 trains**

train length	400 m	
maximum speed	300 km/h	
emergency break	1.0 m/s^2	if speed > 215 km/h
	1.3 m/s^2	if speed $\leq$ 215 km/h

Finally, all trains are supposed to dwell for two minutes at Rotterdam Central Station. This intermediate stop is located between the Groene Hart tunnel and Oude Maas tunnel.

4.3 Results

We start with discussing the outcomes based on deterministic (max,+); see figure 10.6. As we let the hourly train amounts increase, the eigenvalue (upper black graph) decreases until 11 or more trains are processed per hour and in each direction. From this point on, the observed inter–departure time has become constant (equal to 5 minutes and 42 seconds as indicated by dashed line) and thus independent of the regular inter–arrival times at either source (grey graph).

We conclude that the capacity of the line is equal to 10 trains per hour per direction. Consequently, intended traffic loads between 11 and 18 trains per hour per direction are infeasible. Furthermore, although not

shown here, the results point out the Oude Maas tunnel as the bottleneck for HSL South.

Next, define time margin $\Delta(T_{\text{interval}})$ according to

$$\Delta(T_{\text{interval}}) = T_{\text{interval}} - \tau(T_{\text{interval}}),$$

which enables the system to compensate delays.

Examining the remaining (seemingly feasible) alternatives of 8–10 trains per hour, we conclude that the corresponding margins (lower black graph in figure 10.6) are rather small (less than two minutes) for running over a distance of almost 100 kilometers. Consequently, the line is very sensitive to delays which of course is undesirable. In fact, we should expect this delay sensitivity to be even bigger in practice, since our values for τ are based on rather optimistic assumptions. In other words, it is unlikely that these alternative can actually be performed in real–life either without affecting the quality of service, in particular punctuality.

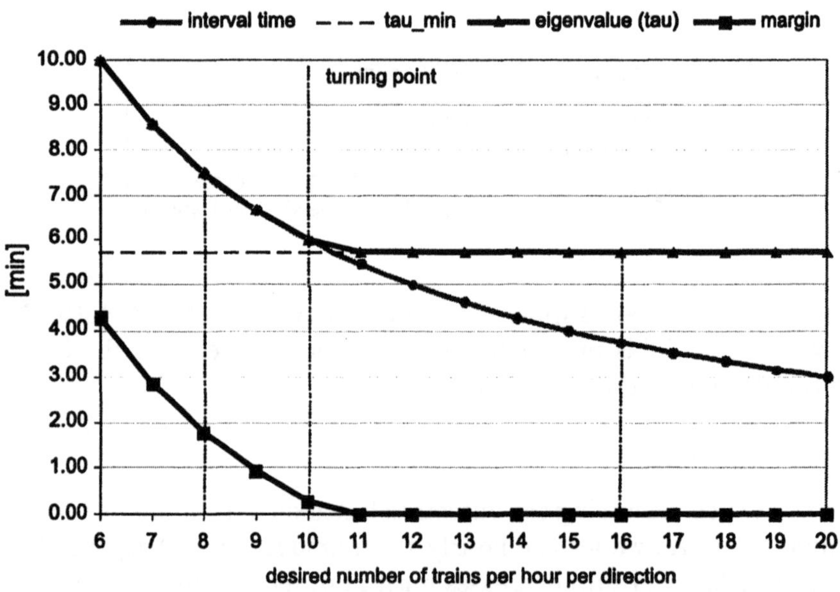

Figure 10.6. Long–run inter–departure times at HSL South for varying train amounts

Alltogether, we conclude that the proposed track layout with two–tube tunnels in combination with a tunnel regime provides insufficient capacity. Hence, either extra tunnel tubes are needed (possibly only for the Oude Maas tunnel, our main bottleneck) or the proposed tunnel access regime has to be abandoned again. In the latter case, alternative measures have to be taken to ensure that passengers can still safely escape to the opposite tunnel tube in case of an emergency. Currently, studies on both options are being carried out. More extensive discussions can be found in [14].

References

[1] F. Baccelli, G. Cohen, G.J. Olsder, and J.P. Quadrat. *Synchronization and Linearity.* Wiley, 1992.

[2] C. Baley (general editor). *European Railway Signalling.* Institution of Railway Signalling Engineers, A&C Black, London, 1995.

[3] J.G. Braker. *Algorithms and applications in timed discrete event systems.* PhD thesis, Delft university of Technology, 1993.

[4] J.G. Braker and G.J. Olsder. The power algorithm in max-algebra. *Linear Algebra and its Applications*, 182:67–89, 1993.

[5] C.G. Cassandras. *Discrete Event Systems, Modelling and Performance Analysis.* Irwin, Boston, 1993.

[6] C.G. Cassandras and S. Lafortune and G.J. Olsder. Introduction to the Modelling, Control and Optimization of Discrete Event Systems. *pp. 217-291 in: Alberto Isidori (ed.), Trends in Control, a European Perspective* Springer Verlag, 1995.

[7] G. Cohen, D. Dubois, J.P. Quadrat, and M. Viot. A linear system-theoretic view of discrete event processes and its use for performance evaluation in manufacturing. *IEEE Transactions on Automatic Control*, AC-30:210–220, 1985.

[8] ECSAG/UNISIG. *ERTMS/ETCS Functional Requirements Specification RFS.* Version 4.29, final, Brussels, 1999.

[9] R.-J. van Egmond and G.J. Olsder. The (max,+) algebra applied to synchronization of traffic light processes *Proceedings WODES, Cagliari, August 1998, pp. 451-456.*

[10] R. Goverde and G. Soto Y Koelemeijer. *Performance Evaluation of Periodic Railway Timetables: Theory and Algorithms.* The Netherlands Trail Research School, Delft University of Technology, 2000.

[11] J. Gunawardena. *Idempotency.* Cambridge University Press, 1998.

[12] Y.C. Ho and X. Cao. *Perturbation Analysis of Discrte Event Dynamic Systems.* Kluwer, Boston, 1991.

[13] Y.C. Ho and C.G. Cassandras. A new approach to the analysis of discrete event dynamic systems. *Automatica*, 19:149–167, 1983.

[14] A.F. de Kort. *A unified approach to capacity assessment for railway infrastructure.* Ph.D. thesis, Delft University of Technology, forthcoming, 2002.

[15] G.J. Olsder. Eigenvalues of dynamic min-max systems. *Journal of Discrete Event Dynamic Systems*, 1:177–207, 1991.

[16] G.J. Olsder. Synchronized continuous flow systems. In S. Balemi, P. Kozàk, and R. Smedinga, editors, *Discrete Event Systems; Modelling and Control*, pages 113–124. Birkhäuser, Basel, 1993.

[17] P.J. Ramadge and W.M. Wonham. Supervisory control of a class of discrete event processes. *SIAM J. Control and Optimization*, 25:206–230, 1987.

[18] Subiono. *On classes of min-max-plus systems and their applications.* Ph.D. thesis, Delft University of Technology, 2000.

Chapter 11

A SENSITIVITY VIEW OF MARKOV DECISION PROCESSES AND REINFORCEMENT LEARNING

Xi-Ren Cao*

Department of Electrical and Electronic Engineering
The Hong Kong University of Science and Technology
Clear Water Bay, Kowloon, Hong Kong

Abstract The goals of perturbation analysis (PA), Markov decision processes (MDPs), and reinforcement learning (RL) are common: to make decisions to improve the system performance based on the information obtained by analyzing the current system behavior. In this paper, we study the relations among these closely related fields. We show that MDP solutions can be derived naturally from performance sensitivity analysis provided by PA. Performance potential plays an important role in both PA and MDPs; it also offers a clear intuitive interpretation for many results. Reinforcement learning, TD(λ), neuro-dynamic programming, etc, are efficient ways of estimating the performance potentials and related quantities based on sample paths. This new view of PA, MDPs and RL leads to the gradient-based policy iteration method that can be applied to some nonstandard optimization problems such as those with correlated actions. Sample path-based approaches are also discussed.

Key words: Potentials, Poisson equations, gradient-based policy iteration, Q-learning, TD(λ)

---

*Supported in part by a grant from Hong Kong UGC. Tel: (852) 2358-7048 Fax: (852) 2358-1485 Email: eecao@ee.ust.hk

1. Introduction

Perturbation analysis (PA) [4, 8, 13, 17] provides performance sensitivities for a discrete event dynamic system (DEDS) by analyzing the dynamical behavior of a single sample path of the DEDS. Performance optimization can be achieved by combining PA with stochastic approximation methods. Markov decision process (MDP) [1, 21, 22, 23] is a general model for performance optimization of DEDSs. Policy iteration, the basic approach in MDPs, can be implemented based on sample paths. The goal of reinforcement learning (RL) [1, 3, 24, 25] is to learn how to make decisions to improve a system's performance by observing its behavior. In this paper, we study the relations among these closely related fields. We show that MDP solutions can be derived naturally from performance sensitivity analysis provided by PA, and that reinforcement learning, TD(λ), neuro-dynamic programming, etc, are sample-path-based efficient ways of estimating the performance potentials and related quantities, which are crucial elements in PA, MDPs and other optimization approaches. This sensitivity-based view of MDPs leads to a new approach, the gradient-based policy iteration, to MDPs with correlated actions at different states.

We first briefly review the basic principles. We show that the effect of a small change in a system parameter on the system performance equals the sum of the effects of all the perturbations (small or large) induced by this small change in the parameter. the effect of a perturbation in a DEDS on its long term performance can be measured quantitatively. From this basic principle, the performance sensitivities can be derived [4, 8].

The above principle can be applied to Markov processes to obtain performance sensitivity [8]. Two types of sensitivities are studied: sensitivities with respect to discrete and continuous parameters. In the discrete case, the difference of the performance of the Markov processes under two different policies is found. In the continuous case, the policies are randomized, with a parameter δ indicating the probability of applying one policy and $1-\delta$ the probability of the other, the derivative of the performance with respect to δ is found; this is the derivative along the direction from one policy to the other in the policy space. With the discrete sensitivity formula, the standard policy iteration optimization algorithm can be very easily derived. Compared with the continuous sensitivity formula, it is clear that at each step the policy iteration algorithm simply chooses the "steepest" direction (with the largest absolute value of the directional derivative) to go for its next policy [5].

The concept of *performance potentials* plays a significant role in sensitivity analysis of Markov processes and MDPs. Just as the potential energy in physics, only the relative values (i.e., the differences) of the performance potentials at different states are meaningful, i.e., the potentials are determined only up to an additive constant. Realization factor is defined as the difference between the potentials at two states i and j, $g(i) - g(j)$, which measures the effort of a jump from state j to state i on the long term (average or discounted) performance. This offers an intuitive explanation for performance sensitivity and MDPs (in view of discrete sensitivity). The performance potential at state i can be approximated by the mean sum of the performance at the first N states on a sample path starting from state i, and hence it can be estimated online. A number of other single sample path-based estimation algorithms for potentials have been derived.

With the potentials estimated, performance derivatives (i.e., PA) can be obtained and policy iteration (i.e., MDPs) can be implemented based on a single sample path. The single sample path-based approach is practically important because it can be applied online to real engineering systems; in most cases, it does not require the system parameters to be completely known (see the example in Section 5). There are two ways to achieve the optimal performance with this approach: by perturbation analysis using performance derivatives, or by policy iteration, both can be implemented online. Stochastic approximation methods have to be used in these two cases to improve the convergence speeds and to reduce stochastic errors. For details, see Section 5 and [12, 19].)

Inspired by the potential-based sensitivity view of MDPs and the single sample path-based approach, a number of new research topics emerge. First, we observe that policy iteration can be implemented by using the performance derivatives. This leads to a new approach, the gradient-based policy iteration, to MDPs with correlated actions at different states. This gradient-based policy iteration can be applied to problems such as MDPs with hidden state components and the distributed control of MDPs, the details are reported in a forthcoming paper [10]. The other topics include the time aggregation of MDPs [9], which was first used in [26] for perturbation analysis. When the number of controllable states is small, this approach may save computations, storage spaces, and/or the number of transitions in sample-path-based approaches.

We first study MDPs with long term average costs. In Section 2, we start with a brief introduction to PA. In Section 2.1, we apply the basic ideas of PA to Markov chains to obtain the performance gradients. *Performance potentials* are introduced here. In Section 2.2, the results are extended to discrete performance sensitivities, i.e., the difference of the performance measures of any two policies. In Section 3, we show that policy iteration in MDP can be easily derived from the discrete sensitivity formula. Gradient-based policy iteration is discussed in Section 3.2. In Section 4, we extend the results to MDPs with discounted performance criteria. All the results for long term average MDPs can be viewed as special cases when the discount factor α goes to one. Thus, potential theory provides a unified framework to MDPs with both long term average and discounted performance criteria. In Section 5, some sample path-based estimation algorithms for potentials are derived. An example is given to illustrate the possible advantages of applying the sample-path-based optimization approach. In Section 6 we give a brief review of Q-learning, TD(λ), and neuro-dynamic programming, which provide efficient ways in estimating the potentials and Q-factors, the latter are used in place of potentials when the system structure is completely unknown. In Section 7, we give a brief summary as a conclusion.

The research of perturbation analysis initiated with Prof. Ho's insightful and creative works in late 70's and early 80's [15, 16, 17, 18]. This paper aims at positioning this research area in the global picture of system optimization. We dedicate this survey on the recent development in this fascinating area to Prof. Y. C. Ho on the occasion of his retirement from teaching.

2. Perturbation Analysis of Markov Chains

The basic principle for PA can be explained by using the concept of *perturbation realization.* A small change in a system parameter (such as the mean service time of a server) induces a series of (small or large) changes on a sample path; each change is called a *perturbation.* The average effect of each perturbation on the system performance can be precisely measured. The total effect of the small change in the parameter on the system performance can then be calculated by adding together the average effects of all the perturbations induced by the parameter change. The sensitivity of the performance with respect to the parameter can then be determined. The idea has been successfully applied to performance sensitivity analysis of queueing networks [18] as well as Markov chains [8, 5] .

2.1 Performance Sensitivity of Markov Chains: the Continuous Case

Consider an irreducible and aperiodic Markov chain $\mathbf{X} = \{X_n : n \geq 0\}$ on a finite state space $\mathcal{S} = \{1, 2, \cdots, M\}$ with transition probability matrix $P = [p(i,j)] \in [0,1]^{M\times M}$. Let $\pi = (\pi_1, \ldots, \pi_M)$ be the vector representing its steady-state probabilities, and $f = (f_1, f_2, \cdots, f_M)^T$ be the performance vector, where "T" represents transpose. We have $Pe = e$, where $e = (1, 1, \cdots, 1)^T$ is an M-dimensional vector whose all components equal 1, and $\pi = \pi P$. The performance measure is the long term average defined as

$$\eta = E_\pi(f) = \sum_{i=1}^{M} \pi_i f_i = \pi f = \lim_{L\to\infty} \frac{1}{L}\sum_{l=0}^{L-1} f(X_l) = \lim_{L\to\infty}\frac{F_L}{L}, \quad (1)$$

where

$$F_L = \sum_{l=0}^{L-1} f(X_l).$$

Let P' be another irreducible transition probability matrix on the same state space. Suppose P changes to $P(\delta) = P + \delta Q = \delta P' + (1-\delta)P$, with $\delta > 0$, $Q = P' - P = [q(i,j)]$. We have $Qe = 0$. The performance measure will change to $\eta(\delta) = \eta + \Delta\eta$. The derivative of η in the direction of Q is defined as $\frac{d\eta}{d\delta} = \lim_{\delta\to 0}\frac{\Delta\eta}{\delta}$.

In this system, a perturbation means that the system is perturbed from one state i to another state j. For example, consider the case where $q(k,i) = -\frac{1}{2}$, $q(k,j) = \frac{1}{2}$, and $q(k,l) = 0$ for all $l \neq i, j$. Suppose that in the original sample path the system is in state k and jumps to state i, then in the perturbed path it may jump to state j instead. Thus, we study two independent Markov chains $\mathbf{X} = \{X_n; n \geq 0\}$ and $\mathbf{X}' = \{X'_n; n \geq 0\}$ with $X_0 = i$ and $X'_0 = j$; both of them have the same transition matrix P. The *realization factor* is defined as [8]:

$$d(i,j) = \lim_{L\to\infty} E\left[\sum_{l=0}^{L-1}\left(f(X'_l) - f(X_l)\right)\middle| X_0 = i,\ X'_0 = j\right], \quad (2)$$
$$i, j = 1, \ldots, M\,.$$

Thus, $d(i,j)$ represents the average effect of a jump from i to j on F_L in (1).

If P is irreducible, then with probability one the two sample paths of $\mathbf{X}$ and $\mathbf{X}'$ will merge together. That is, there is a random number L^* such that $X'_{L^*} = X_{L^*}$. Therefore, from the Markov property, (2) becomes

$$d(i,j) = E\left[\sum_{l=0}^{L^*-1}\left(f(X'_l) - f(X_l)\right)\middle| X_0 = i,\ X'_0 = j\right], \quad i, j = 1, \ldots, M. \quad (3)$$

The matrix $D \in \mathcal{R}^{M\times M}$, with $d(i,j)$ as its (i,j)th element, is called a *realization matrix.* From (3), we can prove that D satisfies the Lyapunov equation [8]

$$D - PDP^T = F, \qquad (4)$$

where $F = fe^T - ef^T$.

(2) indicates that every visit to state i contributes to F_L on the average by the amount of

$$g(i) \approx E[\sum_{l=0}^{L-1} f(X_l)|X_0 = i], \qquad (5)$$

and only the difference between different $g(i)$s are important. Now we consider a sample path consisting of L transitions. Among these transitions, on the average there are $L\pi_i$ transitions at which the system are at state i. After being at state i, the system jumps to state j on the average $L\pi_i p(i,j)$ times. If the transition probability matrix P changes to $P(\delta) = P+\delta Q$, then the change in the number of visits to state j after being at state i is $L\pi_i q(i,j)\delta = L\pi_i[p'(i,j) - p(i,j)]\delta$. This contributes a change of $\{L\pi_i[p'(i,j) - p(i,j)]\delta\}g(i)$ to F_L. Thus, the total change in F_L due to the change of P to $P(\delta)$ is

$$\Delta F_L = \sum_{i,j=1}^{M} L\pi_i[p'(i,j) - p(i,j)]\delta g(i) = \pi[P' - P]g\delta L = \pi Q g \delta L,$$

where $g = (g(1), \cdots, g(M))^T$. Finally, we have

$$\frac{d\eta}{d\delta} = \lim_{\delta\to 0} \frac{1}{\delta}\Delta F_L L = \pi Q g = \pi(P' - P)g. \qquad (6)$$

(5) is only an approximation. When $L \to \infty$, this expression is unbounded. However, because $Qe = 0$, we have $Q(g+ce) = Qg$ for any real constant c. Thus, the sensitivity equation (6) holds when g is replaced by $g + ce$. (This confirms the fact that g is only important up to an additive constant.) Therefore, we may add a constant $-L\eta$ to (5) and obtain

$$g(i) \approx E\{\sum_{l=0}^{L-1}[f(X_l) - \eta]|X_0 = i\}.$$

Letting $L \to \infty$, we obtain the formal definition of the *performance potential* at state i:

$$g(i) = \lim_{L\to\infty} E\{\sum_{l=0}^{L-1}[f(X_l) - \eta]|X_0 = i\}, \qquad (7)$$

which can be proved to be finite for ergodic chains. Again, the above discussion only provides an intuitive explanation; for a rigorous proof, see [8].

We have $d(i,j) = g(i) - g(j)$ and $D = ge^T - eg^T$. From this, we have

$$\frac{d\eta}{d\delta} = \pi Q D^T \pi^T. \tag{8}$$

From (7), it is easy to prove that the potential vector g satisfies the Poisson equation

$$(I - P + e\pi)g = f. \tag{9}$$

2.2 Performance Sensitivity of Markov Chains: the Discrete Case

In sensitivity analysis with discrete parameters, we wish to obtain the difference of the performance measures of two Markov chains with transition probability matrices P and P', respectively, both are ergodic on the same state space. We use prime " $'$ " to denote the values associated with P'.

First, multiplying both side of the Poisson equation with π on the left, we get

$$\pi g = \pi f = \eta. \tag{10}$$

Next, multiplying both side of the Poisson equation with π' on the left yields

$$\pi' Q g = \pi'(P' - P)g = \pi'(I - P)g = \pi' f - \pi g = \pi' f - \eta.$$

That is,

$$\eta' - \eta = \pi' Q g. \tag{11}$$

(11) can be viewed as the discrete version of the performance sensitivity. Note that the discrete sensitivity (11) can be obtained from its continuous countpart (6) by replacing π with π'. This feature serves as the basis for the gradient-base policy iteration for Markov decision processes (MDPs) proposed later in Section 3.2.

For more general cases, we assume that the performance function also changes from f to f'. Let $h = f' - f$. It is easy to check that

$$\eta' - \eta = \pi'(Qg + h). \tag{12}$$

For continuous sensitivity, we set $f(\delta) = f + \delta h$. Together with $P(\delta) = P + \delta Q$, we have

$$\frac{d\eta}{d\delta} = \lim_{\delta \to 0} \frac{1}{\delta} \Delta F_L L = \pi(Qg + h). \tag{13}$$

3. Markov Decision Processes

3.1 Policy Iteration

To know the exact value of the performance difference from (12), one needs to know π'. On the otherhand, if π' is known, one can get η' directly by $\pi' f$. In addition, it is impossible to calculate π' for all the policice since the policy space is usually very large. Fortunately, since $\pi' > 0$ (componentwisely), we can always determine which action is better at each state by using (12). Thus, we can perform policy iteration without knowing the exact value of the performance difference. This leads to the following discussion.

For two M-dimensional vectors a and b, we define $a = b$ if $a(i) = b(i)$ for all $i = 1, 2 \cdots, M$; $a \le b$ if $a(i) < b(i)$ or $a(i) = b(i)$ for all $i = 1, 2 \cdots, M$; $a < b$ if $a(i) < b(i)$ for all $i = 1, 2 \cdots, M$; and $a \preceq b$ if $a(i) < b(i)$ for at least one i, and $a(j) = b(j)$ for other components. The relation $\le$ includes $=$, $\preceq$, and $<$. Similar definitions are used for the relations $>$, $\succeq$, and $\ge$.

Next, we note that $\pi'(i) > 0$ for all $i = 1, 2, \cdots, M$. Thus, from (12), we know that if $Qg + h = (P' - P)g + (f' - f) \succeq 0$ then $\eta' - \eta > 0$. From (12) and the fact $\pi' > 0$, the proof of the following lemma is straightforward.

Lemma 7 If $Pg + f \preceq P'g + f'$, then $\eta < \eta'$.

It is interesting to note that in the lemma, we use only the potentials with one Markov chain, i.e., g. Thus, to compare the performance measures under two policies, only the potentials with one policy is needed.

In an MDP, at any transition instant $n \ge 0$ of a Markov chain $\mathbf{X} = \{X_n, n \ge 0\}$, an action is chosen from an action space $\mathcal{A}$ and is applied to the Markov chain. We assume that the number of actions is finite, and we only consider stationary policies. The actions that are available for $i \in \mathcal{S}$ form a nonempty subset $A(i) \subseteq \mathcal{A}$. A stationary policy is a mapping $\mathcal{L} : \mathcal{S} \to \mathcal{A}$, i.e., for any state i, $\mathcal{L}$ specifies an action $\mathcal{L}(i) \in A(i)$. Let $\mathcal{E}$ be the policy space. If action α is taken at state i, then the state transition probabilities at state i are denoted as $p^{\alpha}(i, j)$, $j = 1, 2, \cdots, M$. With a policy $\mathcal{L}$, the Markov process evolves according to the transition matrix $P^{\mathcal{L}} = [p^{\mathcal{L}(i)}(i, j)]_{i=1}^{M}|_{j=1}^{M}$. We use the superscript $*^{\mathcal{L}}$ to denote the quantities associated with policy $\mathcal{L}$.

For most stable systems in real world applications, the Markov chains under any policy are recurrent. For example, in any stable communication systems, from any state it is always possible to reach the null state where there are no packets in the system. Therefore, in this paper we assume that the Markov chain is recurrent under any policy. In addition, for simplicity we assume that it is aperiodic; although all the results in this paper hold for periodic chains if we replace the steady state value by the corresponding time average value [23]. Therefore, the Markov chains considered are ergodic [11]. This corresponds to the problems classified in [23] as the recurrent case.

The steady-state probabilities corresponding to policy $\mathcal{L}$ is denoted as a vector $\pi^{\mathcal{L}} = (\pi^{\mathcal{L}}(1), \cdots, \pi^{\mathcal{L}}(M))$. Suppose that at each stage with state i and control action $\alpha \in A(i)$, a cost $f(i, \alpha) = f(i, \mathcal{L}(i))$ is incurred. The long term expected value of the average cost per stage corresponding to policy $\mathcal{L}$ is then

$$\eta^{\mathcal{L}} = \lim_{L \to \infty} \frac{1}{L} E\{\sum_{l=0}^{L-1} f[X_l, \mathcal{L}(X_l)]\},$$

For ergodic chains, the above limit exists and does not depend on the initial state. Our objective is to minimize this average cost per stage over the policy space $\mathcal{S}$, i.e., to obtain $\min_{\mathcal{L} \in \mathcal{E}} \eta^{\mathcal{L}}$.

Define $f^{\mathcal{L}} = (f[1, \mathcal{L}(1)], \cdots, f[M, \mathcal{L}(M)])^T$. (9) and (10) becomes

$$(I - P^{\mathcal{L}} + e\pi^{\mathcal{L}})g^{\mathcal{L}} = f^{\mathcal{L}}, \tag{14}$$

and

$$\pi^{\mathcal{L}} g^{\mathcal{L}} = \pi^{\mathcal{L}} f^{\mathcal{L}}.$$

The following optimality theorem follows almost immediately from Lemma 7, which is derived directly from the sensitivity formula (12).

Theorem 1 A policy $\mathcal{L}$ is optimal if and only if

$$P^{\mathcal{L}} g^{\mathcal{L}} + f^{\mathcal{L}} \leq P^{\mathcal{L}'} g^{\mathcal{L}} + f^{\mathcal{L}'} \tag{15}$$

for all $\mathcal{L}' \in \mathcal{E}$.

The optimality condition (15) is, of course, equivalent to the other conditions in the literature. To see this, we rewrite (14) in the following form:

$$\eta^{\mathcal{L}} e + g^{\mathcal{L}} = f^{\mathcal{L}} + P^{\mathcal{L}} g^{\mathcal{L}}. \tag{16}$$

Then Theorem 1 becomes: A policy $\mathcal{L}$ is optimal if and only if

$$\eta^{\mathcal{L}} e + g^{\mathcal{L}} = \min_{\mathcal{L}' \in \mathcal{E}} \{P^{\mathcal{L}'} g^{\mathcal{L}} + f^{\mathcal{L}'}\}. \tag{17}$$

The minimum is taken for every component of the vector. (17) is the optimality equation, or the Bellman equation. From (17), $g^{\mathcal{L}}$ is equivalent to the "differential" or "relative cost vector" in [1], or the "bias" in [23]. In our approach, g is directly related to the long term expected performance, and many results, such as the existence and the uniqueness of the solution to Equation (16), the optimality condition (17), and the convergence of the optimal algorithms, become almost obvious. In addition, as shown in the Section 5, $g^{\mathcal{L}}$ can be easily estimated based on a single sample path. This is an important feature which allows optimization algorithms to be implemented online for real world systems.

Policy iteration algorithms for determining the optimal policy can be easily developed by combining Lemma 7 and Theorem 1. Roughly speaking, at the kth step with policy $\mathcal{L}_k$, we set the policy for the next step (the $(k+1)$th step) as $\mathcal{L}_{k+1} = arg\{\min[P^{\mathcal{L}}g^{\mathcal{L}_k} + f^{\mathcal{L}}]\}$, with $g^{\mathcal{L}_k}$ being the solution to the Poisson equation for $P^{\mathcal{L}_k}$. Lemma 7 implies that performance usually improves at each iteration. Theorem 1 shows that the minimum is reached when no performance improvement can be achieved. We shall not state the details here because they are standard.

Finally, since $P^{\mathcal{L}_k}$ and $f^{\mathcal{L}_k}$ are fixed, we note that $(P^{\mathcal{L}} - P^{\mathcal{L}_k})g^{\mathcal{L}_k} + (f^{\mathcal{L}} - f^{\mathcal{L}_k})$ takes the minimal value (component wise) at $P^{\mathcal{L}_{k+1}}$. Thus, because $\pi^{\mathcal{L}_k} > 0$, from (13) the performance derivative also reaches minimum at $P^{\mathcal{L}_{k+1}}$ (i.e., the largest absolute value, since the derivative is negative). Therefore, we conclude that policy iteration in fact chooses the steepest direction to go in the policy space.

3.2 Gradient-Based Policy Iteration

Because in (11) π' is unknown, one has to compare the effect of the ations on the performance state by state. Therefore, the policy iteration described in Section 3.1 can be used only for problems in which actions at difference state are independent. That is, the approach does not apply to systems in which actions at different states may be related. On the otherhand, the performance gradient (6) or (13) does not depend on π', and all the information except P' can be obtained from the current policy P. Thus, principally one can solve for π and determine the smallest πQg to implement policy iteration. This is the gradient-based policy iteration method, which can be used to solve the optimization problem for systems where the actions at different states are correlated.

For example, consider a simple system where the actions at states 1 ans 2 must be the same. In the gradient-based policy iteration, we simply choose the action that minimizes $\pi(1)\sum_{j=1}^{M} p^{\alpha}(1,j)g(j) + \pi(2)\sum_{j=1}^{M} p^{\alpha}(2,j)g(j)$ as the action for both states 1 and 2 in the next iteration. More details are reported in a recent paper [10].

4. Problems with Discounted Performance Criteria

In this section, we extend the above results to MDPs with discounted performance criteria [7]. We shall see, with performance potentials, both the average and the discounted cost problems can be solved with the same framework, and the average cost problem can be treated as a special case when the discount factor α goes to 1.

Let $f(i)$, $i \in \mathcal{S}$, be a performance function and α, $0 < \alpha \leq 1$, be a discount factor. For $0 < \alpha < 1$, the performance cost is defined as a column vector $\eta_\alpha = (\eta_\alpha(1), \eta_\alpha(2), \cdots, \eta_\alpha(M))^T$ with

$$\eta_\alpha(i) = (1-\alpha)E\{\sum_{n=0}^{\infty} \alpha^n f(X_n)|X_0 = i\}. \tag{18}$$

The factor $(1-\alpha)$ in (18) is used to obtain the continuity of η_α at $\alpha = 1$. In fact, we define

$$\eta_1 = \lim_{\alpha \to 1^-} \eta_\alpha. \tag{19}$$

It is proved in (20) and (24) that the above limit exists.

Lemma 8 $\eta_1 = \eta e$ with η being the average-cost performance:

$$\eta = \pi f = \lim_{N \to \infty} \{E[\frac{1}{N} \sum_{n=0}^{N-1} f(X_n)]\}.$$

Proof. In a matrix form, we have

$$\eta_\alpha = (1-\alpha) \sum_{n=0}^{\infty} \alpha^n P^n f = (1-\alpha)(I-\alpha P)^{-1} f, \qquad 0 < \alpha < 1. \tag{20}$$

The second equation in (20) holds because for $0 < \alpha < 1$ all the eigenvalues of αP are within the unit circle [2]. Next, it is easy to verify

$$(I-\alpha P)^{-1} = (I - \alpha P + \alpha e\pi)^{-1} + \frac{\alpha}{1-\alpha} e\pi, \quad \alpha < 1. \tag{21}$$

Thus,

$$\lim_{\alpha \to 1-} (1-\alpha)(I-\alpha P)^{-1} = e\pi. \tag{22}$$

The lemma then follows directly from (20) and (22). □

Similar to (9), the *α-Poisson equation* is defined as

$$(I - \alpha P + \alpha e\pi) g_\alpha = f. \tag{23}$$

g_α is called the *α-potential*, which is the same as what is defined in the classical potential theory for $0 < \alpha < 1$ (see (30) and [11]). Moreover, (21) and (20) leads to

$$\eta_\alpha = (1-\alpha)g_\alpha + \alpha\eta e. \tag{24}$$

It becomes obvious that (19) does converge.

When $\alpha = 1$, it is the standard Poisson equation. From (23), we have

$$\begin{aligned} g_\alpha &= (I - \alpha P + \alpha e\pi)^{-1} f \\ &= \{\sum_{n=0}^{\infty} \alpha^n (P - e\pi)^n\} f \\ &= \{I + [\sum_{n=1}^{\infty} \alpha^n (P^n - e\pi)]\} f, \qquad 0 < \alpha \le 1. \end{aligned}$$

In particular,

$$g_1 = (I - P + e\pi)^{-1} f = \{I + \sum_{n=1}^{\infty} (P^n - e\pi)\} f.$$

This is the same as the performance potentials defined for the average cost MDPs. We have

$$\lim_{\alpha \to 1-} g_\alpha = g_1,$$

$$\pi g_\alpha = \pi f.$$

Now suppose that P changes to P', then π and η will change to π' and η', respectively. Let $Q = P' - P$. From (20), we get

$$\begin{aligned} \eta'_\alpha - \eta_\alpha &= \alpha(P'\eta'_\alpha - P\eta_\alpha) \\ &= \alpha(P' - P)\eta_\alpha + \alpha P'(\eta'_\alpha - \eta_\alpha). \end{aligned}$$

This leads to

$$\eta'_\alpha - \eta_\alpha = \alpha(I - \alpha P')^{-1} Q \eta_\alpha.$$

Substituting (24) into the right-hand side of the above equation and noting that $Qe = 0$, we obtain

$$\eta'_\alpha - \eta_\alpha = \alpha(1-\alpha)(I - \alpha P')^{-1} Q g_\alpha, \qquad 0 < \alpha < 1, \tag{25}$$

Letting $\alpha \to 1-$, we obtain (11) as a special case.

Similar to Lemma 7, if we assume that f also changes to f', then we have

Lemma 9 If $\alpha P' g_\alpha + f' \preceq \alpha P g_\alpha + f$, then $\eta'_\alpha < \eta_\alpha$, $0 < \alpha \le 1$.

Proof. For $0 < \alpha < 1$, we have

$$(I - \alpha P')^{-1} = I + \alpha P' + \alpha^2 P'^2 + \cdots.$$

Since the Markov chain is positive recurrent, every item in $(I - \alpha P')^{-1}$ is positive. The lemma for $0 < \alpha < 1$ follows directly from (25); for $\alpha = 1$, it is the same as Lemma 7. □

Now, policy iteration algorithms for discounted cost can be derived easily. At the kth step, it simply chooses the policy that minimizes $\alpha P^{\mathcal{L}} g_\alpha^{\mathcal{L}_k} + f^{\mathcal{L}_k}$, componentwisely, as the policy for the next step.

Next, we study the performance sensitivity. let $P(\delta) = P + Q\delta$. Thus, $P(1) = P'$, $P(0) = P$. Then from (25), we have

$$\frac{d\eta_\alpha}{d\delta} = \alpha(1-\alpha)(I - \alpha P)^{-1} Q g_\alpha, \qquad 0 < \alpha < 1. \tag{26}$$

From (25) and (26), it is clear that policy iteration for discounted problems also chooses the next policy along the steepest direction. We have

$$\frac{d\eta_1}{d\delta} = e\frac{d\eta}{d\delta} = \lim_{\alpha \to 1} \frac{d\eta_\alpha}{d\delta}.$$

5. Single Sample Path-Based Implementations

5.1 Estimating Performance Potentials

In this section, we present some sample-path-based estimation algorithms for potentials and realization factors [5] [6].

First, from (7), we can choose a fixed L and use $g_L(i) = E[\sum_{l=0}^{L-1} f(X_l) \, |X_0 = i] - L\eta$ as an estimate of $g(i)$. Define $\epsilon_i(x) = 1$, if $x = i$, and $\epsilon_i(x) = 0$, otherwise. Given a long sample path $(X_0, X_1, \cdots, X_K)$, we can prove

$$g_L(i) = \lim_{K \to \infty} \left\{ \frac{\sum_{k=0}^{K-L+1} \{\epsilon_i(X_k)[\sum_{j=0}^{L-1} f(X_{k+j})]\}}{\sum_{k=0}^{K-L+1} \epsilon_i(X_k)} - \frac{L}{K} \sum_{k=0}^{K-1} f(X_k) \right\}, \tag{27}$$

$w.p.1.$

For realization factor, algorithms can be developed from (3). However, such algorithms require two sample paths, one starting with $X_0 = i$ and the other with $X_0' = j$. Another algorithm overcomes this difficulty and is based on a finite sample path. On a Markov chain $\mathbf{X} = \{X_n, n \geq 0\}$ with $X_0 = i$, we define $L_i(j) = min\{n : n \geq 0, X_n = j\}$; i.e., at $n = L_i(j)$, the Markov chain reaches state j for the first time. We have

$E[L_i(j)|X_0 = i] < \infty$ [11], and from [8]

$$d(j,i) = E\{ \sum_{l=0}^{L_i(j)-1} [f(X_n) - \eta] | X_0 = i \}. \tag{28}$$

(28) relates $d(i,j)$ to a finite portion of the sample paths of $\mathbf{X}$. To develop an algorithm based on (28), we define $u_0 = 0$, and $u_{k+1} = min\{n : n > u_k, X_n = i\}$, $k \geq 0$, where i is a fixed state. u_k, $k \geq 0$ are regenerative points. For any $j \neq i$, define $v_k(j) = min\{n : u_{k+1} > n > u_k, X_n = j\}$ and $\chi_k(j) = 1$, if $\{u_{k+1} > n > u_k, X_n = j\} \neq \emptyset$; and $\chi_k(j) = 0$, otherwise. From (28), we have

$$\begin{aligned} d(i,j) &= \lim_{K\to\infty} \frac{1}{\sum_{k=0}^{K-1} \chi_k(j)} \left\{ \left[\sum_{k=0}^{K-1} \{\chi_k(j) \sum_{n=v_k(j)}^{u_{k+1}-1} f(X_n) \right] \right. \\ &\quad \left. - \left[\sum_{k=0}^{K-1} \chi_k(j)[u_{k+1} - v_k(j)] \right] \eta \right\}, \quad w.p.1 \end{aligned} \tag{29}$$

where η can be simply estimated by

$$\eta = \lim_{N\to\infty} \frac{1}{N} \sum_{n=0}^{N-1} f(X_n), \quad w.p.1.$$

Given a sample path of a system run under the current policy, we can estimate the potentials using (27) or (29). Then the performance derivatives in a continuous parameter space can be obtained by (6) or (13). The derivatives can be used together with stochastic approximation techniques to reach the optimal performance. Algorithms and proofs of convergence and other results in this direction is reported in [19]. We can also use the potentials estimated on a sample path to implement policy iteration. When the new policy is found, we run the system under the new policy for another sample path and update the policy again until some criterion is met. We shall refer to this procedure as the *sample path-based MDP algorithm.* Stochastic approximation techniques can also be used in this procedure to provide fast algorithms. Research in this direction is reported in [12]. The main advantage of the sample path-based approach is that it can be applied online to real systems without estimating the system parameters, and the inversion of large matrices is avoided. In addition, the sample path-based approach offers flexibility in obtaining sub-optimal policies with less computation. For more details, see [6] and the example in Section 5.2.

The above discussion applies to discounted MDPs as well. In particular, ignoring the constant term, the α potential can be written as

$$g_\alpha(i) = E\{\sum_{n=0}^{\infty} \alpha^n f(X_n)|X_0 = i\}, \qquad 0 < \alpha < 1. \tag{30}$$

This is the same as the α potential defined in the standard potential theory [11]. When $\alpha = 1$, the above expression goes to infinity. Thus, (7) can be viewed as the extension of the α potential to the case of $\alpha = 1$.

5.2 An Example

We use an example to illustrate the application of the single sample path-based approach.

Example 1 A manufacturing system consists of two machines and N pieces of works, which are circulating between the two machines, as shown in Figure 11.1. Each work piece has to undertake three consecutive operations at machine 1; thus machine 1 is illustrated by three circles in the figure. The service rates of these three operations are λ_1, λ_2, and λ_3, respectively. Machine 2 has only one operation with service rate λ_4. The system can be modeled as a Markov process with state denoted as (n, i), where n is the number of pieces in machine 1 and $i = 1, 2, 3$ denotes the operation that the piece at machine 1 is undertaking. A work piece, after the completion of service at machine 1, goes to machine 2 with probability $p^\alpha(n)$ and feeds back to machine 1 with probability $1 - p^\alpha(n)$. The subscript "α" represents an action with $\alpha \in \mathcal{A}$. For any $\alpha \in \mathcal{A}$ and $n = 1, 2, \cdots, N$, $p^\alpha(n) \in [0, 1]$. The performance to be optimized is the weighted sum of the two machines' throughputs. (The cost function f does not depend on the actions.) Some transition probabilities of the Markov chain embedded at the operation completion times are (for $0 < n < N$):

$$\begin{aligned}
p[(n,1),(n+1,1)] &= \frac{\lambda_4}{\lambda_1 + \lambda_4}, \\
p[(n,1),(n,2)] &= \frac{\lambda_1}{\lambda_1 + \lambda_4}, \\
p^\alpha[(n,3),(n+1,3)] &= \frac{\lambda_3}{\lambda_3 + \lambda_4}, \\
p^\alpha[(n,3),(n-1,1)] &= \frac{\lambda_3}{\lambda_3 + \lambda_4} p^\alpha(n), \\
p^\alpha[(n,3),(n,1)] &= \frac{\lambda_3}{\lambda_3 + \lambda_4} [1 - p^\alpha(n)],
\end{aligned}$$

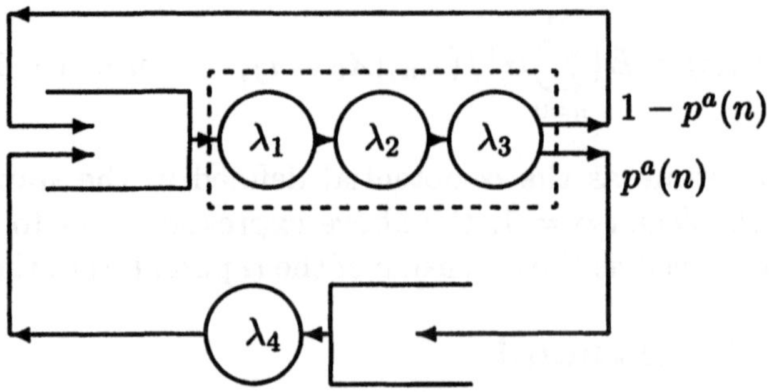

Figure 11.1. The Example

Other nonzero transition probabilities have a similar form.

The transitions from states $(n, 1)$ and $(n, 2)$ do not depend on actions. For state $(n, 3)$, the optimality equation is

$$\begin{aligned} & \frac{\lambda_4}{\lambda_3+\lambda_4} g^\alpha(n+1,3) + \frac{\lambda_3}{\lambda_3+\lambda_4} p^\alpha(n) g^\alpha(n-1,1) \\ & + \frac{\lambda_3}{\lambda_3+\lambda_4}[1-p^\alpha(n)] g^\alpha(n,1) \\ \leq\ & \frac{\lambda_4}{\lambda_3+\lambda_4} g^\alpha(n+1,3) + \frac{\lambda_3}{\lambda_3+\lambda_4} p^{\alpha'}(n) g^\alpha(n-1,1) \\ & + \frac{\lambda_3}{\lambda_3+\lambda_4}[1-p^{\alpha'}(n)] g^\alpha(n,1), \end{aligned}$$

for all $\alpha' \in \mathcal{A}$. This is equivalent to

$$[p^\alpha(n) - p^{\alpha'}(n)] g^\alpha(n-1,1) - [p^\alpha(n) - p^{\alpha'}(n)] g^\alpha(n,1) \leq 0. \qquad (31)$$

In (31), only the action related probabilities are involved; the system parameters λ_1, λ_2, λ_3, and λ_4 do not appear. □

In the above example, the service rates govern the evolution of the system, which runs autonomously. The control action can effect only some of the system transitions. The transition probabilities corresponding to the uncontrolled transitions are the same under all policies; they cancel each other in the Bellman equation and hence do not appear in the final form. Since we can estimate g^αs on a sample path, we can implement policy iteration without knowing these transition probabilities by observing a sample path.

Many practical systems have the same feature as the above example. Indeed, for many such systems control can be exercised only at a very limited region (e.g., the admission control can be applied only at the access points of a high-speed communication network); the rest part of the systems simply evolves by its own nature. The dashed box in Figure 11.1 can also be viewed as a machine whose service time has an Erlangian distribution, in such cases the transitions between the three stages are not controllable. This type of service distribution and the more general forms, such as the coxian distribution and the phase-type distributions, are very common in practical systems.

In summary, Example 1 and the above discussion illustrate that the sample-path-based approach has the following advantages.

1 Given a sample path, policy iteration can be implemented without knowing the whole transition matrix; only those items related to control actions have to be known. Matrix inversion is not required.

2 The approach saves memory spaces required for implementing MDPs. Generally speaking, only the M potentials, not the $M \times M$ transition matrix, have to be stored. This can be further reduced when there are some states which cannot be reached by controlled states. As shown in (31), in Example 1 only $g^{\alpha}(n, 1)$, $n = 0, 1, \cdots, N$, have to be estimated and stored; $g^{\alpha}(n, 2)$ and $g^{\alpha}(n, 3)$, $n = 0, 1, \cdots, N$, do not even need to be estimated.

3 Different from the standard computational approach where all the potentials are obtained altogether by a matrix inversion, in the sample path based approach potentials can be estimated one by one. This feature brings in a number of possible applications.

   (a) The computational efforts and memory spaces of each iteration may be further reduced at the cost of the convergence rate. The idea is, if the state space is too large, at each iteration we can estimate the potentials of only a subset of the state space and update the actions for the states in this subset. For instance, in Example 1 we may set $0 = n_0 < n_1 < n_2 < \cdots < n_{k-1} < n_k = M$. Then in the ith iteration, $i = 1, 2, \cdots, k$, we may estimate $g^{\alpha}(n, 1)$ only for $n_{i-1} - 1 \leq n \leq n_i$, $i = 1, 2, \cdots, k$. Of course, it may need more iterations to reach the optimal policy; however, at each iteration the computation and the menory requirements may be reduced to fit the capacity of the available computing equipments. This feature may be important for on-line optimization using special designed hardwares which may have limited capacities. In high speed communication networks, the effect of a slow convergence rate in terms of iterations may be compensated by the fast speed in system evolution.

(b) For many practical systems, we may have some a priori knowledge about which states are more important than others. Then we can estimate only the potentials for the states that are needed for updating the actions on these important states. This may reduce the computation and memory at the cost of the best performance achieved.

6. Reinforcement Learning

As shown in the last section, potentials can be estimated on a single sample path. More efficient algorithms can be developed with stochastic approximation methods. For example, from (7), we have

$$g(i) = E\{\sum_{l=0}^{\infty}[f(X_l) - \eta] | X_0 = i\}.$$

Given a sample path $\{X_0, \cdots, X_n, X_{n+1}, \cdots,\}$, at the nth transition, we have

$$g(X_n) = E\{\sum_{l=0}^{\infty}[f(X_{n+l}) - \eta] | X_n\} = [f(X_n - \eta] + E[g(X_{n+1})] \quad (32)$$

Applying stochastic approximation methods to (32) leads to the algorithm

$$g(X_n) := g(X_n) + \gamma[f(X_n) - \eta + g(X_{n+1}) - g(X_n)], \qquad 0 < \gamma < 1,$$

where γ is the step size. Set

$$d_n = [f(X_n) - \eta + g(X_{n+1}) - g(X_n)].$$

We have

$$g(X_n) := g(X_n) + \gamma d_n. \quad (33)$$

More generally, we have

$$g(X_n) := g(X_n) + \gamma \sum_{m=n}^{\infty} \lambda^{m-n} d_n, \qquad 0 \le \lambda \le 1. \quad (34)$$

This algorithm is known as *TD(λ)* [3, 24, 25]. Setting $\lambda = 0$ (using the convention $0^0 = 1$) leads to TD(0) shown in (33). Different variants of TD(λ) exist and convergence can be proved.

When the system structure, i.e., the transition probabilities, $p(i,j)s$, are completely unknown, we cannot apply policy iteration directly by using the potentials. (As shown in Example 1, we need to know at least the transition probabilities related to the actions, i.e., $p^{\alpha}(n)$) to implement policy iteration with estimated potentials.) In this case, we can estimate the Q-factor defined as [3, 25]

$$Q(i,\alpha) = \{\sum_{j=1}^{M} p^{\alpha}(i,j)g(j)\} + f^{\alpha}(i) - \eta, \tag{35}$$

for every state-action pair (i,α). From Theorem 1 we can choose $\alpha = arg\{min_{\alpha' \in \mathcal{A}}[Q(i,\alpha')]\}$ as the action at state i in the policy iteration. However, there is one problem for this approach: it is impossible to estimate $Q(i,\alpha')$ on the current sample path with $\alpha \neq \alpha'$ being the action taken at state i, because the pair (i,α') does not appear on the sample path. One way to obtain the estimation of $Q(i,\alpha')$, $\alpha' \neq \alpha$, is to apply the important sampling technique. This requires to know the ratios of $p^{\alpha'}(i,j)$ and $p^{\alpha}(i,j)$. In Example 1, this ratio can be obtained if $p^{\alpha'}(n)$ and $p^{\alpha}(n)$ are known. This shows that policy iteration based on Q-factors is almost the same as that based on potentials.

The value iteration of Q-factor, however, leads to the optimal Q-factor and can be implemented when the system structure is completely unknown. This is briefly explained as follows. From the definition (35), we have

$$\eta + Q(i,\alpha) = \{\sum_{j=1}^{M} p^{\alpha}(i,j)g(j)\} + f^{\alpha}(i), \tag{36}$$

Taking minimum on both sizes, we get

$$\eta + min_{\alpha \in \mathcal{A}} Q(i,\alpha) = min_{\alpha \in \mathcal{A}}\{\sum_{j=1}^{M} p^{\alpha}(i,j)g(j) + f^{\alpha}(i)\},$$

Comparing with the Bellman equation

$$\eta + g(i) = min_{\alpha \in \mathcal{A}}\{\sum_{j=1}^{M} p^{\alpha}(i,j)g(j) + f^{\alpha}(i)\},$$

we conclude that at the optimal policy $g(i) = min_{\alpha \in \mathcal{A}} Q(i,\alpha)$. Substituting this into (36), we have at the optimal policy

$$\eta + Q(i,\alpha) = \{\sum_{j=1}^{M} p^{\alpha}(i,j)[min_{\alpha \in \mathcal{A}} Q(j,\alpha)]\} + f^{\alpha}(i). \tag{37}$$

This is the Bellman equation for Q-factors. Based on (37), the Robbins-Monro stochastic approximation method leads to

$$Q(i,\alpha) := (1-\gamma)Q(i,\alpha)+\gamma[min_{\beta\in\mathcal{A}}Q(j,\beta)]+f^{\alpha}(i)-\eta], \qquad 0<\gamma<1. \tag{38}$$

This is the Q-learning algorithm, which can be applied to any sample path: when the system jumps from state i to j, the Q-factor is updated according to (38). It can be proved that (38) converges to the optimal $Q^*(i,\alpha)$ and $\alpha^* = arg\{min_{\alpha'\in\mathcal{A}}[Q^*(i,\alpha')]\}$ is the optimal action at state i.

Compared with the approach based on potentials, Q-learning can be applied to systems where the structure is completely unknown. However, from (38) it is clear that Q-learning requires the sample path visit every state-action pair. In addition, the number of Q-factors increases to $M \times K$ (where K is the number of possible actions at each state).

In addition to the above approaches, *Neuro-Dynamic programming* is proposed to overcome the difficulty of the so-called "curse of dimensionality". Roughly speaking, in neuro-dynamic programming, we try to approximate the potential function $g(i)$ by $g(i,r)$, with a continuous parameter r. This generally involves with two steps:

1 develop an approximation architecture, e.g., a neuro-network, to represent $g(i,r)$,

2 find a training algorithm for updating the parameter vector r, based on the information observed on a sample path.

After training, the parameter r reaches a proper value. The neuro-network will output an approximate value of $g(i)$ for an input integer i. For details and successful examples, see [3].

7. Discussions and Conclusions

Figure 11.2 illustrates the relation among PA, MDPs, and RL. All these areas are related to performance improvement and optimization. PA provides performance sensitivity with respect to both continuous and discrete parameters. Policy iteration in MDPs can be viewed as a direct consequence of discrete performance sensitivity provided by PA, This new view of PA and MDPs leads to the gradient-based policy iteration method that can be applied to some nonstandard problems such as those with correlated actions at different states. Performance potential plays an important role in both PA and MDPs; it also offers a clear intuitive interpretation for many results. Reinforcement learning, TD(λ), neuro-dynamic programming, etc, are efficient ways of estimating the

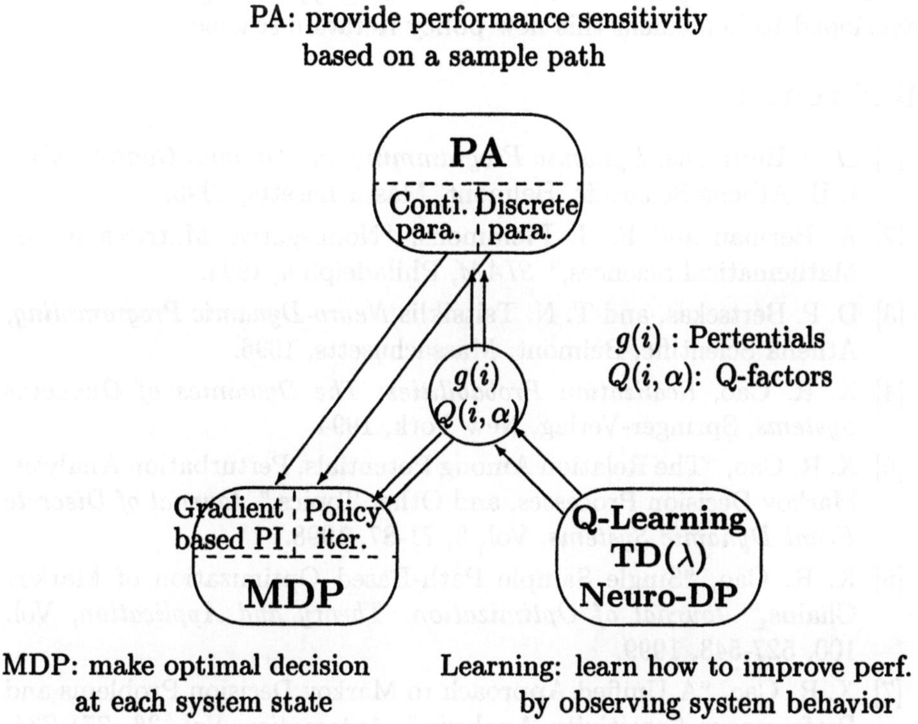

Figure 11.2. The Relations Among PA, MDP and RL

performance potentials and related quantities based on sample paths. In particular, Q-learning can be used when the system structure is unknown. The potentials and Q-factors can be implemented on a single sample path; the sample path-based approach leads to on-line optimization schemes that are of practical importance to real world engineering problems such as the optimization of manufacturing and communication systems.

The gradient-based policy iteration method is a new topic for further research. Sample-path-based and Q-learning-type of algorithms can be developed to implement this new policy iteration scheme.

References

[1] D. P. Bertsekas, *Dynamic Programming and Optimal Control,* Vols. I, II, Athena Scientific, Belmont, Massachusetts, 1995.

[2] A. Berman and R. J. Plemmons, "Nonnegative Matrices in the Mathematical Sciences," *SIAM*, Philadelphia, 1994.

[3] D. P. Bertsekas, and T. N. Tsitsiklis,*Neuro-Dynamic Programming,* Athena Scientific, Belmont, Massachusetts, 1996.

[4] X. R. Cao, *Realization Probabilities: The Dynamics of Queueing Systems,* Springer-Verlag, New York, 1994.

[5] X. R. Cao, "The Relation Among Potentials, Perturbation Analysis, Markov Decision Processes, and Other Topics," *Journal of Discrete Event Dynamic Systems,* Vol. 8, 71-87, 1998.

[6] X. R. Cao, "Single Sample Path-Based Optimization of Markov Chains," *Journal of Optimization: Theory and Application*, Vol. 100, 527-548, 1999.

[7] X. R. Cao, "A Unified Approach to Markov Decision Problems and Performance Sensitivity Analysis," *Automatica*, Vol. 36, 771-774, 2000.

[8] X. R. Cao and H. F. Chen, "Potentials, Perturbation Realization, and Sensitivity Analysis of Markov Processes," *IEEE Transactions on AC*, Vol. 42, 1382-1393, 1997.

[9] X. R. Cao, Z. Y. Ren, S. Bhatnagar, M. Fu, and S. Marcus, "A Time Aggregation Approach to Markov Decision Processes," *Automatica*, to appear, 2001.

[10] X. R. Cao and H. T. Fang, "Gradient-Based Policy Iteration: An Example," submitted to 2002 IEEE Conference on Decision and Control.

[11] E. Çinlar, *Introduction to Stochastic Processes,* Prentice Hall, Englewood cliffs, NJ, 1975.

[12] H. T. Fang and X. R. Cao, "Single Sample Path-Based Recursive Algorithms for Markov Decision Processes," *IEEE Trans. on Automatic Control*, submitted.

[13] P. Glasserman, *Gradient Estimation Via Perturbation Analysis,* Kluwer Academic Publisher, Boston, 1991.

[14] P. W. Glynn and S. P. Meyn, "A Lyapunov Bound for Solutions of Poisson's Equation," *Ann. Probab.,* 916-931, Vol. 24, 1996.

[15] Y. C. Ho, A. Eyler, and T. T. Chien, "A Gradient Technique for General Buffer Storage Design in a Serial Production Line," *International Journal on Production Research,* Vol. 17, 557-580, 1979.

[16] Y. C. Ho (editor), *Discrete-Event Dynamic Systems: Analyzing Complexity and Performance in the Modern World,* IEEE Press, New York, 1992.

[17] Y. C. Ho and X. R. Cao, *Perturbation Analysis of Discrete-Event Dynamic Systems,* Kluwer Academic Publisher, Boston, 1991.

[18] Y. C. Ho and X. R. Cao, "Perturbation Analysis and Optimization of Queueing Networks," *Journal of Optimization Theory and Applications,* Vol. 40, 4, 559-582, 1983.

[19] P. Marbach and T. N. Tsitsiklis, "Simulation-based optimization of Markov reward processes," *IEEE Transactions on Automatic Control* Vol. 46, 191-209, 2001.

[20] J. G. Kemeny and J. L. Snell, *Finite Markov Chains,* Van Nostrand, New York, 1960.

[21] S. P. Meyn, "The Policy Improvement Algorithm For Markov Decision Processes with General State Space," *IEEE Transactions on Automatic Control*, Vol. 42, 1663-1680, 1997.

[22] S. P. Meyn and R. L. Tweedie, *Markov Chains and Stochastic Stability,* Springer-Verlag, London, 1993.

[23] M. L. Puterman, *Markov Decision Processes: Discrete Stochastic Dynamic Programming*, Wiley, New York, 1994.

[24] R. S. Sutton, "Learning to Predict by the Methods of Temporal Differences," *Machine Learning*, Vol. 3, 835-846, 1988.

[25] R. S. Sutton and A. G. Barto, *Reinforcement Learning: An Introduction*, MIT Press, Cambridge, MA, 1998.

[26] B. Zhang and Y. C. Ho, "Performance Gradient Estimation for Very Large Finite Markov Chains," *IEEE Transactions on AutomaticControl*, Vol. 36, 1218-1227, 1991.

Chapter 12

OPTIMAL CONTROL FOR DISCRETE EVENT AND HYBRID SYSTEMS*

Christos G. Cassandras and Kagan Gokbayrak
Dept. of Manufacturing Engineering
Boston University
Boston, MA

Dedicated to Yu-Chi (Larry) Ho

Abstract As time-driven and event-driven systems are rapidly coming together, the field of optimal control is presented with an opportunity to expand its horizons to these new "hybrid" dynamic systems. In this paper, we consider a general optimal control problem formulation for such systems and describe a modeling framework allowing us to decompose the problem into lower and higher-level components. We then show how to apply this setting to a class of switched linear systems with a simple event-driven switching process, in which case explicit solutions may often be obtained. For a different class of problems, where the complexity lies in the nondifferentiable nature of event-driven dynamics, we show that a different type of decomposition still allows us to obtain explicit solutions for a class of such problems. These two classes of problems illustrate the differences between various sources of complexity that one needs to confront in tackling optimal control problems for discrete-event and hybrid systems.

1. Introduction

The theory and extensive applications of optimal control for dynamic systems are documented in several books, including "Applied Optimal Control" [1], co-authored by Bryson and Ho. Since this paper, along with

*This material is based on work supported in part by the National Science Foundation under Grants ACI-98-73339 and EEC-00-88073, by AFOSR under contract F49620-01-0056, and by the Air Force Research Laboratory under contract F30602-99-C-0057.

the rest in this collection, is dedicated to the second author, it is most appropriate that its topic covers two areas to which Larry Ho has made fundamental contributions: optimal control and discrete event systems. The emergence of *hybrid* systems, which has largely motivated the work presented here, arguably owes its rapid growth to the foundations laid by the development of modeling frameworks and analytical techniques for discrete event systems.

Optimal control methodologies have been based on a modeling framework for dynamic systems which revolves around the differential equation

$$\dot{x} = f(x, u, t) \tag{1}$$

describing the dynamics of a system whose state at time t is captured by a vector $x(t)$ and whose behavior is dependent upon some input or control vector $u(t)$. We refer to this class of systems as *time-driven* because their state continuously changes as time evolves. Typical examples include mechanical systems whose state consists of variables representing physical position and velocity, or chemical systems whose state includes the temperature or pressure for certain physical processes. In contrast, Discrete Event Systems (DES) are characterized by *event-driven* dynamics. In this case, state variables are usually discrete and change only as a result of the occurrence of events; they do not continuously change over time. One encounters DES in modern technological settings such as automated manufacturing, communication networks, and computer operating systems. As an example, let $x(t)$ be the number of packets waiting in a buffer to be processed by some network switch at time t. The only way this quantity can change is if a packet is removed from the buffer and transmitted (in which case the state changes by -1) or if a new packet is added to the buffer (in which case the state changes by $+1$). Both changes occur instantaneously when the event "packet processed" or the event "new packet arrives" respectively takes place. Clearly, (1) is neither natural nor appropriate for describing the dynamics of such an event-driven system: the state in between events is unchanged and the derivative of $x(t)$ is not defined at event times (at least not in a strict sense that can make it useful for analysis purposes). Modeling frameworks for DES were developed over the 1980s along several different directions, including untimed and timed automata and Petri nets (for a comprehensive overview, see [2]). Because these modeling frameworks are drastically different form (1), optimal control methodologies developed on the basis of time-driven dynamics are simply not transferable to the DES setting, although fundamental principles (e.g., dynamic programming) obviously still apply. To complicate matters, many interesting optimal control problems one encounters in DES involve some form of uncertainty (e.g., in the example of a packet buffer, the processes that describe arrival and processing events are generally stochastic); this implies that ideas and techniques from *stochastic* optimal control need to be invoked.

The advent of hybrid systems makes the development of optimal control beyond classical time-driven systems an even greater challenge. The term "hybrid" is used to characterize systems that combine time-driven and event-driven dynamics. Broadly speaking, two categories of modeling frameworks have been proposed to study hybrid systems: Those that extend event-driven models to include time-driven dynamics; and those that extend the traditional time-driven models to include event-driven dynamics; for an overview, see [3]. A simple way to think of a hybrid system is as one characterized by a set of operating "modes", each one evolving according to time-driven dynamics of the form (1). The system switches between modes through discrete events which may be controlled or uncontrolled. Controlling the switching times, when possible, and choosing among several feasible modes, whenever such choices are available, gives rise to a rich class of optimal control problems. This has motivated efforts to extend classical optimal control principles [4],[5],[6] and to apply dynamic programming techniques [7],[8]. While in principle this is possible, the computational complexity involved becomes prohibitive: not only does one have to deal with the well-known "curse of dimensionality" in such problems, but there are at least two additional sources of complexity to deal with, i.e., the presence of switching events causing transitions from one mode to another (which introduces a combinatorial element into the control), and the presence of event-driven dynamics for the switching times (which introduce nondifferentiabilities). Therefore, keys to the successful development of optimal control methods for hybrid systems are: (*i*) seeking structural properties that allow the decomposition of such systems into simpler components, and (*ii*) making use of efficient numerical techniques. Along these lines, progress has been reported for classes of hybrid systems whose structure may be exploited. For example, in [9] a Mixed Logical Dynamical (MLD) system framework is proposed, which allows the use of efficient methods developed for piecewise affine systems, and in [10] optimal controllers are presented for the class of autonomous switched linear systems.

The hybrid system modeling framework we consider in this paper is motivated by the fact that it is often natural to hierarchically decompose systems into a lower-level component representing physical processes characterized by time-driven dynamics and a higher-level component representing discrete events related to these physical processes (e.g., switching from one mode of operation to another, as in shifting gears in an automotive system). Our objective is to formulate and solve optimal control problems associated with trade-offs between the operation of physical processes and timing issues related to the overall performance of the system. For a class of such optimal control problems, a hierarchical decomposition method was introduced in [11]. This method enables us to design a controller which has the task of communicating with both components and jointly solving coupled optimization problems, one for each component. The same basic idea is also independently proposed in [12]. The explicit solution of the lower and higher-level problems depends on the specifics of the time-driven and event-driven dynamics involved.

In the remainder of this paper, we first describe a convenient modeling framework for hybrid systems allowing us to decompose them into lower (time-driven) and higher (event–driven) level components. Subsequently, we formulate an optimal control problem and present a solution approach based on this decomposition. In Section 3 we describe how to apply this setting to a class of hybrid systems where all operating modes have linear time-driven dynamics and the event-driven switching process is very simple. In this case, we find that the hierarchical decomposition approach requires solving a nonlinear parametric optimization problem coupled with a number of standard optimal control problems subject to purely time-driven dynamics; explicit solutions of such problems are occasionally possible, as illustrated by an example. In Section 4, we consider a different class of problems where the complexity lies in the event-driven dynamics which are nondifferentiable. In this case, we show that a different type of decomposition allows us to obtain explicit solutions for a class of such problems. While the approaches presented for solving these two classes of problems help to lay the foundations for extending classical optimal control theory, they also serve to identify the differences between various sources of complexity that one needs to understand and address in order to handle the challenges of new technological environments where these complexities manifest themselves.

2. Optimal Control Problem Formulation

In the hybrid systems we consider, the state of the system consists of *temporal* and *physical* components. The temporal components keep track of the time information for events that may cause switches in the operating mode of the system. Let $i = 1, 2, \ldots$ index these events. We denote the *physical state* of the system after the ith event by $z_i(t)$ with dynamics:

$$\dot{z}_i = g_i(z_i, u_i, t), \quad z_i(x_{i-1}) = z_i^0, \quad t \in [x_{i-1}, x_i) \tag{2}$$

where u_i is the control applied over an interval $[x_{i-1}, x_i)$ defined by two event occurrences at times x_{i-1} and x_i. In what follows, we shall write u_i to denote a function $u_i(t)$ defined over $[x_{i-1}, x_i)$; similarly for $z_i(t)$. We shall assume that $u_i(t)$ is allowed to be piecewise continuous and is in general an n-dimensional vector.

In the case of a *single* event process in the system, the event-driven dynamics characterizing the *temporal states* x_i are given by

$$x_i = x_{i-1} + \gamma_i(z_i, u_i) \tag{3}$$

for $i = 1, 2, \ldots$, where $\gamma_i(\cdot)$ represents the amount of time between switches, which generally depends on the physical state z_i and control u_i over $[x_{i-1}, x_i)$.

In the case of *multiple* asynchronous event processes in the system indexed by $j = 1, \ldots, M$, we need to introduce a *Timed Automaton* which determines which of the M events in the system triggers the next switch and at what precise time. The exact structure of a timed automaton is described in [2]. We limit ourselves here to a brief review of its definition and basic operation. An *Automaton*, denoted by G, is a five-tuple $G = (Q, E, f, \Gamma, x_0)$ where Q is the set of *states*; E is the finite set of *events* associated with the transitions in G; $f : Q \times E \to Q$ is the *transition function*, where $f(q, e) = r$ means that there is a transition labeled by event e from state q to state r (in general, f is a *partial* function on its domain); $\Gamma : Q \to 2^E$ is the *active event function* (or feasible event function), where $\Gamma(q)$ is the set of all events e for which $f(q, e)$ is defined and it is called the *active event set* (or feasible event set) of G at q; and q_0 is the *initial* state.

The automaton G operates as follows. It starts in the initial state q_0 and upon the occurrence of an event $e \in \Gamma(q_0) \subseteq E$ it makes a transition to state $f(q_0, e) \in Q$. This process then continues based on the transitions for which f is defined. Note that an event may occur without changing the state, i.e., it is possible that $f(q, e) = q$.

As it stands, this model is based on the premise that a given event sequence $\{e_1, e_2, \ldots\}$ is provided, so that, starting at state q_0, we can generate a state sequence $\{q_0, f(q_0, e_1), f(f(q_0, e_1), e_2), \ldots\}$. We extend our modeling setting to *Timed* automata by incorporating a *Clock Structure* associated with the event set E which now becomes the input from which a specific event sequence can be deduced. This clock structure is a set $\mathbf{V} = \{\mathbf{v}_i : i \in E\}$ of clock (or lifetime) sequences

$$\mathbf{v}_i = \{v_{i,1}, v_{i,2}, \ldots\}, \;\; i \in E, \;\; v_{i,k} \in \mathbb{R}^+, \;\; k = 1, 2, \ldots$$

With this in mind, the original automaton now becomes a six-tuple $(Q, E, f, \Gamma, x_0, \mathbf{V})$ where $\mathbf{V} = \{\mathbf{v}_i : i \in E\}$ is a clock structure. The timed automaton generates a state sequence

$$q' = f(q, e') \tag{4}$$

driven by an event sequence $\{e_1, e_2, \ldots\}$ generated through

$$e' = \arg \min_{i \in \Gamma(q)} \{y_i\} \tag{5}$$

with the clock values y_i, $i \in E$, defined by

$$y_i' = \begin{cases} y_i - y^* & \text{if } i \neq e' \text{ and } i \in \Gamma(q) \\ v_{i,N_i+1} & \text{if } i = e' \text{ or } i \notin \Gamma(q) \end{cases} \qquad i \in \Gamma(q') \tag{6}$$

where the *interevent time* y^* is defined as

$$y^* = \min_{i \in \Gamma(q)} \{y_i\} \tag{7}$$

and the *event scores* N_i, $i \in E$, are defined by

$$N_i' = \begin{cases} N_i + 1 & \text{if } i = e' \text{ or } i \notin \Gamma(q) \\ N_i & \text{otherwise} \end{cases} \qquad i \in \Gamma(q') \tag{8}$$

In addition, initial conditions are: $y_i = v_{i,1}$ and $N_i = 1$ for all $i \in \Gamma(q_0)$. If $i \notin \Gamma(q_0)$, then y_i is undefined and $N_i = 0$.

In this setting, the clock structure $\mathbf{V}$ is assumed to be fully specified in a deterministic sense. A *Stochastic Timed Automaton* is obtained when the clock sequences are specified only as stochastic sequences $\{V_{i,k}\} = \{V_{i,1}, V_{i,2}, \ldots\}$, in which case each $\{V_{i,k}\}$ is characterized by a distribution function $G_i(t) = P[V_i \leq t]$.

In the sequel, the details of the timed automaton that controls the mode switches are not required and are suppressed by simply representing the event-driven dynamics in the form

$$x_i = x_{i-1} + \gamma_i(y_{i,1}, \ldots, y_{i,M}, z_i, u_i) \tag{9}$$

where $y_{i,1}, \ldots, y_{i,M}$ are the event clocks of the timed automaton (through which the triggering event and its occurrence time for the next switch are determined) after the $(i-1)$th switch. We note, that $\gamma_i(\cdot)$ above generally involves the min operation seen in (5), introducing nondifferentiabilities which can significantly complicate the analysis. Looking at (2) and (9), note also that the choice of control u_i affects both the physical state z_i and the temporal state x_i. Thus, the switches at times $x_1, x_2, \ldots$ are generally *not* exogenous events that dictate changes in the state dynamics, but rather temporal states intrically connected to the control of the system; this is one of the crucial elements of a "hybrid" system.

The optimal control problem we now consider has the general form

$$\min_{\mathbf{u}} \; J = \sum_{i=1}^{N} [\phi_i(x_{i-1}, x_i) + \psi_i(x_i)] \tag{10}$$

subject to (2) and (9), where

$$\mathbf{u} = \begin{cases} u_1(t) & t \in [x_0, x_1) \\ \vdots & \vdots \\ u_N(t) & t \in [x_{N-1}, x_N) \end{cases} \tag{11}$$

The function $\phi_i(x_i, x_{i-1})$ is the cost of operating the system with control u_i over the interval $[x_{i-1}, x_i)$, which results in the physical state $\{z_i(t),\, t \in [x_{i-1}, x_i)\}$. This is generally expressed as

$$\phi_i(x_{i-1}, x_i) = h(z_i^f) + \int_{x_{i-1}}^{x_i} L_i(z_i(t), u_i(t))dt$$

where $h(z_i^f)$ is a terminal cost associated with the physical state $z_i^f = z_i(x_i)$ and $L_i(z_i(t), u_i(t))$ is a cost function dependent on the physical state and control during the ith mode. Note that $\phi_i(\cdot)$ is expressed as a function of the starting and ending times for the ith mode, but it obviously depends on the choice of control used over $[x_{i-1}, x_i)$.

The function $\psi_i(x_i)$ is the cost associated with the occurrence time x_i of the ith event. The intent is to penalize switching times that occur later rather than earlier, which creates a trade-off with the requirement that a certain desired physical state $z_i(x_i)$ be attained, which may not be possible if x_i is too short.

In (10), the time horizon is determined by a given number of switches N and the control does not include selecting the next mode. Thus, it is assumed that the sequence of modes to be used is prespecified (e.g., as in changing gears in an automobile). Moreover, note that in this formulation the only control is $\mathbf{u}$, through which the physical and tempoiral states are affected. In some variations of this problem, it is possible that the switching times $x_1, \ldots, x_N$ may be exogenous controllable variables as well.

Let $s_i = x_i - x_{i-1} \geq 0$, $i = 1, 2, \ldots$ Assuming stationarity of the cost $\phi_i(\cdot)$ in the sense that $\phi_i(x_{i-1}, x_{i-1} + s_i) = \phi_i(0, s_i) \equiv \phi_i(s_i)$, we can write

$$\phi_i(s_i) = h(z_i^f) + \int_0^{s_i} L_i(z_i(t), u_i(t))dt \tag{12}$$

Thus, the optimal control problem we consider can be written as

$$\min_{\mathbf{u}} \quad J = \sum_{i=1}^{N} [\phi_i(s_i) + \psi_i(x_i)] \tag{13}$$

subject to (2) and (9) with $\mathbf{u}$ as defined in (11) and $\phi_i(s_i)$ as defined in (12).

2.1 Hierarchical Decomposition

Consider an interval $[x_{i-1}, x_i)$ and suppose the initial and final physical states, $z_i^0 = z_i(x_{i-1})$ and $z_i^f = z_i(x_i)$, as well as the length of the time interval $s_i = x_i - x_{i-1}$, are fixed. If this were the case and $u_i(t)$, $t \in [x_{i-1}, x_i)$, is given, then $\{z_i(t),\ t \in [x_{i-1}, x_i)\}$, is specified through (2). Thus, with z_i^0, z_i^f, and s_i fixed, each mode can be individually analyzed and an optimal control for it can be sought, which will be parameterized by z_i^0, z_i^f, and s_i. Since actually z_i^0, z_i^f, and s_i are also unknown and part of the desired solution, it follows that we can rewrite (13) as

$$\min_{\mathbf{z}^0, \mathbf{z}^f, \mathbf{s}} \sum_{i=1}^{N} \left[\min_{u_i(z_i^0, z_i^f, s_i)} \quad \phi_i(s_i) + \psi_i(x_i) \right]$$

where $\mathbf{s} = [s_1, \ldots, s_N]$, $\mathbf{z}^0 = [z_1^0, \ldots, z_N^0]$, and $\mathbf{z}^f = [z_1^f, \ldots, z_N^f]$. This imposes a decomposition that gives rise to a collection of inner minimization problems subject to (2), and an outer minimization problem subject to (9). At the lower level of this decomposition we first seek to determine the cost

$$\theta_i(z_i^0, z_i^f, s_i) \equiv \min_{u_i} \phi_i(s_i) \tag{14}$$

subject to (2) for all $i = 1, 2, \ldots$, which we view as the minimal cost for a *given* time interval s_i and boundary conditions z_i^0 and z_i^f for the physical state. Accordingly, the optimal control is

$$u_i^*(z_i^0, z_i^f, s_i) \equiv \arg\min_{u_i} \phi_i(s_i) \tag{15}$$

Note that $u_i^*(z_i^0, z_i^f, s_i)$ is in general time-varying over $[x_{i-1}, x_i)$. It is the control that minimizes (12) when applied to $z_i^0 = z_i(x_{i-1})$ with a target state $z_i^f = z_i(x_{i-1} + s_i)$, given s_i. Once $u_i^*(z_i^0, z_i^f, s_i)$ and $\theta_i(z_i^0, z_i^f, s_i)$ are determined, we can proceed with the higher-level optimization problem:

$$\min_{\mathbf{z}^0, \mathbf{z}^f, \mathbf{s}} \sum_{i=1}^{N} [\theta_i(z_i^0, z_i^f, s_i) + \psi_i(x_i)] \tag{16}$$

subject to (9) where we try to determine the optimal event times and physical states at these times. Once these are known, the relationship (15) is used to determine the optimal controls for the N time intervals involved in the operation of the system.

The hybrid controller for coordinating the two problems above operates as follows (see Figure 12.1):

1 *System identification*: the physical dynamics, $\mathbf{g}$, the costs associated with the physical dynamics, ϕ, the temporal dynamics, $\mathbf{f}$, and the costs associated with the temporal dynamics, ψ, are all input to the controller.

2 *Lower-level problem, parameterized by s_i*: The lower level controller solves (14) to determine $\theta_i(z_i^0, z_i^f, s_i)$ and $u_i^*(z_i^0, z_i^f, s_i)$ for all $i = 1, \ldots, N$.

3 *Higher-level problem*: The higher level controller solves (16) to determine the optimal values s_i^*, $(z_i^0)^*$, and $(z_i^f)^*$ for all $i = 1, \ldots, N$.

4 *Lower-level problem, given s_i^**: The lower level controller evaluates $u_i^* = u_i^*((z_i^0)^*, (z_i^f)^*, s_i^*)$ for all $i = 1, \ldots, N$.

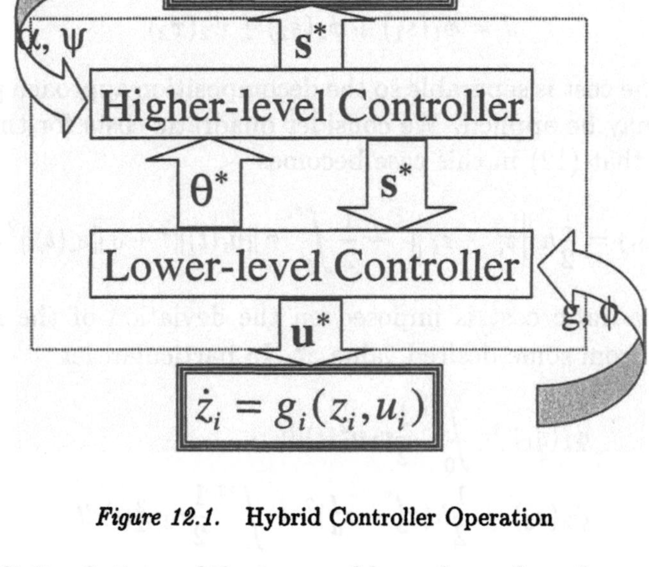

Figure 12.1. Hybrid Controller Operation

The explicit solutions of the two problems depend on the nature of the respective cost functions and dynamics. In some cases, the higher-level problem is greatly simplified when the event-driven dynamics involve a single exogenous event process, while in other cases the event-driven dynamics introduce nondifferentiabilities that prohibit the use of standard optimization methods. In the next two sections, we consider two different classes of hybrid systems with different characteristics, illustrating the major differences between the two problem levels (see also [13]).

3. Switched Linear Systems

Let us consider a class of hybrid systems consisting of linear time-invariant time-driven dynamics and associated quadratic criteria for all modes. In order to illustrate the hierarchical decomposition approach and obtain some explicit numerical results, we limit ourselves to a specific two-mode example as follows:

$$\dot{z}_1 = u_1, \qquad z_1(x_0) = z_0$$
$$\dot{z}_2 = \alpha z_2 + u_2, \qquad z_2(x_1) = z_1(x_1)$$

The event-driven dynamics determining how to switch between modes have the simple form given in (3):

$$x_1 = x_0 + s_1(z_1, u_1)$$
$$x_2 = x_1 + s_2(z_2, u_2)$$

The cost to minimize is

$$J = \phi_1(s_1) + \phi_2(s_2) + \psi_2(x_2)$$

Note that the cost is separable so the decomposition approach previously described may be applied. We consider quadratic costs for the physical process, so that (12) in this case becomes

$$\phi_i(s_i) = \frac{1}{2}h\left\|z_i^f - z_i^d\right\|^2 + \frac{1}{2}\int_0^{s_i} r\left\|u_i(t)\right\|^2 + q\left\|z_i(t)\right\|^2 dt$$

where a quadratic cost is imposed on the deviation of the final state $z_i^f = z_i(x_i)$ from some desired value z_i^d. In particular, let

$$\phi_1(s_1) = \int_0^{s_1} \frac{1}{2} r_1 u_1^2(t)dt$$
$$\phi_2(s_2) = \frac{1}{2}h(z_2^f - z_d^f)^2 + \int_0^{s_2} \frac{1}{2} r_2 u_2^2(t)dt$$

We also consider a quadratic cost on the final time x_2:

$$\psi_2(x_2) = \beta x_2^2$$

Let us now consider the lower-level problem which, for each mode, is a standard Linear Quadratic (LQ) problem [1]. The Hamiltonian for the first stage $[x_0, x_0 + s_1)$ of the physical process is

$$H_1(t) = \frac{1}{2} r_1 u_1^2(t) + p_1(t) u_1(t)$$

where $p_1(t)$ is the costate for the first stage, hence the optimality conditions for the first stage are

$$\begin{aligned} \dot{z}_1(t) &= u_1(t) \\ \dot{p}_1(t) &= 0 \\ r_1 u_1(t) + p_1(t) &= 0 \end{aligned}$$

Therefore, the optimal control in the first stage is a constant $u_1$and we get

$$u_1 = \frac{z_1(x_1) - z_1(x_0)}{s_1} \equiv \frac{z_1^f - z_1^0}{s_1}$$

and

$$\begin{aligned} \theta_1(z_1^0, z_1^f, s_1) &= \min_{u_1} \phi_1(s_1) = \frac{1}{2} r_1 u_1^2 s_1 \\ &= \frac{1}{2} \frac{r_1}{s_1} \left[z_1^f - z_1^0 \right]^2 \end{aligned}$$

Similarly, for the second stage the Hamiltonian is

$$H_2(t) = \frac{1}{2} r_2 u_2^2(t) + p_2(t) u_2(t) + \alpha p_2(t) z_2(t)$$

where $p_2(t)$ is the costate for the second stage, hence the optimality conditions for the second stage are

$$\begin{aligned} \dot{z}_2(t) &= \alpha z_2(t) + u_2(t) \\ \dot{p}_2(t) &= -\alpha p_2(t) \\ r_2 u_2(t) + p_2(t) &= 0 \end{aligned}$$

Therefore,

$$u_2(t) = -\frac{2\alpha \left(z_2^f - z_2^0 e^{\alpha s_2} \right)}{(e^{-\alpha s_2} - e^{\alpha s_2})} e^{-\alpha t}$$

and

$$\begin{aligned} \theta_2(z_2^0, z_2^f, s_2) &= \min_{u_2} \phi_2(s_2) \\ &= \frac{1}{2} h (z_2^f - z_d^f)^2 + \frac{r_2 \alpha \left(z_2^f - z_2^0 e^{\alpha s_2} \right)^2}{(e^{\alpha s_2} - e^{-\alpha s_2})} e^{-\alpha s_2} \end{aligned}$$

The higher-level optimization problem (16) becomes

$$\min_{\substack{s_1,s_2,z_1^0,\\ z_2^0,z_1^f,z_2^f}} \left[\frac{1}{2}\frac{r_1}{s_1}\left(z_1^f - z_1^0\right)^2 + \frac{1}{2}h\left(z_2^f - z_d^f\right)^2 + \frac{\left(z_2^f - z_2^0 e^{\alpha s_2}\right)^2 \alpha r_2}{(e^{2\alpha s_2}-1)} + \beta x_2^2 \right]$$

subject to $z_1^0 = z_0$, $z_2^0 = z_1^f$, $s_1, s_2 \geq 0$, and $x_2 = s_1 + s_2$. The resulting reduced optimization problem is

$$\min_{\substack{s_1,s_2,\\ z_1^f,z_2^f}} \left[\frac{1}{2}\frac{r_1}{s_1}\left(z_1^f - z_0\right)^2 + \frac{1}{2}h\left(z_2^f - z_d^f\right)^2 \right. \tag{17}$$
$$\left. + \frac{\left(z_2^f - z_1^f e^{\alpha s_2}\right)^2 \alpha r_2}{(e^{2\alpha s_2}-1)} + \beta\left(s_1+s_2\right)^2 \right]$$

where $s_1, s_2 \geq 0$.

The optimality conditions are given in terms of four algebraic equations which can be solved to yield s_1, s_2, z_1^f, and z_2^f, assuming $s_1 > 0$ and $s_2 > 0$ (if we allow $s_1 = 0$ and $s_2 = 0$, we get $z_1^f = z_0$ and $z_2^f = z_1^f$, respectively):

$$\frac{r_1}{s_1}(z_1^f - z_0) = \frac{2\left(z_2^f - z_1^f e^{\alpha s_2}\right)\alpha r_2}{(e^{\alpha s_2} - e^{-\alpha s_2})}$$

$$h(z_2^f - z_d^f) = -\frac{2\left(z_2^f - z_1^f e^{\alpha s_2}\right)\alpha r_2}{(e^{2\alpha s_2} - 1)}$$

$$r_1(z_1^f - z_0)^2 = 4\beta s_1^2(s_1 + s_2)$$

$$\beta(s_1+s_2) = \frac{\alpha^2 r_2\left(z_2^f - z_1^f e^{\alpha s_2}\right)\left(z_1^f e^{\alpha s_2}\right)}{(e^{2\alpha s_2}-1)} + \frac{\alpha^2 r_2 e^{2\alpha s_2}\left(z_2^f - z_1^f e^{\alpha s_2}\right)^2}{(e^{2\alpha s_2}-1)^2}$$

For a numerical example, setting $r_1 = 2$, $r_2 = 10$, $h = 10$, $z_d = 10$, $z_0 = 0$, $\alpha = 1$ and $\beta = 10$, yields the following solution: It is optimal to start operating in the first mode with constant control $u_1(t) = 5.72$ for $t \in [0, x_1)$ and switch to the second mode at time $x_1 = 0.4$ when $z_1^f = 2.29$. The system operates in the second mode with control $u_2(t) = 1.66e^{-t}$ until time $x_2 = 1.64$ when $z_2^f = 9.67$.

In this type of a simple switched linear system, the main complexity lies in the fact that the LQ problems involved are coupled through the continuity requirements $z_i^0 = z_{i-1}^f$ associated with the ith mode switch. This leads to a higher-level nonlinear optimization problem such as (17).

Obtaining an analytic form for this nonlinear optimization problem depends on the availability of a closed-form solution for the LQ problems associated with each mode at the lower level of the decomposition. Except for relatively simple cases such as the one considered above, this is not possible, and numerical techniques must be invoked. This has motivated recent work by Xu and Antsaklis [12] making use of the sensitivities of the optimal cost at the lower level with respect to the switching times $x_1, \ldots, x_N$ involved. On the other hand, the presence of more challenging event-driven dynamics shifts the main complexity burden to the higher-level problem, as discussed in the next section.

4. Switched Systems with Nondifferentiable Event-Driven Dynamics

In the problem of the previous section, the switching time dynamics are determined through

$$x_i = x_{i-1} + s_i(z_i, u_i) \tag{18}$$

which is a simple linear relationship. Thus, as already mentioned, the complexity is concentrated in determining the optimal amount of time spent in mode i, $s_i(z_i, u_i)$, whereas the event driven dynamics yielding the switching time sequence $\{x_1, x_2, \ldots\}$ in this case are extremely simple.

The situation is very different when switching times are dependent upon two or more event processes, in which case we must resort to (9) and analyze the timed automaton that coordinates these processes. In what follows, we shall discuss a class of systems where the event-driven switching time dynamics are described by

$$x_i = \max(x_{i-1}, a_i) + s_i(u_i) \tag{19}$$

where $\{a_i\}$, $i = 1, \ldots, N$, is a given sequence of event times corresponding to an asynchronous event process operating independently of the physical processes $\{z_i(t),\ t \in [x_{i-1}, x_i)\}$. In fact, the "max-plus" recursive equation (19) is the Lindley equation, well-known in queueing theory [2], describing the times at which "customers" depart from a simple queueing system; in this case, a_i is the ith customer's arrival time and $s_i(u_i)$ is the time required to process the ith customer, dependent upon some control u_i. In comparing (19) to (18), note that the key difference is the presence of the max function, which introduces a nondifferentiable component into the solution of the overall problem.

Thus, the problem we consider here is (13), subject to (2) and (19) and $\mathbf{u}$ as defined in (11). This problem is largely motivated by the structure of many manufacturing systems: Discrete entities (referred to as *jobs*) move through a network of workcenters which process the jobs so as to change their physical characteristics according to certain specifications. Associated with each job is a *temporal* state and a *physical* state. The temporal state of a job evolves according to event-driven dynamics and includes information such as the waiting time or departure time of the job at the various workcenters. The physical state evolves according to time-driven dynamics which, depending on the particular problem being studied, describe changes in such quantities as the temperature, size, weight, chemical composition, or some other measure of the "quality" of the job. The interaction of time-driven with event-driven dynamics leads to a natural trade-off between temporal requirements on job completion times and physical requirements on the quality of the completed jobs. For example, while the physical state of a job can be made arbitrarily close to a desired "quality target", this usually comes at the expense of long processing times resulting in excessive inventory costs or violation of constraints on job completion deadlines. Our objective, therefore, is to formulate and solve optimal control problems of the form (13) which capture such trade-offs.

In the context of manufacturing systems, the mode switches correspond to jobs that we index by $i = 1, \ldots, N$. We shall limit ourselves to a single-stage process modeled as a single-server queueing system. The objective is to process N total jobs. The server processes one job at a time on a first-come first-served nonpreemptive basis (i.e., once a job begins service, the server cannot be interrupted, and will continue to work on it until the operation is completed). Jobs arriving when the server is busy wait in a queue whose capacity is $\geq N$. As job i is being processed, its *physical* state z_i, evolves according to time-driven dynamics of the general form (2), where x_{i-1} is the time when processing begins. The control variable u_i will be assumed here to be scalar and not time-dependent for simplicity; it is used to attain a final desired physical state corresponding to a target "quality level". Specifically, if the service time for the ith job is $s_i(u_i)$ and $\Gamma_i(u_i)$ is a given set (e.g., a threshold above which z_i satisfies a desired quality level), then the control u_i is chosen to satisfy the stopping rule

$$s_i(u_i) \tag{20}$$
$$= \min \left\{ t \geq 0 : \left[z_i(x_{i-1} + t) = \int_{x_{i-1}}^{x_{i-1}+t} g_i(z_i, u_i, \nu) d\nu + \zeta_i \right] \in \Gamma_i(u_i) \right\}$$

where u_i takes a fixed constant value during the interval $[x_{i-1}, x_{i-1} + t)$, and the "min" is assumed to exist. On the other hand, the *temporal* state of the ith job, x_i, represents the time when the job completes processing and departs from the system. Letting a_i be the arrival time of the ith job, the event-driven dynamics describing the evolution of the temporal state are given by (19). A typical state trajectory if this type of system is shown in Fig. 12.2. Note that the interval $(x_1, a_2]$ in this example corresponds to an "idle period" for the workcenter and causes the second switching time to become, from (19), $x_2 = \max(x_1, a_2) + s_i(u_i) = a_2 + s_i(u_i)$.

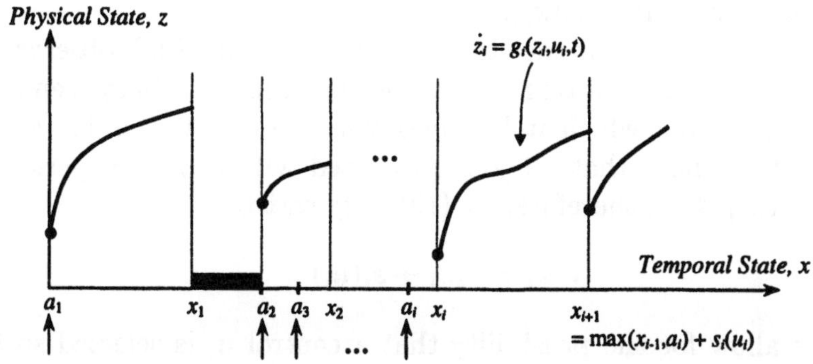

Figure 12.2. Typical state trajectory

To summarize, the optimization problem of interest here is

$$\min_{u_1,\ldots,u_N} \sum_{i=1}^{N} [\phi_i(u_i) + \psi_i(x_i)] \tag{21}$$

$$\text{s.t. } x_i = \max(x_{i-1}, a_i) + s_i(u_i)$$

where $a_1, \ldots, a_N$ is a given input sequence of event times and $s_i(u_i)$ is specified through (20). This is similar in form to classical discrete-time optimal control problems commonly found in the literature (e.g., [1]) except for two issues. First, the index $i = 1, \ldots, N$ does not count time steps, but rather asynchronously departing jobs. Second, the presence of the "max" function in the state equation (19) prevents us from using standard gradient-based techniques, since it introduces a nondifferentiability at the point where $a_i = x_{i-1}$. In effect, (21) formulates an optimal control problem for a DES with dynamics given by (19), where the control variables regulate interevent times.

In order to overcome the nondifferentiability introduced through the max in (19), one can resort to nonsmooth optimization techniques, as in [14]. Under some reasonable assumptions on $\phi_i(u_i)$ and $\psi_i(x_i)$, it is shown that a unique solution exists and can be obtained explicitly through an algorithm based on calculating generalized gradients and solving a Two Point Boundary Value Problem (TPBVP). However, it turns out that one can exploit the structure of state trajectories, as seen in Fig. 12.2, to decompose the overall problem (21) into a collection of simpler nonlinear constrained optimization problems. A key observation is that, unlike the problem considered in the previous section, there is no coupling of physical states when a mode switch occurs: after a job is processed and departs at time x_i with physical state $z_i(x_i)$, the new mode corresponds to a completely new job with physical state $z_{i+1}(x_i^+)$ which is independent of $z_i(x_i)$.

Taking a closer look at the state trajectory in Fig. 12.2, observe that it consists of periods during which the workcenter is busy, separated by intervals during which it becomes temporarily idle, as in (x_1, a_2). Recalling (19), note that all jobs processed within a busy period are such that $x_{i-1} \leq a_i$, therefore, by (19) they satisfy

$$x_i = x_{i-1} + s_i(u_i)$$

where we allow for the possibility that a control u_i is selected so that $x_{i-1} = a_i$; we refer to jobs with the property that they depart exactly when the next job arrives as "critical", a manifestation of the intuitive fact that processing jobs on a "just in time" basis is occasionally (but not always) optimal. On the other hand, if a job initiates a busy period (equivalently, terminates an idle period), then

$$x_i = a_i + s_i(u_i)$$

To formalize this partition of the state trajectory into "busy" and "idle" periods, we make the following definitions :

Definition 1. An *idle* period is a time interval (x_k, a_{k+1}) such that $x_k < a_{k+1}$ for any $k = 1, ..., N-1$.

Definition 2. A *Busy Period* (BP) is a set of contiguous jobs $\{k, ..., n\}$, $1 \leq k \leq n \leq N$ such that the following conditions are satisfied

1. $x_{k-1} < a_k$
2. $x_n < a_{n+1}$
3. $x_i \geq a_{i+1}$ for every $i = k, ..., n-1$.

Definition 3. A *busy-period structure* is a partition of the jobs $1, ..., N$ into busy periods.

Obtaining an explicit solution to problem (21) is tantamount to identifying the BP structure of the optimal state trajectory (sample path) and then solving a nonlinear optimization problem within each BP. If a BP is defined by the job indices (k, n), then we denote this problem by

$Q(k,n)$:

$$Q(k,n): \min_{u_1,\ldots,u_N} \sum_{i=k}^{N} \{\phi_i(u_i) + \psi_i(a_k + \sum_{j=k}^{i} s_j(u_j))\} : s_i(u_i) \geq 0\} \quad (22)$$

$$\text{s.t.} \quad a_k + \sum_{j=k}^{i} s_j(u_j) \geq a_{i+1}, \quad i = k, ..., n-1$$

Note that we have set $\psi_i(x_i) = \psi_i(a_k + \sum_{j=k}^{i} s_j(u_j))$ since, within a BP, $x_j = a_j + s_j(u_j)$ for all $i = k, ..., n-1$. The constraint represents the requirement $x_i \geq a_{i+1}$ for any job $i = k, ..., n-1$ belonging to the BP.

Let us also impose some basic conditions on the cost functions $\phi_i(u_i)$ and $\psi_i(x_i)$:

Assumption 1: For each $i = 1, \ldots, N$, $\phi_i(\cdot)$ is strictly convex, twice continuously differentiable, and monotonically decreasing with $\lim_{u_i \to 0^+} \phi_i(u_i) = -\lim_{u_i \to 0^+} \frac{d\phi_i}{du_i} = \infty$ and $\lim_{u_i \to \infty} \phi_i(u_i) = \lim_{u_i \to \infty} \frac{d\phi_i}{du_i}$ $= 0$.

Assumption 2: For each $i = 1, \ldots, N$, $\psi_i(\cdot)$ is strictly convex, twice continuously differentiable, and its minimum is obtained at a finite point δ_i.

In addition, let restrict ourselves to controls u_i that affect the processing time of the ith job linearly:

Assumption 3: For each $i = 1, \ldots, N$, $s_i(\cdot)$ is monotonically increasing and linear.

The latter assumption allows us, for simplicity, to replace $s_i(u_i)$ by u_i and directly control all processing times. Moreover, note that under the first two assumptions problem $Q(k,n)$ is a convex constrained optimization problem which is readily solved using standard techniques. We denote the (unique) solution to this problem by $u_i^*(k,n)$, $i = k, ..., n$ and the corresponding event times by $x_i^*(k,n)$, $i = k, ...$ Under these conditions, the following result provides a crucial necessary and sufficient condition for identifying a BP on the optimal state trajectory making use of solutions.

Theorem 1 *(Cho et al. [15]) Jobs $k, \ldots, n$ constitute a single busy period on the optimal sample path if and only if the following conditions are satisfied:*

1. $a_k > x_{k-1}^*$
2. $x_i^*(k,i) \geq a_{i+1}$ for all $i = k, \ldots, n-1$
3. $x_n^*(k,n) < a_{n+1}$

The significance of this necessary and sufficient condition is best understood as follows. Given N jobs, the total number of possible BPs is given 2^{N-1}. Thus, the complexity of exploring all BP structutres in order to determine the optimal one is combinatorial in nature. However, this complexity is reduced to linear using the theorem above: One proceeds forward in time sequentially solving a problem of the form $Q(k, n)$, with $k = 1$ initially, until the condition $x_n^*(k, n) < a_{n+1}$ is satisfied for some $n \geq k$, at which point a BP defined by (k, n) is identified. The process then repeats for a new problem $Q(n+1, m)$ with $m = n+1, n+2, \ldots$ until a new BP is similarly identified. This gives rise to the *Forward Algorithm* (FA) presented in [15]. It is easy to see that the complexity of the FA is of the order of N convex constrained optimization problems of the form (22). In fact, one can improve the efficiency of the FA even further, as described in [16].

5. Conclusions

As time-driven and event-driven systems are rapidly merging, giving rise to a class of "hybrid" dynamic systems, the field of optimal control is presented with an opportunity to expand its horizons, combining the fundamental principles on which it was originally conceived with new ideas that must tackle new forms of complexities. Some of these complexities are purely combinatorial in nature, the result of discrete elements in the problem such as switching events and modes to choose form, which enlarge an already large state space. Others are the result of nonlinear dynamics, including the type of nondifferential behavior one encounters in DES. Yet another form of complexity, which was only briefly mentioned in Section 1, is due to the stochastic nature of many state and/or input variables of interest.

In order to deal with these unavoidable and sometimes new types of complexities, it is clear that we have to go beyond existing methods and to try and exploit any features present in the structure of a system or the optimization problem itself. In this paper, we aimed to show how a natural hierarchical decomposition of at least some classes of hybrid systems can simplify the task of solving optimal control problems. Still, in most cases, one must resort to numerical techniques in order to obtain explicit solutions, and this is before even considering the issue of selecting over a set of feasible modes when a switch occurs or seeking solutions that are in some sort of feedback form useful in practice.

In the case of the optimal control problem (21), we have been able to take advantage of a different type of decomposition (over time, as opposed to over the system structure). However, we limited ourselves to a scalar problem and it is unclear whether a similar efficient decomposition can be used in a vector setting (e.g., a manufacturing system consisting of multiple workcenters).

Finally, an interesting issue that this new line of research has brought up is that of applying Perturbation Analysis (PA) methods to hybrid systems in a way similar to the successful development of PA for DES (see [2],[17]) that has contributed to the solution of some complex optimization problems. Looking at problems in the fields of manufacturing, communication networks, transportation, and command-control systems, one is struck by the natural way in which hybrid systems manifest themselves and optimization problems arise which the PA framework can help solve through on-line techniques that exploit the structure of sample paths and the data one can readily extract from them. There are promising signs in this direction (e.g., see [18]), a fact that should be gratifying to Larry Ho, since Perturbation Analysis for DES is also an area that he pioneered in the early 1980s.

References

[1] A. E. Bryson and Y. C. Ho, *Applied Optimal Control.* Hemisphere Publishing Co., 1975.

[2] C. G. Cassandras and S. Lafortune, *Introduction to Discrete Event Systems.* Kluwer Academic Publishers, 1999.

[3] P. J. Antsaklis, ed., *Proceedings of the IEEE*, vol. 88, 2000.

[4] M. S. Branicky, V. S. Borkar, and S. K. Mitter, "A unified framework for hybrid control: Model and optimal control theory," *IEEE Trans. on Automatic Control*, vol. 43, no. 1, pp. 31–45, 1998.

[5] B. Piccoli, "Hybrid systems and optimal control," in *Proceedings of 37th IEEE Conf. On Decision and Control*, pp. 13–18, Dec. 1998.

[6] H. J. Sussmann, "A maximum principle for hybrid optimal control problems," in *Proceedings of 38th IEEE Conf. On Decision and Control*, pp. 425–430, Dec. 1999.

[7] S. Hedlund and A. Rantzer, "Optimal control of hybrid systems," *Proceedings of 38th IEEE Conf. On Decision and Control*, pp. 3972–3977, 1999.

[8] X. Xu and P. J. Antsaklis, "A dynamic programming approach for optimal control of switched systems," in *Proceedings of 39th IEEE Conf. On Decision and Control*, pp. 1822–1827, Dec. 2000.

[9] A. Bemporad, F. Borelli, and M. Morari, "Optimal controllers for hybrid systems: Stability and piecewise linear explicit form," in *Proceedings of 39th IEEE Conf. On Decision and Control*, pp. 1810–1815, Dec. 2000.

[10] A. Giua, C. Seatzu, and C. V. D. Mee, "Optimal control of switched autonomous linear systems," in *Proceedings of 40th IEEE Conf. On Decision and Control*, pp. 2472–2477, Dec. 2001.

[11] K. Gokbayrak and C. G. Cassandras, "Hybrid controllers for hierarchically decomposed systems," in *Proceedings of 3rd Intl. Workshop on Hybrid Systems: Computation and Control*, pp. 117–129, March 2000.

[12] X. Xu and P. J. Antsaklis, "Optimal control of switched systems: New results and open problems," *Proceedings of the ACC*, pp. 2683–2687, 2000.

[13] K. Gokbayrak and C. G. Cassandras, "A hierarchical decomposition method for optimal control of hybrid systems," in *Proceedings of 39th IEEE Conf. On Decision and Control*, pp. 1816–1821, December 2000.

[14] C. G. Cassandras, D. L. Pepyne, and Y. Wardi, "Optimal control of a class of hybrid systems," *IEEE Trans. on Automatic Control*, vol. AC-46, no. 3, pp. 398–415, 2001.

[15] Y. C. Cho, C. G. Cassandras, and D. L. Pepyne, "Forward decomposition algorithms for optimal control of a class of hybrid systems," *Intl. Journal of Robust and Nonlinear Control*, vol. 11, no. 5, pp. 497–513, 2001.

[16] P. Zhang and C. G. Cassandras, "An improved forward algorithm for optimal control of a class of hybrid systems," in *Proceedings of 40th IEEE Conf. On Decision and Control*, pp. 1235–1236, Dec. 2001.

[17] Y. C. Ho and X. Cao, *Perturbation Analysis of Discrete Event Dynamic Systems.* Dordrecht, Holland: Kluwer Academic Publishers, 1991.

[18] C. G. Cassandras, Y. Wardi, B. Melamed, G. Sun, and C. G. Panayiotou, "Perturbation analysis for on-line control and optimization of stochastic fluid models," *IEEE Transactions on Automatic Control*, 2002. To appear.

MIX
Papier aus verantwortungsvollen Quellen
Paper from responsible sources
FSC® C105338
FSC
www.fsc.org

If you have any concerns about our products,
you can contact us on
ProductSafety@springernature.com

In case Publisher is established outside the EU,
the EU authorized representative is:
Springer Nature Customer Service Center GmbH
Europaplatz 3, 69115 Heidelberg, Germany

Printed by Libri Plureos GmbH
in Hamburg, Germany